# Spatio-Temporal Databases

Rita de Caluwe • Guy de Tré •
Gloria Bordogna (Editors)

# Spatio-Temporal Databases

## Flexible Querying and Reasoning

With 112 Figures

Springer

Editors:

**Professor Dr. Rita de Caluwe**
**Dr. Guy de Tré**
Ghent University
Dept. of Telecommunications
and Information Processing
Computer Science Laboratory
Sint-Pietersnieuwstraat 41
9000 Ghent, Belgium
E-mail: *rita.decaluwe@ugent.be*
and *guy.detre@ugent.be*

**Dr. Gloria Bordogna**
Italian National Research Council
CNR IDPA - Institute for
Environmental Process Dynamics
Via Pasubio 5
24044 Dalmine (BG), Italy
E-mail: *gloria.bordogna@idpa.cnr.it*

ISBN 3-540-22214-6 Springer Berlin Heidelberg New York

Library of Congress Control Number: 2004106677



Springer is a part of Springer Science+Business Media
springeronline.com

Printed in Germany



Cover design by Erich Kirchner, Heidelberg
Map reproduction by courtesy of Ghent University Library
Map photography by Jörg Verstraete
Typesetting: Camera-ready by the editors
Production: Luisa Tonarelli
Printing: Mercedes Druck, Berlin

Printed on acid free paper 30/3141/LT – 5 4 3 2 1 0

# Foreword

"Spatio-Temporal Databases: Flexible Querying and Reasoning," or STD for short, is an important work. Edited by prominent researchers in database theory and applications, Professor Rita de Caluwe, Dr. Guy de Tré and Dr. Gloria Bordogna, STD addresses issues and problems which lie on the frontiers of data knowledge and engineering. The principal thrust of STD is directed at spatial-temporal database systems–systems which are rapidly growing in visibility and importance, in reflection of (a) the advent of geographical information systems (GIS); and (b) the exponential growth in satellite imagery, as well as dynamic imagery in the Web, multimedia, medicine and other fields.

Spatial-temporal database systems represent a natural evolution of conventional database systems. But the growing variety and complexity of data types in spatio temporal systems are transcending the capability of conventional database theory to deal with them. In particular, in the context of spatio-temporal systems the issues of imprecision, uncertainty and fuzziness move to the center of the stage and call for methods of analysis that are not a part of standard theories. In this perspective, STD may be viewed as a thrust into new territories with new tools, new ideas and new objectives.

The contributed chapters in STD range over a wide variety of issues and problems. What stands out in STD is the strong, but not exclusive, emphasis on concepts and techniques drawn from fuzzy set theory, possibility theory and, more broadly, fuzzy logic. This is a significant paradigm shift, reflecting a realization that conventional bivalent-logic-based methods do not have the capability to come to grips with the pervasive imprecision, uncertainty, incompleteness and partiality of truth that are the norm in spatio-temporal systems. What are the boundaries of a mountain, a forest or a desert? What is the definition of an edge? These are elementary but by no means trivial examples of questions which come into view when we attempt to translate perceptions into measurements.

What is the rationale for the use of fuzzy logic in dealing with spatio-temporal database systems? Spatial-temporal systems are intrinsically complex and call for methods which can cope with partiality of likelihood, partiality of possibility and partiality of truth. This is what fuzzy logic is designed to do. In fact, in fuzzy logic everything is, or is allowed to be, a matter of degree. Furthermore, variables are allowed to be granular, with a granule being a clump of values drawn together by indistinguishability, similarity, proximity or functionality. The concept of granularity underlies the concepts of a linguistic variable and fuzzy if-then rules. The calculus of fuzzy if-then rules plays an important role in the analysis of spatio-temporal systems and is an object of considerable attention in STD.

STD deserves high marks for organization, coverage, clarity of exposition and, above all, up-to-dateness. I was impressed to see, the recently developed methodology of computing with words and generalized constraints, employed by Guy de Tré, Rita de Caluwe, Jörg Verstraete and Alex Hallez in Chapter 6 to enrich database models and address the issue of imperfect information. I may add that it is my personal conviction that the concepts of a generalized constraint and generalized constraint propagation should have a position of centrality in knowledge representation, deduction and search. In fact, in fuzzy logic, deduction may be viewed as a form of generalized constraint propagation.

A point of semantics. In STD some of the authors employ the term "vague" in the same sense as "fuzzy." As I have pointed out on a number of occasions, "vague" and "fuzzy" are not synonymous. More specifically, "vagueness" is a property of propositions rather than predicates. Thus, a proposition, $p$, is vague if it is insufficiently specific. On the other hand, a predicate, $q$, is fuzzy if its denotation has unsharp boundaries, and a proposition, $p$ is fuzzy if it contains fuzzy predicates. For example, the proposition, "I will be back sometime," is vague and fuzzy, while the proposition, "I will be back in about five minutes," is fuzzy but, for most purposes, is not vague.

In conclusion, STD contributes so much and in so many important ways to the advancement of our understanding of the theory and applications of spatio-temporal database systems. In STD, one finds many concepts, methods and techniques that are original with the authors and reflect their high level of expertise and expository skill. STD points to new directions in database theory and applications, and is of great value as a reference. The editors, the authors and the publisher have produced a book that is highly informative, authoritative and reader-friendly. They deserve our applause.

Lotfi A. ZADEH<br>Berkeley, California<br>April 2004

# Contents

# Basic Notions and Rationale of the Handling of Imperfect Information in Spatio-Temporal Databases

Rita de Caluwe[1], Guy de Tré[1], and Gloria Bordogna[2]

[1] Department of Telecommunications and Information Processing
Ghent University
Sint-Pietersnieuwstraat 41, B-9000 Gent (Belgium)
{rita.decaluwe,guy.detre}@ugent.be

[2] Institute for Environmental Process Dynamic (IDPA)
Dep. of Milan – Lab. of Georesources, National Research Council (CNR)
via Pasubio 5, 24044 – Dalmine (BG), Italy
gloria.bordogna@idpa.cnr.it

## Basic Notions

**Space and time**, both are familiar notions, although they strangely do not let capture themselves in strict definitions. Everything people do, is experienced in space and time. Space and time are, therefore, the logical prerequisites of all human experience. As such they are used as a framework for querying and reasoning. The perspectives in this book are limited to querying and reasoning about information kept in databases and that carries time and space aspects. The underlying time and space models themselves are only presented in as far as they are needed for that purpose. In the same way as models able to deal with imprecision and uncertainty can be conceived, these notions can find their way in the querying and reasoning formalisms too, whether applied to traditional database models or to the more sophisticated database models.

With respect to databases and how to find out and use the information contained in them, it makes sense to distinguish clearly between the data model and the database model, strata on top of which querying and reasoning can be built. Database systems exist by the virtue of database models and thus of the underlying data models; it has to be seen as a term with the flavour of computer applications, hence coloured by implementation aspects and offerings on the software market. To come to a more comprehensive meaning of flexible querying and reasoning as dealt with in this book, it can be helpful to first have a closer look to the meaning of data, data model, database model and database system.

**Data** are factual information used as a basis for reasoning, discussion and computation. The are what is known in the real world, facts and figures carrying semantics, which is in the end most of the time represented in a very clumsy way in a computer environment. It is noticeable that data can be related among them.

A **database** is any collection of data that is specially structured and organised for rapid search and retrieval by a computer. In a database data are represented according to a data model and is related one another according to a database model.

A **data model** is defined by a **data type**, being a set of legal values, together with a (sufficient) number of (a non-conflicting set of) assorted operators on these values. In traditional systems, only a very limited number of basic types is made available, but sometimes the systems are granted type generators, which converts them into open ended type systems. The most common examples of data types are of the primitive number types integer, real, character type, ..., which can be arranged in arrays, list structures and so on, sometimes in more complicated structures turned into new types by using type generators; appropriate operations are assorted to them as arithmetic operations on numbers. In general, very few of the real world semantics can be formally captured in the models, but database models and database systems built on this kind of strata have been used for years in the absence of more sophisticated data models. As far as there is no need for more powerful systems, there is nothing wrong with using these models: their availability simply is a blessing. However, many times there is a real lack of expressing imprecise and uncertain data, especially of data that only can be described in linguistic terms supported by common understanding; the provided data models are not of sufficient, support, for instance to qualify an item as tall or ephemeral or quick or to indicate the possibility of one event or another, happened or to happen. Even when one is able to cope in an acceptable way with failing semantics at the level of the data model, it can be very helpful to be offered the possibility to leave a number of details to the system, thanks to the use of underpinning more advanced models. A number of such models have been proposed in literature. These models, known in literature as fuzzy models and uncertainty models, rely in essence on fuzzy set theory, on probability theory and on possibility theory and can be easily implemented in the context of open type systems. In general they are seen as generalizations of traditional models, sometimes as extensions of them. A very straightforward way of dealing with them, is to see them as the basic models of which the traditional ones are special cases, for particular parameter values.

The notion of **database model** defined on top of a data model takes into account and incorporates into the model the expression of the relationships between the data, network, hierarchical, relational and object database systems, being the best known at present. In up to date settings, these relationships are interpreted as constraints, which more and more play a role in database theory.

Databases are therefore instantiations of database models. **Spatio-temporal databases** are databases in which space and time are types of data, at least conceptually, implementation issues sometimes preventing their direct instantiation as basic types.

The spatio-temporal database application field is wide and **GIS (Geographical Information Systems)** is an important part of it. In this context the management of imprecise and /or uncertain spatio-temporal information is particularly important since any observation of geographic phenomena is affected more or less by some kind of imperfection. Many of the contributions in this book are directly centred on GIS, some others are more general or focus on related application fields, presenting theoretical viewpoints and techniques that are inspiring or can be used mutatis mutandis for GIS.

**Flexible querying** tries to enhance querying expressiveness as offered in traditional database systems in many different ways, always aiming thereby to ease the extraction of data and information from the database relevant to the user. It can be applied to databases based on traditional data models, as well as to databases based on other, more advanced data models, such as the ones offering the possibility to deal with fuzzy and uncertain data. In this book the focus is on flexible querying related to the spatial and temporal components of the information that is represented in a database. Spatio-temporal querying implies the specification of spatial and/or temporal constraints on the entities. The logics used to build the expressions that make up the core of the query can be the traditional binary logic, a ternary logic —as Kleene's logic—, a four-valued logic —as Belnap's logic— or even more general ones. In particular, fuzzy and uncertainty models mostly rely on fuzzy logic and possibilistic logic. Much work has also been done to bridge the gap between the use of stringent formal forms of queries, as imposed by most of the systems at present, and query formulations that allow to express more naturally the information needs of the user, as by natural language sentences. From the foregoing discussion it should be clear that flexible querying constitutes a desirable plus-value in the hands of the user, particularly useful in the cases in which has to be dealt with spatial and temporal information that is affected by imprecision and uncertainty. For historical reasons, the development of the research on flexible querying took off with applications on traditional databases, obviously the easiest part to explore in the long way to go.

**Flexible reasoning** is not so much heard as a stand-alone term. Most of the time it is used in combination with flexible querying. The techniques for formal reasoning rely on logic as query formulation does, but additionally exploit the deductive capabilities drawn from the inference rules that are valid in the logic. Reasoning about data and information that resides in a database thus can bring up in a straightforward way more information than simple querying can. Spatio and temporal reasoning can exploit specific logics such as temporal logic. By flexible spatio and temporal reasoning is meant qualitative reasoning about the dynamic changes in the spatial domain, derived by either imprecision or uncertainty (or both), in case of incomplete knowledge of the available information.

## Rationale of the Handling of Imperfect Information in Spatio-temporal Databases

This book on flexible querying and reasoning in spatio-temporal databases is divided into two parts. The first part bundles the contributions in which the focus is rather on advances at the theoretical level, thereby discussing examples and opening further perspectives. The second part presents contributions of which the core is a well-developed application.

Though not the focus of the book, representation of the information is an important component in the majority of the chapters, as querying and reasoning are mostly developed on top of one or another specific data and/or database model for the representation of points, lines and objects, the properties associated with them and their intrinsic position, mutual position in aggregations of objects, or their behaviour during the course of time. Most of the authors recognise the need to handle imprecision and uncertainty, ranging from dealing with the static aspects of imprecise or uncertain positions to the dynamic aspects of sequences of changes during time course and possible trajectories. The volume shows definitely how the use of the more advances techniques that enable to deal with imprecision and uncertainty, can add value to the solution of a number of problems related to GIS.

### THEORETICAL PART

Chapter 1 by Suzana DRAGICEVIC proposes an approach to represent geographic entities with high spatial variability. Geographic objects are classified according to the characteristics of crispness and vagueness of their body and boundary and are represented in the database by a snapshot model that records distinct states of a dynamic phenomenon at known time instants. The status of the phenomenon at an intermediate spatio-temporal location is derived by a fuzzy temporal interpolation of the known states. To this end probable trajectories of transitions between classes are defined which can model fast, slow or even cyclic changes. An example of fuzzy spatio-temporal interpolation to model land-use changes is presented.

Chapter 2 by Martin ERWIG treats the querying of the spatial dynamic phenomena, by introducing spatio-temporal patterns as a useful means to express repeating or cyclic constraints. The focus is on deriving constraints that allow spatio-temporal patterns to become well-designed composable abstractions that can be smoothly integrated into spatio-temporal query languages. The outline of a language with graphical notation is presented, suited to address problems in GIS and in the related fields of geo- and eco-sciences.

Chapter 3 by Cindy CHEN, Haixun WANG and Carlo ZANIOLO presents a customisable spatio-temporal extension of SQL by enriching a small set of its built-in primitives with User Defined Aggregates that can be programmed directly through the extended SQL itself, hence in a very easy way for the experienced SQL user. The obtained spatio-temporal objects are modelled by counter clockwise directed triangles. A satisfactory overall performance is achieved through the use of the language.

Chapter 4 by Mirco NANNI, Alessandra RAFFAETA, Chiara RENSO and Franco TURINI offer a constraint logical language, STACLP, that provides a set of spatial and temporal primitive operators allowing the user to perform temporal reasoning on spatial data. The chosen formalism can be used as an advanced spatio-temporal query language on geographical data, but as well can be exploited as a deductive rule-based approach to represent domain knowledge on such data. Moreover, it is well suited to represent trajectories of moving objects. STACLP allows to tackle several analysis tasks requiring the integration of deductive and inductive capabilities. A case study in the field of behavioural ecology is added.

Chapter 5 by Gloria BORDOGNA and Sergio CHIESA presents, in an object-based framework, an ontology for imperfect spatio-temporal information, based on fuzzy set theory and possibility theory A dual representation of dynamic spatial phenomena is proposed in which both uncertainty and imprecision can be modelled as orthogonal concepts, affecting either the spatial or the temporal dimensions. Operations going along with the kind of imperfection and methods for the evaluation of spatio-temporal constraints are studied. Abstract spatio-temporal data types are introduced, to guarantee efficient retrieval.

Chapter 6 by Guy DE TRE, Rita DE CALUWE, Jörg VERSTRAETE and Axel HALLEZ discusses the advantage of using generalized constraints as introduced by Lotfi ZADEH to enrich spatio-temporal database models in order to cope with imperfect information. To handle spatio-temporal information, a new data type is introduced. Relying on a many-valued logic based on extended possibilistic truth values, missing information can also be dealt with. It is shown how the generalized constraints enable to specify the semantics of the database and to impose selection criteria for flexible querying. Both data modelling and querying aspects are addressed in this chapter.

Chapter 7 by Fangju WANG introduces new parsing techniques employing approximate rule matching based on a fuzzy grammar and on possibility theory, for a natural language interface to a GIS, in which can be dealt with uncertainty. This parse strategy aims at minimal computational cost.

Chapter 8 by Adnan YAZICI, Öznur YAVUZ and Roy GEORGE highlights all the static and dynamic aspects of spatio-temporal data objects in the settings of the management of a video database, conform to the MPEG-7 standard. Querying based on the XML Path Language is a well developed part of the chapter.

Chapter 9 by Eliseo CLEMENTINI studies in depth objects with fuzzy boundaries, defined as broad boundaries i.e. delimited by their minimum and maximum extent. New interpretations of the fuzzy boundaries and new application fields for them are investigated. An application is presented in which aggregations of spatial objects are considered, either as tessellations or as networks; it is studied how uncertainty can affect the definition of the position of their common borders and common endpoints.

Chapter 10 by Maria SOMODEVILLA and Fred PETRY too studies vague regions; these are defined by a core and boundary, approximated by their minimum bounding rectangles. It is shown how an expert through an interactive process can achieve a more precise representation of the vague region. Thematic and temporal dimensions have been incorporated in the model to complement spatial reasoning. Both location-based querying and feature-based querying are supported by the model and illustrated in the chapter.

Chapter 11 by Markus SCHNEIDER pays attention to the definition of fuzzy spatial data types and predicates to suit integration into query languages. The 9-intersection model for topological operations is considered as a basis for defining fuzzy topological operators for fuzzy regions. To integrate fuzzy topological predicates into an SQL-like query language, qualitative linguistic descriptions of nuances of topological relationships are introduced, such as "nearly completely inside", "somewhat inside", "quite overlap", etc.

## APPLICATION PART

Chapter 12 by Paola CARRARA, Gabriella PASI, Monica PEPE and Anna RAMPINI discusses a possibility-based model to index remote sensing images. This study contributes to the efficient exploitation of image archives, as could be crucial in emergency and disaster management. The capability of spectral signatures as indexes of images from remote sensing is investigated. The management of uncertainty is faced at distinct levels: in the image classification phase in associating the pixels to one or more classes, in the indexing phase in associating images with spectral signatures and in the retrieval phase to compute the relevance of the images. The proposed method has been evaluated in a database of images representing heterogeneous landscapes.

Chapter 13 by Josef BENEDIKT, Sebastian REINBERG and Leopold RIEDL explores how to evaluate and use linguistic representations for spatial data. Spatial information analysis is carried out with the help of flow-charts based on an easy-to-use programming language. It is shown that sensitivity analysis can be simplified by adopting a fuzzy representation of geographic entities characterized by unsharp boundaries.

Chapter 14 by Ben DERUDDER and Frank WITLOX studies the classification of the major cities in the world city network. This chapter provides evidence that a fuzzy set approach can aid in an exploratory data mining analysis of these inherently vague spatial data. Central to the approach is the generalization of a clustering method using a parameterised fuzzy c-means algorithm. This provides more flexibility in classifying city objects and better supports the analysis of (identified patterns in) world city networks than with a traditional approach.

Chapter 15 by Sidharta GAUTAMA, Johan D'HAEYER and Wilfried PHILIPS addresses image analysis and shows how spatial constraints can help to automatically detect changes in geographic information. The system uses error-tolerant graph matching to find correspondences between the detected image information and the road vector data. Spatial relations between objects play here an important role in finding reliable object-to-object mappings. Flexible querying is realized through the use of relaxation labelling, which in combination with the taking into account of acceptable data inconsistencies, leads to more reliable results for detecting change than the use of global transformation schemes.

Chapter 16 by Tom DE MUER, Andy VERKEYN and Dick BOTTELDOOREN studies noise annoyance and uses a fuzzy rule base to model the vague and complex interactions between its triggers. Large parts of them are related to geographical information, as source location (roads and railway lines), land use, building location and size etc. Many of the input data are imperfect. Imprecision and uncertainty are identified as different flavours of imperfection that need to be treated using different conceptual frameworks. A final issue addressed in this chapter, is the presentation of the vague and uncertain outcome and the aggregation to regions of interest for local policy makers (neighbourhoods, cities, regions).

This book originates from an acknowledgement by the editors of a lack on a basic document in the field of dealing with spatio-temporal databases, offering an overview of the state of the art thus far. So many people have put their efforts in trying to resolve spatio-temporal problems in a computer environment, ranging from modelling information, over querying, mining and reasoning, to say exploiting at best this kind of information, that bringing their efforts together in a book has been felt to be beneficial as well for the further research of those already involved, as for providing a starting platform to those who wish to step into this fast developing, promising, useful field. This volume, companion to the volume Fuzzy Modelling of Spatial Data for Geographic Information Systems, edited by Maria COBB, Fred PETRY and Vincent ROBINSON, focuses on the aspect of exploiting the information which is kept in databases, on querying and reasoning, incorporating the modelling aspects only as far as necessary, but of course highlighting the ultimate purpose of the research, as application is, illustrated by some stringent examples.

Well-known researchers that pioneered the fields of uncertainty and imprecision management in databases and geographical information systems have contributed to the book. Contacts with new researchers have been established made through the process of elaboration of this volume. We are grateful to all of them for the efforts they have put into their contribution and not in the least for their spirit of cooperation, which was utmost encouraging to us and turned our part of the job into a truly joyful and exciting one. For the text editing in Latex and figure layouts, we could account with the valuable help of Jörg VERSTRAETE. While putting this book together, despite the increasing workload never estimated at its real value at the time of the initiative, we not only never came to doubt about the meaningfulness of our enterprise, but rushed towards the end with even greater enthusiasm than ever. We hope the reader will share it in welcoming this book.

Ghent
February 2004

Rita DE CALUWE
Guy DE TRE
Ghent University

Gloria BORDOGNA
National Research Council
Institute for Environmental Process Dynamic

# Part I

# THEORY

# 1

# Fuzzy Sets for Representing the Spatial and Temporal Dimensions in GIS Databases

Suzana Dragićević

Spatial Analysis and Modeling Laboratory
Department of Geography
Simon Fraser University
8888 University Drive
Burnaby, BC, Canada
suzana_dragicevic@sfu.ca

## 1.1 Dynamic geographic phenomena and GIS

Geographic phenomena are dynamic and any effort to study, analyze, and predict them must consider both the space and time dimension. Space can be defined based on an absolute or a relative framework. According to the absolute view, space is a container in which objects are located; its structure is best described by the familiar and common three-dimensional Euclidean geometry. The relative view focuses on objects as the subject matter, and space is measured as relationships between objects. While the absolute view involves measurement referenced to a constant base with non-judgmental observations, the relative view implies interpretation of patterns and processes within specific contexts [47, 50]. Time can be described as the absolute linear line on which events occur and on which intervals between events can be measured. Space is three-dimensional and physically perceptible while time is understood as a unidirectional continuous flow. Both geography and cartography deal with space and time, and their relationships with various geographic phenomena. If there is no understanding of time and its impacts, geographers could not analyze and understand spatial changes nor cartographers able to represent these changes on maps [61].

The concept of geographic space and time provides different metrics and scales in their representations of large-scale environments, relationships, and interactions between humans and environment [33]. In the late 20th century, geographers began to systematically analyze the concept and representation of time as well as its effects on geographic phenomena. Time is used as a measure of human activity in studies within time geography and was introduced as a necessary variable in other areas such as historical, quantitative or cultural geography [29, 45, 60].

Measuring geographic space at a particular time gives an indication of changes that occur to geographic entities, their attributes and dynamic process [58, 59, 66]. As time progresses, all measurements of change in geographic space becomes relative to previous measurements. Relative geographic space is hence linked to relative time and nothing in the physical world is independently spatial or temporal, everything is a continual process [6]. As a consequence, dealing with geographic phenomena highlights the issues of how to represent spatio-temporal change. In recent times, problems related to the temporal dimension of geographic data within geographic information science and systems (GIS) have generated great interest in the issues of geographic representation of the space-time dimensions and dynamics in GIS databases [50].

GIS are flexible tools designed to handle, visualize and analyze geographic data and are intended to deal with the representation of dynamic spatio-temporal process. However, current GIS suffer from several limitations. These are: the atemporal nature of GIS databases; the paradigm of spatial data representations, which is the static characteristics of vector or raster data models; and the inherent fuzzy nature of geographical entities and boundaries that is approximated using crisp representation in GIS databases. This chapter focuses on examining the role of fuzzy set theory in the representation of the spatial and temporal dimensions in GIS databases when dealing with dynamic geographic phenomena.

## 1.2 Inherent nature of geographic entities

Geographic entities from the real world and measured data are represented in a GIS databases by traditional cartographic abstract objects such as points, lines and polygons. They can be represented in an aspatial manner as attributes related to the space. Moreover, spatial data and objects can be represented in a GIS with vector and raster models. The information is usually separated into different GIS thematic layers. The vector GIS data model is the traditional cartographic choropleth map of geographic entities transposed unto a digital layout. Geographic objects are usually represented by entities such as points, lines, and polygons. The vector approach allows an understanding of topology, i.e. the relationship between entities in space. The raster data model provides information on space as a continuum and is usually divided into square cells. This results in neither boundaries nor information on spatial topology between the geographic objects. The traditional isopleth maps represent geographic objects or their attributes as continuous variations over space [9].

Vector and raster models are static and only depict the information captured in one instant of time. Therefore their representation in a GIS database is in the form of snapshots. The time component of GIS data and objects is just an attribute of space [36]. Vector and raster models cannot maintain the topological relations between geographic objects when a change event is introduced [41, 44]. Information on change is a calculation of a difference between the states of the objects in two different snapshots. Snapshots may not be captured at the moment of the change, so if the interval between two consecutive snapshots is too long, the entire information related to the change may be lost [13]. The complete analysis of change should be focused on the process itself, and the events that induce the changes [35, 48].

Geographic objects represented as points, lines and polygons (Figure 1.1) are defined by spatial coordinates in two $(x, y)$ or three $(x, y, z)$ dimensions. When the time dimension is introduced, a geographical object is moving - varying its position (2D or 3D) with time, or changing - its position by changing the shape [51]. [15] further observes that object(s) can appear, disappear, succession, split, merge, expand, or shrink thereby making spatial and temporal relationships more complex to model in spatio-temporal GIS databases.

Another issue is that historically geographic data and objects were represented with sharply defined boundaries. Crisp sets and Boolean logic was used to represent spatial information on the map and then later in the GIS [25]. Many researchers have argued that the traditional crisp approach is not the most appropriate for representing geographic data [9, 24, 37, 52]. In the real world, geographic entities are inherently characterized by vagueness or fuzziness [56]. The indeterminate boundaries of geographic entities can be those whose position and shape are not known or cannot be measured precisely, and those with boundaries that are ill-defined [17]. For example, the natural boundaries of shorelines, soil classes, or forest are spatially and temporarily delimited in a gradual manner. The boundaries between urban and rural land-use is another example, and contains fuzzy transition zones with a mixture of for example low-density residential areas, green spaces and farmlands.

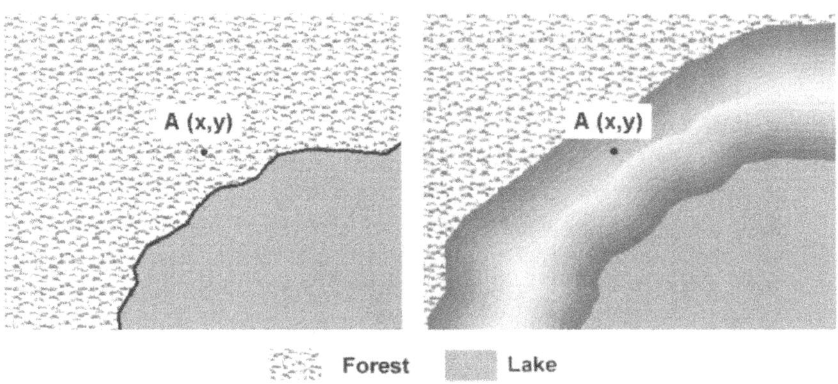


**Figure 1.1.** Crisp versus fuzzy representation of geographical entities

Representing this inherent vagueness of geographical features in GIS databases can be modeled by using fuzzy set theory [52]. Fuzzy set theory has been developed by [70] and is a generalization of classic set theory. It deals with classification of objects or phenomena that have continuous values and where the classes of objects and phenomena are represented with gradual transitions. A fuzzy set, because of its ability to represent degrees of membership, is a more suitable approach for modeling geographical entities. In a GIS database, real world objects can be represented by the degrees of belonging to multiple classes or objects. The degree of belonging to a particular set — geographic class or object is determined through the membership function. The fuzzy set provides a way to express the degree of membership to a particular class or a set.

In crisp set theory an element belongs or does not belong to a certain set — a geographical class or object. In fuzzy set theory an element $x$ can belong to a certain set partially and its affiliation can be represented by a real number within the range of $[0,1]$. Therefore a membership function $f(x)$ determine the degree of belonging $\mu(x)$ to class or object $C$ can be defined as follows:

$$\mu_C(x) = \begin{cases} f(x), & x \in C \\ 0, & x \notin C \end{cases}$$

The value of $\mu(x)$ is within the bounded range of $[0,1]$. Figure 1.1 illustrates a particular point $A$ with $(x_A, y_A)$ location on a map such that according to the crisp set theory point $A$ fully belongs to the forest class type and does not belong to the lake class type. According to the fuzzy set theory, point $A$ is partially belonging to the forest land-use and partially to the lake, and this is expressed with the degrees of memberships to both the forest or lake classes.

## 1.3 Fuzzy sets in representing space in GIS

Fuzzy set theory has been used in geography in the areas of remote sensing, GIS, spatial decision-making among others for handling the inherent fuzziness of geographical data from a spatial perspective. [53] provides a comprehensive review on the application of the fuzzy sets in GIS.

When dealing with the spatial dimension, fuzzy set theory can be applied for more realistic representation of geographical boundaries. Further, geographic objects can be represented by entities but if the content of the object or the boundary or both are indeterminate then the concept of fuzzy objects is needed [10, 11]. When representing geographic objects from the real world there are many that can be identified (Figure 1.2). Figure 1.2 shows a combination of objects and contents: A is an object with crisp content and crisp boundaries, CcCb; B is an object with crisp content and fuzzy boundaries, CcFb; C is an object with fuzzy content and crisp boundaries, FcCb; and D is an object with fuzzy content with fuzzy boundaries, FcFb respectively.

In order to achieve realistic modeling of geographical objects and boundaries with high spatial variability, a geographical boundary fuzzy set is defined [65]. A fuzzy membership function is derived such that the membership grade denotes the degree of sharpness at a boundary of the object of interest. A membership grade of 1 represents a sharp geographical boundary. A partial membership is indicated by a grade value between 0 and 1 and it models a boundary of intermediate sharpness. Thus, the sharper changes of the entity at the boundary are characterized by membership grades closer to 1 depicting more differentiable adjacent polygons for the entities of interest.

The fuzzy function associated with the membership grade that represents "change sharpness" can be defined in terms of a change rate at the boundary of the entity of interest. According to [65], since a geographic entity can be represented as a function of spatial locations $f(l)$ where $f$ denotes a value of the content of the object, such as a geographic class, related to the entity at any location $l$ the corresponding fuzzy membership function can be expressed as a function of the magnitude of the rate of change $|df/dl|$. This form of membership function is appropriate for determining the boundary membership grades for mainly vector-based GIS data and their attributes.

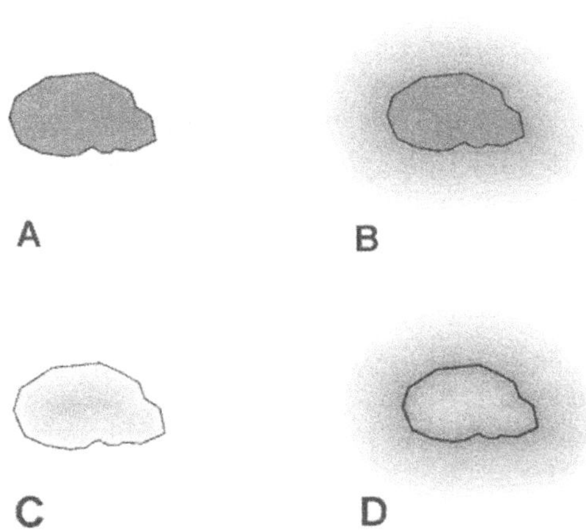


**Figure 1.2.** Different type of geographic objects: a) Crisp content with crisp boundaries (CcCb); b) Crisp content with fuzzy boundaries (CcFb); c) Fuzzy content with crisp boundaries (FcCb); d) Fuzzy content with fuzzy boundaries (FcFb)

The corresponding fuzzy membership function for the spatial dimension can be defined by using a generalized gaussian function (Figure 1.3e) described as:

$$\mu(x) = \begin{cases} 1 & b \le x \le c \\ f_1(x) & a < x < b \\ f_2(x) & c < x < d \\ 0 & x < a \wedge x > d \end{cases}$$

By setting parameters $a$, $b$, $c$ and $d$ it is possible to vary the shape of the fuzzy membership function. For example, if $a = b$ and $c = d$ the equation will represent the Boolean membership function that describes geographic object with crisp content and crisp boundary CcCb (Figure 1.3a). Other type of geographic objects CcFb, FcCb, FcFb can be represented by simplified fuzzy membership functions that are presented in Figures 1.3b, 1.3c, and 1.3d respectively. Figure 1.4 shows an example of the different land use classes comprising different type of geographic objects. At time $T1$ (Figure 1.4a) for example, forest is an object with fuzzy content and fuzzy boundary, road and agricultural land-use are represented as crisp objects with crisp boundaries, and lake and urban land-use are objects with crisp contents and fuzzy boundaries. At time $T2$ (Figure 1.4b) the change occurs and part of the forest land becomes grassland. It can be noticed that some objects contains fuzzy boundaries (e.g. urban, forest, lake, grassland) but by being adjacent to crisp object (e.g. road, agricultural) which boundaries can be measured, their type of boundaries change. [12] partially treated the fuzzy objects and their spatial change but there is potential for more work in this area.

The fuzzy membership function that models spatial boundaries between individual objects can be derived from the Similarity Relation Model (SR) or Semantic Import Model (SI) [7, 52]. The SR model is based on clustering theory, and is also known as the fuzzy $k$-mean clustering technique. Suitable application areas are physical geography, mostly in classification procedures in climatic [42] and soil science [43], or remote sensing [27, 64, 71, 72].

The Semantic Import Model uses information on data and their grouping known a priori by the experts and end-users that play a key role in determining the fuzzy membership functions. This approach is adopted in applications for soil survey, land suitability and evaluation among others [7, 8, 18, 30].

Fuzzy set theory can be used for spatial representation in 3 dimensions, such as Digital Elevation Models (DEM) and Digital Terrain Models (DTM). The fuzzy terrain models that incorporate uncertainty and fuzziness in predicting the variability of the landscape, topographic surface, ocean floor has been developed and presented in the literature [4, 40, 55].

## 1.4 Fuzzy sets for representing time in GIS

There are different types of times and various ways of understanding time by geographers [16, 28]. The development of temporal capabilities in GIS databases has been extensively explored in last two decades [14, 34, 36, 54, 58, 63]. But most of this development has been theoretical in nature and full

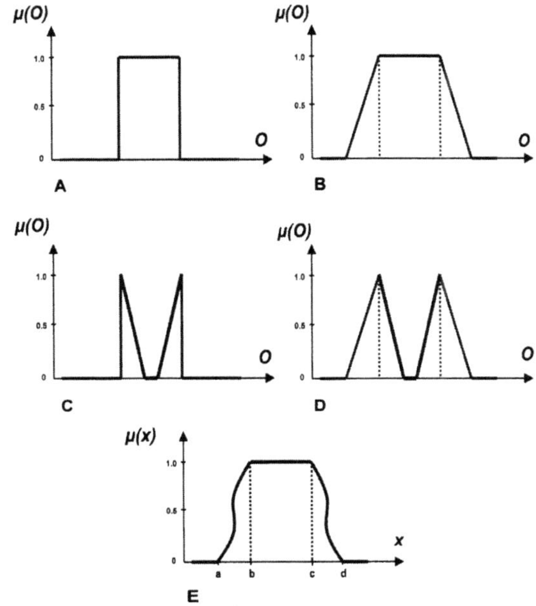


**Figure 1.3.** Different type of simplified linear fuzzy functions describing geographic objects with: a) CcCb; b) CcFb; c) FcCb; d) FcFb; e) generalized gaussian curve

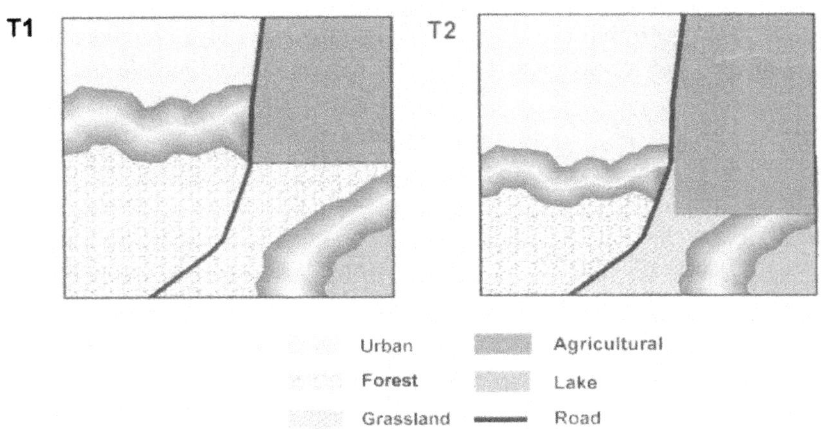


**Figure 1.4.** Snapshots of different land use classes in time T1 and T2 that depict the changes of different geographic objects with composite boundaries

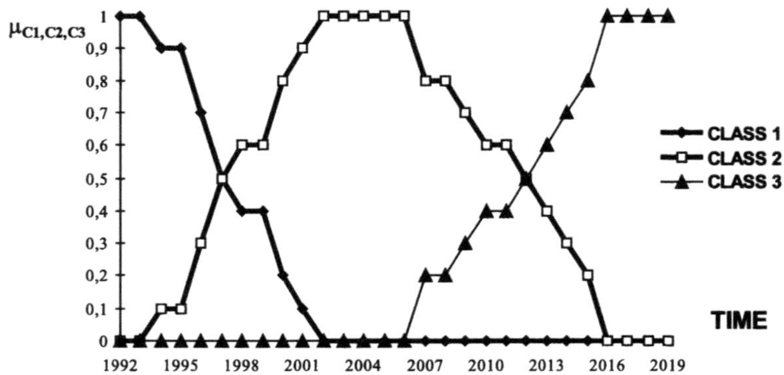


**Figure 1.5.** Temporal fuzzy functions for different land-use class transitions at the same location: from Class 1 to Class 2 in the period 1992-2004, and then from Class 2 to Class 3 in the period 2005-2019

implementation of the temporal dimension in current GIS still represents a challenge [49]. At an operational level, [21] used fuzzy set theory to represent time in a GIS database. A traditional GIS database is used where information is stored in a series of snapshot layers. The known states describe geographic objects and their boundaries in a crisp manner. The information about the change that has occurred in the interval between two consecutive snapshots is not available. Hence, one possible solution is in performing a fuzzy temporal interpolation between known states. Using probable trajectories of gradual progressing from one class to another, the degrees of membership in a specific class at a particular intermediate space-time location are calculated using fuzzy set membership functions. The membership function of the temporal transition $f(t)$ determine the degree of belonging $\mu(x)$ of an element $x$ to a class or object $C$ at a particular time $t$ and can be defined as follows:

$$\mu_C(x,t) = \begin{cases} f(t), & x \in C \\ 0, & x \notin C \end{cases}$$

Further, for an element $x$ in transition between two temporarily adjacent classes or objects, $C_1$ and $C_2$ at time $t_i$, where $i \in [1, N]$, the membership functions can be extended to:

$$\mu_{C_1 \to C_2}(x,t) = \begin{cases} f_{C_1}(x,t_1) = 1 \wedge f_{C_2}(x,t_1) = 0 \\ \quad \text{where } x \in C_1 \wedge x \notin C_2 \\ f_{C_1}(x,t_i) = \mu_{C_1}(x,t_i) \wedge f_{C_2}(x,t_1) = \mu_{C_2}(x,t_i) \\ \quad \text{where } x \in C_1 \to C_2 \\ f_{C_1}(x,t_N) = 0 \wedge f_{C_2}(x,t_N) = 1 \\ \quad \text{where } x \notin C_1 \wedge x \notin C_2 \end{cases}$$

Figure 1.5 illustrate these concepts by using fuzzy membership functions modeling the gradual transition of land use classes from Class type 1 to Class type 2 in the period 1992–2001. Different membership functions based on expert knowledge are developed for different possible class changes, e.g. forest to agricultural, forest to urban, agricultural to road etc, to depict realistic estimation of time needed for a change to take place and possible belonging to a particular class type. As an example, in the year 1995, the GIS database provides the information that the particular geographical object is in transition from Class1 to Class 2, where possible degrees of belonging are 0.7 for Class1 and 0.3 for Class 2. In the year 2001 the geographic object is not anymore belonging to Class 1 but is fully belonging to Class 2.

## 1.5 Modeling past and future of the change process

Representing only spatial or just the temporal dimension in a GIS is not sufficient to analyze or model dynamic and complex geographic phenomena. To model change involves incorporating both dimensions simultaneously. The research efforts of a large number of GIScientists dealing with spatio-temporal issues of GIS databases have been evident in several books and comprehensive reviews on the state of the art in the area [1, 3, 23, 50]. The different approaches developed are: entity-based [35], event-based [14, 48], object-oriented [62], interpolation-based [19, 57, 73], voronoi-based [44].

Incorporating a fuzzy set approach in spatio-temporal modeling of change has been dealt with by [19]. The model of temporal change is linked to the model of spatial distribution of the change. Fuzzy set theory is used for modeling the temporal dimension and standard spatial interpolation technique such as inverse distance weighing method is used for modeling the spatial dimension of the change process.

The simulation of a gradual change between two consecutive snapshot layers is accomplished in three steps. The first step is to determine a number of intermediate layers between two consecutive snapshot layers captured at distinct time instances $T1$ and $T2$. These layers correspond to the shortest possible transition time for a cell to change from one geographic class to another. The second step is to establish the generic layer based on inverse distance weighting method in order to create the information about the change of spatial boundaries between the two explicitly recorded snapshot states. The inverse distance weighting interpolation method estimate values at each cell as a function of their distance from the known cells in the initial snapshot layer by assigning greater weights to the closer cells than those further away. The third step consists of implementing the fuzzy membership functions in order to generate the missing information about the change of geographic entities in the intermediate layers by analysis of the generic and two basic snapshot layers. Fuzzy membership functions are developed for simple transition such as rural to urban land-use classes. A functional module was developed to handle the modeling procedures and fuzzy membership functions in GRASS GIS software [20].

The change process is characterized by its temporal length, start and end in time sometimes unknown, and amplitude of mutations that can be measurable in space. The concept of branching time [28] is suitable given that it permits building scenarios of alternative sequences of events and their consequences that can occur in the past or in the future. Scenarios simulating the different possible outcomes of geographic entity change that happen in the past are developed with a fuzzy spatio-temporal approach: scenario with the shortest duration, longest duration and variable duration of transitions. Further, measures of the dynamics of change were proposed that describe different spatio-temporal speed and mechanisms of change such as diffusive and road-influenced [22]. Diffusive mechanism of change relies on the generic layer that is created from the spatial inverse distance interpolation method. The road-influenced mechanism of change uses the generic layer with buffer zones created around the main roads in the initial layer to simulate the spatial urbanization process influenced by the main transportation network.

Figure 1.6 represents a composite map of different scenarios of rural-urban transition generated with fuzzy spatial-interpolation method for City of Terrebonne, Canada for periods 1956–1966, 1966–1976, and 1976–1986. In the first period the diffusive technique is used, in the second period the road influenced, and in the third a combined technique is used to model the spatial dimension of the process. The combined technique is based on the GIS-based overlay of generic layers depicting the diffusive and road-influenced mechanisms. Each class is characterized in addition with the values of the degree of fuzzy memberships in the particular moments in time. However, these spatial mechanisms used to represent the stages of spatial change of urban land-use transitions are a simplification of a much more complex process. This approach could be enhanced if the modeling of urban land-use change is determined by functions that model spatial change more realistically then spatial interpolation or buffer operations. As example the cell neighborhood could be considered in the class transition rules or the use of a development density function related to population growth [69] would make the model of urban land-use change more realistic.

The advantage of the proposed approach is its capability to develop different models of spatial evolution based on different speed, mechanisms, rate of change, and temporal granularity. Many spatial phenomena also occur as cyclic changes [32]. The presented method mostly uses the linear fuzzy membership functions but however it is possible to expand this concept with cyclic, exponential, gaussian or any other type of curves that is already used to fuzzy spatial dimension.

Furthermore, the fuzzy set approach can model spatio-temporal changes in the future (Figure 1.5). Class 3 appears as a succession of change that happen in the past (period 1992–2001) when Class 1 mutate in Class 2, and then in the future Class 2 will mutate in Class 3 (period 2006–2016). [38] explore

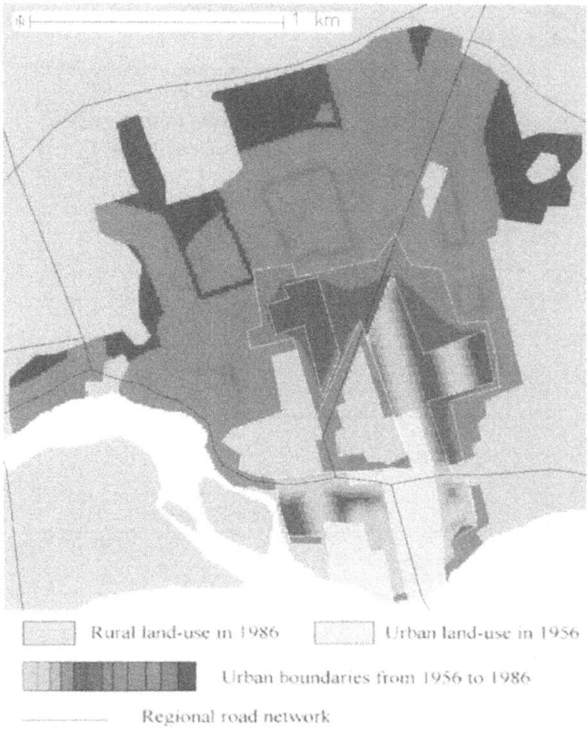


**Figure 1.6.** Composite map of different modeling scenarios for periods of 1956–1966, 1966–1976, and 1976–1986 of rural-urban change in City of Terrebonne

the impact of the variation of temporal dynamics of the time dimension in transition-based and process-based models of land-use change. Variation of temporal dynamics strongly affects the spatial dimension of the model output. When modeling the past events, the known states of the geographical objects at particular instants in time (years 1992 and 2001 on the Figure 1.5) are taken into consideration. When modeling the future, only the initial spatial state and moment in time is known (year 2006 on the Figure 1.5) and then the ending state can vary.

The transitions between two known states are generated from the initial conditions in the spatial dimension that is determined in a crisp manner in a GIS database. The fuzzy functions developed use geographic objects that are represented with crisp contents and crisp boundaries (CcCb). However many possible situations can arise when different types of geographic objects start their mutation over time. Table 1.1 provides the possible outcomes of transitions of four types of geographical objects. Development of fuzzy membership functions that depict the change from fuzzy state toward crisp or fuzzy represents a challenge.

**Table 1.1.** Possible transitions of different type of geographic objects and boundaries over two position in time

| Type of geographic objects at time $T_1$ | Possible type of transitions of geographic objects at time $T_2$ |
|---|---|
| $C_C C_B$ | $C_C C_B$<br>$C_C F_B$<br>$F_C F_B$<br>$F_C C_B$ |
| $F_C C_B$ | $C_C C_B$<br>$C_C F_B$<br>$F_C F_B$<br>$F_C C_B$ |
| $C_C F_B$ | $C_C C_B$<br>$C_C F_B$<br>$F_C F_B$<br>$F_C C_B$ |
| $F_C F_B$ | $C_C C_B$<br>$C_C F_B$<br>$F_C F_B$<br>$F_C C_B$ |
| $C_C$ = Crisp Content<br>$C_B$ = Crisp Boundary | $F_C$ = Fuzzy Content<br>$F_B$ = Fuzzy Boundary |

In reality geographic entities can be represented with objects that contains patches (islands of different class types) or are composite of different type of boundaries that can be partially fuzzy and partially crisp. As an example (Figure 1.4b) forest land-use is represented as object with fuzzy content and fuzzy boundary but only part of the boundary that is towards urban land use. Forest boundary with road is crisp. Boundaries within agricultural and road land-use or urban and road land-use are fully crisp because their location is measurable in space and time. Spatio-temporal modeling in GIS databases of complex objects with composite types of boundaries is another challenge.

The proposed fuzzy spatio-temporal approach can be extended for inclusion of finite neighborhood effects for the consideration of different adjacent land use types and other factors that may influence the change process. In such case land-use change can be modeled with fuzzy controlled cellular automata. [67] used fuzzy logic to capture the feature of land conversion behavior to control the cellular automata process. [31] propose fuzzy urban sets, and further [39] modeled urban growth with fuzzy membership function influenced by urban population density but constrained with crisp topographic slope map.

## 1.6 Conclusion

Fuzzy set theory plays an important role in the spatial and temporal representation of geographic data and objects in a GIS framework. This chapter provides some insights on how fuzzy set theory is used to handle the inherent fuzzy nature of geographic objects. Moreover, the theory is implemented for fuzzy spatio-temporal modeling of the dynamic geographic phenomena that occurred in the past. It can be used for the prediction of changes and can be integrated with for example cellular automata for more realistic modeling procedures.

The use of alternative sets such as rough sets is another related approach that can treat vagueness, fuzziness and uncertainty in geographic information remains a promising avenue of research in GIScience [26, 46]. Rough membership function is a special type of fuzzy function [68], and the exploration of its spatial context in GIS databases is already evident in the literature [2, 5].

*Acknowledgements.* This study is supported by the Natural Sciences and Engineering Research Council (NSERC) of Canada under grant 250518-02. The anonymous referee comments on an earlier draft of the paper is gratefully acknowledged.

## References

1. Abraham T., Roddick J. (1999) Survey of spatio-temporal databases. GeoInformatica 3:61–69.
2. Ahlqvist O., Keukelaar J., Oukbir K. (2000) Rough classification and accuracy assessment. International Journal of Geographical Information Science 14(5):475–496.
3. Al-Taha K.K., Snodgrass R., Soo M.D. (1994) Bibliography on spatiotemporal databases. International Journal of Geographical Information Systems 8(1):95–103.
4. Anile M.A.P., Furno P., Gallo G., Massolo A. (2003) A fuzzy approach to visibility maps creation over digital terrains. Fuzzy Sets and Systems 135:63–80.
5. Bittner T., Stell J. (2002) Vagueness in rough location. GeoInformatica, 6:99–121.
6. Blaut J.M. (1961) Space and Process. The Professional Geographer 13(4):1–7.
7. Burrough P.A. (1989) Fuzzy mathematical methods for soil survey and land evaluation. Journal of Soil Science 40:477–492.
8. Burrough P.A., MacMillan R.A., Deursen W.V. (1992) Fuzzy classification methods for determining land suitability from soil profile observations and topography. Journal of Soil Science 43:193–210.
9. Burrough P.A. (1996) Natural objects with indeterminate boundaries. In: Burrough P.A., Frank A. (Eds.) Geographic objects with indeterminate boundaries, Taylor and Francis, London, pp. 3–28.

10. Burrough P.A., Frank A. (1996) Geographic Objects with Intermediate Boundaries. Innovations in GIS 2. Taylor and Francis, London.
11. Cheng T., Molenaar M., Lin H. (2001) Formalizing fuzzy objects from uncertain classification results. International Journal of Geographical Information Sciences 15(1):27–42.
12. Cheng T. (2002) Fuzzy objects: their changes and uncertainties. Photogrammetric Engineering and Remote Sensing 98(1):41–49.
13. Chrisman N.R. (1998) Beyond the snapshot: changing the approach to change, error and process. In: Egenhofer M.J., Golledge R.G. (Eds.) Spatial and Temporal Reasoning in Geographical Information Systems. Oxford University Press, New York, pp. 85–93.
14. Claramunt C., Theriault M. (1995). Managing time in GIS: an event-oriented approach. In: Recent Advances in Temporal Databases: Proceedings of the International Workshop on Temporal Databases. Zurich, Switzerland.
15. Claramunt C., Theriault M. (1997). A qualitative representation of evolving spatial entities in two-dimensional spaces. In: Carver S. (Ed.) Innovations in GIS V. Taylor and Francis, London, pp. 119–129.
16. Couclelis H. (1999) Space, time, geography. In: Longley P., Goodchild M., Maguire D., Rhind D. (Eds.) Geographical information systems: principles and technical issues. John Wiley and sons, New York, pp. 29–38.
17. Couclelis H. (1999) Towards an operational typology of geographic entities with ill-defined boundaries. In: Burrough P.A., Frank A. (Eds.) Geographic objects with indeterminate boundaries. Taylor and Francis, London, pp. 45–55.
18. Davidson D.A., Theocharopoulos S.P., Bloksma R.J. (1994) A land evaluation project in Greece using GIS and based on Boolean and fuzzy set methodologies. International Journal of Geographical Information Systems 8(4):369–384.
19. Dragićević S., Marceau D.J. (1999) Spatio-temporal interpolation and fuzzy logic for GIS simulation of rural-to-urban transition. Cartography and Geographical Information Science 26(2):125–137.
20. Dragićević S., Marceau D.J. (2000) An application of fuzzy logic reasoning for GIS temporal modeling of dynamic processes. Fuzzy Sets and Systems 113:69–80.
21. Dragićević S., Marceau D.J. (2000) A fuzzy set approach for modelling time in GIS. International Journal of Geographical Information Science 14(3):225–245.
22. Dragićević S., Marceau D.J., Marois C. (2001) Space, time, and dynamics modeling in historical GIS databases: a fuzzy logic approach. Environment and Planning B: Planning and Design 28:545–562.
23. Egenhofer M., Golledge R.G. (1998) Spatial and temporal reasoning in geographic information systems. Oxford University Press, Oxford.
24. Fisher P. (1996) Boolean and fuzzy regions. In: Burrough P.A., Frank A. (Eds.) Geographic objects with indeterminate boundaries. Taylor and Francis, London, pp. 87–94.
25. Fisher P. (1997) Concepts and paradigms of spatial data. In: Craglia M., Couclelis H. (Eds.) Geographic Information Research: Bridging the Atlantic. Taylor & Francis, London, pp. 297–307.
26. Fisher P. (2001) Alternative set theories for uncertainty in spatial information. In: Hunsaker C., Goodchild M., Friedl M., Case T. (Eds.) Spatial Uncertainty in Ecology. Springer, New York, pp. 351–362.

27. Foody G.M. (1996) Approaches for the production and evaluation of fuzzy land cover classifications from remotely-sensed data. International Journal of Remote Sensing 17(7):1317–1340.
28. Frank A.U. (1998) Different types of 'times' in GIS. In: Egenhofer M.J., Golledge R.G. (Eds.) Spatial and Temporal Reasoning in Geographical Information Systems. Oxford University Press, New York, pp. 40–62.
29. Hagerstrand T. (1970) What about people in regional science? Papers of the Regional Science Association 14(7):7–21.
30. Hall G.B., Wang F., Subaryono (1992) Comparison of Boolean and fuzzy classification methods in land suitability analysis using geographical information systems. Environment and Planning A 24:497–516.
31. Heikkila E.J., Shen T.-Y., Yang K.-Z. (2003) Fuzzy urban sets: theory and application to desakota regions in China. Environment and Planning B: Planning and Design 30:239–254.
32. Hornsby K., Egenhofer M., Hayes P. (1999) Modeling cyclic change. In: Chen P., Embley D., Koulondjan J., Liddle S., Reddinck J. (Eds.) Advances in Conceptual Modelling. Springer-Verlag Lecture Notes in Computer Science, Berlin, pp. 98–109.
33. Kelmelis J.A. (1998) Process dynamics, temporal extent, and casual propagation as the basis for linking space and time. In: Egenhofer M.J., Golledge R.G. (Eds.) Spatial and Temporal Reasoning in Geographical Information Systems. Oxford University Press, New York, pp. 94–103.
34. Langran G. (1992) Time in geographic information systems. Taylor and Francis, London.
35. Langran G. (1993) Issues of implementing a spatiotemporal system. International Journal of Geographical Information Systems 7(4):305–314.
36. Langran G., Chrisman R. (1988) A framework for temporal geographic information. Cartographica 25(3):1–14.
37. Leung Y. (1987) On the imprecision of boundaries. Geographical Analysis 19:125–151.
38. Liu X.H., Andersson C. (2004) Assessing the impact of temporal dynamics on land-use change modeling. Computers, Environment and Urban Systems 18(1–2):107–124.
39. Liu Y., Phinn R. (2003) Modelling urban development with cellular automata incorporating fuzzy-set approaches. Computers, Environment and Urban Systems 27(6):637–658.
40. Lodwick W.A., Santos J. (2003) Constructing consistent fuzzy surfaces from fuzzy data. Fuzzy Sets and Systems 135:259–277.
41. Marceau D.J., Guindon L., Bruel M., Marois C. (2001) Building temporal topology in a GIS database to study the land-use changes in a rural-urban environment. Professional Geographer 53(4):546–558.
42. McBratney A., Moore A. (1985) Application of fuzzy sets to climatic classification. Agricultural and Forest Meteorology 35:165–185.
43. McBratney A., Odeh I. (1997) Application of fuzzy sets in soil sciences: Fuzzy logic, fuzzy measurements, and fuzzy decision. Geoderma 77:85–113.
44. Pang M.Y.C., Shi W. (2002) Development of a process-based model for dynamic interaction in spatio-temporal GIS. Geoinformatica 6(4):323–344.
45. Parkes D., Thrift N. (1980). Times, spaces, and places. John WIley and Sons, New York.

46. Pawlak Z. (1982). Rough sets. International Journal of Computer and Information Sciences 11:241–356.
47. Peuquet D.J. (1994) Its about time: a conceptual framework for the representation of temporal dynamics in geographical information systems. Annals of the Association of American Geographers 84(3):441–461.
48. Peuquet D.J., Duan N. (1995) An event-based spatiotemporal data model (ESTDM) for temporal analysis of geographical data. International Journal of Geographical Information Systems 9(1):7–24.
49. Peuquet D.J. (2001) Making space for time: issues in space-time data representation. Geoinformatica 5(1):11–32.
50. Peuquet D.J. (2002) Representations of Space and Time. The Guilford Press, New York.
51. Pfoser D., Tryfona N. (1999) Capturing fuzziness and uncertainty of spatiotemporal objects. In: Proc. of the 6th International Symposium on the Advances in Spatial Databases, Hong Kong, pp. 111–132.
52. Robinson V.B. (1988) Some implications of fuzzy set theory applied to geographical databases. Computers, Environment and Urban Systems 12:89–97.
53. Robinson V.B. (2003) A perspective on the fundamentals of fuzzy sets and their use in geographical information systems. Transitions in GIS 7(1):3–30.
54. Roddick J., Patrick J. (1992) Temporal semantics in geographic information systems - a survey. Information Systems 17:249–267.
55. Santos J., Lodwick W.A., Neumaier A. (2002) A new approach to incorporate uncertainty in terrain modeling. In: Egenhofer M.J., Mark D.M. (Eds.) Geographic Information Science: Second International Conference, GIScience 2002, Boulder, Colorado, USA, September 2002. Springer, Germany, pp. 291–299.
56. Schneider M. (1999) Uncertainty management for spatial data in databases: fuzzy spatial data types. In: Guting R.H., Papadias D., Lochovsky F. (Eds.) SSD'99. Springer-Verlag, Germany, pp. 330–351.
57. Shibasaki R., Huang S. (1996) Spatio-temporal interpolation by integrating observation data and a behavioural model. In: Kraak M., Molenaar M. (Eds.) SDH'96, Advances in GIS Research II. Taylor and Francis, Delft, The Netherlands, pp. 655–666.
58. Snodgrass R.T. (1992) Temporal databases. In: Frank A.U., Campari J., Fozmentini U. (Eds.) GIS – From Space to Territory: Theories and Methods of Spatio-Temporal Reasoning. Theory and Methods of Spatio-Temporal Reasoning in Geographical Space, Pisa, Italy, pp. 22–64.
59. Stead S.D. (1998) Temporal dynamics and geographical information systems. In: Egenhofer M.J., Golledge R.G. (Eds.) Spatial and Temporal Reasoning in GIS. Oxford University Press, New York, pp. 214–219.
60. Thrift N. (1977) An introduction to time geography. Concepts and Techniques in Modern Geography 3:3–36.
61. Vasilev I.R. (1997) Mapping Time. Cartographica 34(2):1–51.
62. Wachowicz M. (1999) Object-oriented design for temporal GIS. Taylor and Francis, London.
63. Wachowicz M., Healey R.G. (1994) Towards temporality in GIS. In: Worboys M.F. (Ed.) Innovations in GIS. Taylor and Francis, UK, pp. 105–115.
64. Wang F. (1990) Improving remote sensing image analysis through fuzzy information representation. Photogrammetric Engineering and Remote Sensing 56(8):1163–1169.

65. Wang F. and Hall G.B. (1996) Fuzzy representation of geographical boundaries in GIS. International Journal of Geographical Information Systems 10(5):573–590.
66. Worboys M. (1994) A unified model for spatial and temporal information. The Computer Journal 37:26–33.
67. Wu F. (1998) Simulating urban encroachment on rural land with fuzzy-logic-controlled cellular automata in a geographical information system. Journal of Environmental Management 53:293–308.
68. Yao Y. (1998) A comparative study of fuzzy sets and rough sets. Information Sciences 109:227–241.
69. Yeh A.G.-O., Li X. (2002) A cellular automata model to simulate development density for urban planning. Environment and Planning B: Planning and Design 29:431–450.
70. Zadeh L.A. (1965) Fuzzy Sets. Information and Control 8:338–353.
71. Zhang J., Foody G.M. (1998) A fuzzy classification method of sub-urban land cover from remotely sensed imagery. International Journal of Remote Sensing 19:21-38.
72. Zhang J., Stuart N. (2001) Fuzzy methods for categorical mapping with image-based land cover data. International Journal of Geographical Information Science 15(2):175–195.
73. Zhang W., Hunter G. (2000) Temporal interpolation of spatially dynamic objects. GeoInformatica 4(4):403-418.

# 2

# Toward Spatio-Temporal Patterns

Martin Erwig

School of Electrical Engineering and Computer Science
Oregon State University
Corvallis, Oregon 97331, U.S.A.
erwig@cs.orst.edu

## 2.1 Introduction

Our world is changing at an increasing pace causing the validity intervals of information to shrink. Therefore, the *change* of information becomes important information itself—an important information resource that should be exploitable by query languages. In particular, information about the position and extent of objects is of high interest in many areas of science and elsewhere (vehicle tracking, history, security, health care, ecology, weather prediction, urban planning, and so on). In all these areas relevant information has to be extracted from huge data sets. Beyond data mining that often serves the purpose of identifying interesting objects in spatial or spatiotemporal data sets, queries are of interest that find out about relationships between objects once they have been identified. In the realm of spatio*temporal* databases this means to find out, in particular, about the *changes* that objects and their relationships undergo. Moreover, many applications require not only the detection of one single change, but rather look for sequences of changes describing particular *developments*.

Existing spatiotemporal database systems and query languages offer only basic support to query changes of data. Most of these systems allow the formulation of queries that ask for changes at particular time points. However, it is often very difficult, if not impossible, to express queries for sequences of such changes. In other words, existing query languages do not offer a systematic, scalable concept to query developments of objects and relationships; instead they require the user to encode these by a number of individual conditions. This method is awkward, in particular, because it also requires the formulation of additional side conditions. Moreover, this approach does not work in queries for arbitrary numbers of changes.

A concept of *spatiotemporal patterns* and their integration into spatiotemporal query languages can help to close this gap. Spatiotemporal patterns enable the formulation of queries about complex object developments; their integration into query languages allows the formulation of queries that are currently not possible. Moreover, spatiotemporal patterns can also simplify the formulation of queries that are possible with existing languages.

Beyond the integration of spatiotemporal patterns into a textual query language we will also discuss the design of a visual query language and a corresponding user interface to support the formulation of spatiotemporal patterns and queries. This is of particular importance because most users of spatiotemporal data (for example, scientists) do not have a formal computer education and do not know how to use a query language, in particular, query languages for complex data like spatiotemporal data. Moreover, the fast growing set of available spatial and spatiotemporal data and its wide dissemination through the Internet increases the class of possible users for these data. Offering ordinary users access to spatiotemporal data is therefore becoming a more and more important issue that should be addressed by developing a visual query language and a corresponding user interface.

In the remainder of this chapter we first motivate the need for spatiotemporal patterns in Section 2.2. We discuss related work on spatiotemporal data models and query languages in Section 2.3. We review the notion of spatiotemporal objects and their integration into databases in Section 2.4. Spatiotemporal predicates are described in Section 2.5. Spatiotemporal patterns will be discussed in Section 2.6, and the design of a visual query language is sketched in Section 2.7. Finally, we present some conclusions in Section 2.8.

## 2.2 The Need for Spatiotemporal Patterns

Queries about changing objects are becoming more and more important. One reason might be that the world is changing at a constantly increasing pace so that information about a situation at a particular instant in time has a decreasing half-life. In other words, the intervals for which certain pieces of information remain valid become shorter and shorter. Therefore, the *change* of information becomes important information itself—an important information resource that should be exploitable by query languages. In the area of temporal databases this issue has been addressed for quite some time now [4, 60, 63].

### 2.2.1 Spatiotemporal Data

A particularly important kind of information change is that of objects' locations and/or spatial extensions. Many phenomena in all areas of natural sciences involve the change of spatial information. A few example are given below.

- Meteorology: all kinds of weather data, moving storms, tornados, developments of high pressure areas, movement of precipitation areas, changes in freezing level, droughts, El Niño effects, ...

- Biology: animal movements, mating behavior, species relocation and extinction, ...
- Crop sciences: seasonal grasshopper infestation, harvesting, soil quality changes, land usage management, ...
- Forestry: forest growth, forest fires, hydrology patterns, canopy development, planning tree cutting, planning tree planting, ...
- Medicine: patients' cancer developments, supervising developments in embryology, ...
- Geophysics: earthquake histories, volcanic activities and prediction, ...

A characteristic feature of most of these examples is that changes are of continuous nature. Moreover, many human-related activities are related to changes in spatial information as well:

- People: movements of terrorists/criminals/spies, movement of people in emergency situations, pedestrian patterns/habits, ...
- Cars/trucks/taxis: tracking, rerouting, fleet management, ...
- Urban planning: parcel management, development of social areas, urban economics, tourism planning, bus routes, ...
- Crime/disaster prevention: risk area analyses, resource allocations (police, health care, fire stations), ...
- History: country expansions, reunifications, tribe movements, ...
- Military: missile tracking, troop movements, ...
- Planes/ships: routes, detours, ...
- Ecology: causal relationships in environmental changes, tracking down pollution incidents, ...

In particular, since mobility is constantly increasing, we can observe an increase in the time dependency of spatial information related to human activities. Many of the second group of examples represent continuous changes, some are discrete in nature.

In addition to the shown individual areas, combinations of phenomena are also of interest, for example, what changes in forests can be linked to which kind of animal behavior, which weather developments are responsible for grasshopper infestation, etc. Moreover, some combinations pose particular planning challenges, for example, extreme weather events require rerouting of cars, planes, and ships.

### 2.2.2 Spatiotemporal Patterns

The shown examples reveal one important characteristic of change in spatial data that makes it interesting for data processing by scientists and others: interesting phenomena are those that are not random but rather follow certain rules. In other words, applications are interested in spatial data changes that exhibit a certain regular structure.

Regular structures in space and time, in particular, repeating structures, are often called *patterns*. Patterns that describe changes in space and time are referred to as *spatiotemporal patterns* [47].

Not surprisingly, the notion of spatiotemporal patterns occurs in many different areas of nature [51] and sciences [17], in particular, in geosciences [48] and geophysics [57], meteorology [45, 68], ecology [73] and environmental studies [70], and even embryology [65]. A collection of applications in ecology and related sciences and the challenges they pose for computer science has been gathered during a recent NSF workshop [44]. An area of growing importance is that of crime detection and prevention, in which spatiotemporal patterns also play an important role, for example, in crime-pattern analysis with the help of GIS [72, 14, 53], tracking criminals [52], and crime prevention [37].

To make the following discussion concrete, we use a couple of example queries that could arise in different application areas. Here we will comment on the requirements that a data model and query language must have to support these applications. In Section 2.3 we will discuss how most of the existing approaches fall short of supporting these requirements in some way or another. The first example relates to the environment.

> **Q1.** Find all satellites that were able to observe an oil spill at least four times for at least 5 minutes each time.

This innocent looking query demands a lot from a data model and query language. For example, in addition to moving points the data model must also support evolving *regions*. Moreover, the movement of the satellite and the oil spill are *continuous*.

Next we consider some examples of spatiotemporal queries that can be relevant for crime prevention. Suppose we want to find out suspicious movements of, say cars. With regard to security of the public mail system we might want to pose the following query.

> **Q2.** Find all cars that stopped at two or more mailboxes in a short period of time.

Why would anyone *not* put all their mail into the first mailbox? Therefore, the above query gives hints for suspicious behavior. To answer this query a language must be able to express *repetitions* of *patterns* of *changes*, for example, in speed or topological relationships (here: meeting a mailbox). Moreover, the language must be able to express *time conditions* on the duration of certain predicates. Another example is to identify suspicious behavior near sites of high public interest (like Disneyland):

> **Q3.** Find cars that surrounded an area of interest multiple times or at a significantly less than average speed.

In addition to the need for a repeated pattern, this query requires the possibility to express a constraint on an object's movement by another object's spatial attribute (here, boundary).

A spatiotemporal database system should be able to answer queries like these. However, as we will discuss next, most existing systems and language proposals are rather limited in their support for querying spatiotemporal patterns and cannot express queries as the ones shown.

### 2.2.3 Limitations of Existing Approaches

One goal of spatiotemporal databases research is to enable the intelligent use of the collected data. In addition to the challenges of efficient storage and indexing spatiotemporal data [49, 1], a main problem is that of extracting relevant and useful information from the data. Two complementing areas of database research address two different aspects of this problem.

(1) *Data mining* [16], which is also sometimes referred to as *knowledge discovery* [32], is concerned with the discovery of patterns in large data sets. Data mining often uses statistical analyses and methods to extract interesting constellations/arrangements of points in a usually very huge data space. In particular, spatiotemporal data mining [69, 6, 50] has the goal of identifying interesting structures and patterns in spatiotemporal data sets. In many cases, spatiotemporal data mining is used to identify spatiotemporal objects. Once found, these objects of interest can the be stored in a database where they can be processed further. In GIS systems the so-called *field view* [55, 36] of spatial and spatiotemporal data is used. Data mining techniques can be used to move from this low-level field view of raw data to the higher-level view of objects. This conception was prevailing at a recent NSF workshop on spatiotemporal data models for scientific applications [44].

(2) *Query Languages* work on data that is well structured and that is stored in rather fixed formats obeying schema definitions and possibly additional integrity constraints. The distinctive feature of query languages (compared to data mining) is the assumption that the knowledge in a database is already present, whereas the goal of data mining is to discover it. Query languages are used to find out further relationships among objects.

Although considerable research has been performed in recent years on data models and query languages, most of them are not capable of expressing the example queries shown above.

What are the reasons for this situation? We believe that there are two major problems of data models and query languages that are responsible for the gap between research and applications: (1) Lack of expressiveness and (2) high complexity, which has a particularly bad impact on the usability.

**Problem 1: Lack of Expressiveness.** There have been quite a few proposals for spatiotemporal query languages. However, most languages are able to express only a small fraction of the queries that are possible on spatiotemporal data and that are needed to solve application problems like the ones indicated above. For example, data models that are only concerned with moving objects, such as [62, 71], cannot express queries that involve changing regions, such as Q1. Models that are restricted to discretely changing objects, such as [10, 74, 54, 9, 41, 18], cannot express queries about continuous objects; again, query Q1 relies on exploiting the continuous nature of the involved objects. None of the existing data models can express queries about changing relationships of objects in a satisfactory way, as it is required for query Q2. In particular, queries like Q3 that ask for a repetition of certain patterns are either impossible to express or lead to incredibly complex ad hoc solutions. Although constraint databases promise a declarative and expressive approach to spatiotemporal databases [11], actual approaches to spatiotemporal query languages have been described only in [40, 41, 12]. The DEDALE system [40, 41] can due to its embedding of constraint objects into relations only deal with discretely changing spatiotemporal objects. The proposal of Chomicki [12] is based on constraints in the $n + 1$-dimensional space ($n$ space dimension plus one time dimension) and corresponds more closely to the field view, so that it is not really clear whether applications involving many attributes (spatial and others) can be modeled well. In any case, the constraint approach like all other proposals lacks abstractions to express patterns of changes.

A possible route for a solution to the last problem is given by the concept of *developments* [24] and *spatiotemporal predicates* [27], described in Section 2.5, which form the basis for spatiotemporal patterns, discussed in Section 2.6.

**Problem 2: Complexity and Lack of Usability.** A data model and query language for spatiotemporal data must be necessarily more complex than, for example, SQL, because SQL is a language for querying flat relations over atomic data, whereas spatiotemporal databases have to deal with spatial data (point, lines, and regions) and their possible changes. It seems that the goal of simplicity contradicts the goal of expressiveness because one might think in order to express more advanced queries one needs more and complex operations in the query language. Indeed, extending, for example, relational algebra by spatial data types, increases the number of operations significantly because the added data types require their own operations [42]. This is even more the case when considering more complex data types for spatiotemporal objects [22]. Apparently, the constraint approach does not suffer from an inflation of operations. However, it is not immune to an increase in complexity because the domains over which constraints are formulated become more complex [31]. Although the vocabulary of constraint languages remains very much unaffected, queries grow complex, too, because without a mechanism to express patterns of changes, the individual changes have to be spelled out in detail, which leads to an inflation of conditions comparable to the inflation of joins found in relational approaches.

The problem of usability is of growing importance because the fast growing number of data sources and the raising availability through the Internet causes the target group of possible users also to grow. For many of these possible users languages like SQL are already too complex since most of them do not have formal computer science training at all; the requirement to learn a formal language with keywords, binding rules, etc. is in many cases too high a barrier.

Visual query languages and visual query interfaces to spatiotemporal databases can help solving this problem. Visual query languages are particularly promising for spatial databases because spatial data has a direct and intuitive visual representation that can be exploited for formulating queries in a more direct way. As we will discuss in Section 2.7, we can also obtain a representation of spatiotemporal data that is effective for expressing spatiotemporal queries in a simple and direct way.

## 2.3 Related Work

Some early models for spatiotemporal data have been proposed starting in the mid-90s. In [74] a spatial data model has been generalized to become spatiotemporal. Spatiotemporal objects are defined as so-called *spatio-bitemporal complexes* whose spatial features are described by simplicial complexes and whose temporal features are given by bitemporal elements attached to all components of simplicial complexes. On the other hand, temporal data models have been generalized to become spatiotemporal and include variants of Gadia's temporal model [34] which can be found in [10, 7]. In [66] a discrete snapshot model is described. The main drawback of all these approaches is that they are incapable of modeling *continuous* changes of spatial objects over time. An extension of the snapshot model toward capturing continuous changes has been described by Yeh and Cambray [75, 76] who introduce the notion of *behavioral time sequences.* In this approach, which is based on the work described in [59], each element of such a sequence contains a geometric value, a date, and a behavioral function, which describes the evolution up to the next element of the sequence.

The constraint-based approach to modeling spatiotemporal data considers spatiotemporal objects as point sets in a multi-dimensional space [39]. The description of objects is given by logical formulas (constraints). Moreover, queries are also expressed by logical formulas. An embedding of constraints into a relations is performed in [40] where also a SQL-like query language is proposed. In other work [11] spatiotemporal objects are represented by spatial objects plus affine geometric transformations to describe their evolution. Although the logical/constraint-based approach is very general, it is not clear, for example, whether it can be efficiently implemented. Other possible problems are the handling of objects that require non-linear constraints and the treatment of metric operations (for example, computing distances).

A logic-based approach to model predicates on (spatio)temporal data is offered through temporal logic. In particular, Allen's work [3] is centered around the definition of a predicate $Holds(p, i)$ that asserts that a property $p$ is true during a time interval $i$. Combinations of predicates can be constructed through logical predicates and quantifiers. The work is based on his earlier work [2] where he has identified thirteen possible relationships between intervals, such as *before*, *equal*, *meets*, *overlaps*, and *during*. Allen's temporal logic is solely based on time intervals and does not include time points. Galton [35] has extended Allen's approach to the treatment of temporally changing topological relationships. Topological relationships are based on the RCC model [15, 56]. In contrast to Allen, Galton also takes time points into account. In specifying changes of spatial situations he uses the notion of a *fluent* or *state*, which can have different values at different times. Whereas Galton takes into account time points, he does not consider geometric points. We believe that it is important to have spatial as well as time points available in a data model. First, spatial points are needed in many applications. Second, having also spatial points provides some form of symmetry to the existence of points in the time domain. Third, spatiotemporal values can be projected to spatial values by function application at time points. This feature is absolutely essential and much needed in spatiotemporal query languages.

Our approach is based on (1) the notion of *temporal objects* [30] and (2) the concept of *spatiotemporal data types* [21, 22, 43].

The notion of a *temporal object* is based on the observation that anything that changes over time can be expressed as a function over time. A temporal version of an object of type $\alpha$ is then given by a function from *time* to $\alpha$. Spatiotemporal objects like *moving points* and *evolving regions* are regarded as special instances of temporal objects where $\alpha$ is a spatial data type like *point* or *region*. Spatiotemporal objects are encapsulated in data types that offer tailor-made operations for them, which facilitates their integration as complex values into databases [64]. The ADT approach has several advantages. First, employing ADTs is more expressive than relying on attribute-timestamped models, let alone tuple-timestamped models, since continuous changes can be modeled [22]. Second, a definition of ADT values is valid regardless of a particular DBMS data model and query language. The reason is that ADT values are *not* modeled by concepts of a DBMS data model and that they therefore do not depend on them. This facilitates an easy and elegant conceptual integration of ADT values into relational, complex-object, object-oriented, or other data models and query languages.

ADT operations can be used within queries to construct new, derived spatiotemporal objects. In particular, boolean-valued operations can be used as predicates to formulate conditions in queries. However, like for all other existing query language approaches, this approach is basically suited to express queries about changes at single time points, so that a combination of conditions is required to express multiple changes. What makes this worse is the fact that, in general, additional side conditions are needed in such queries to express temporal ordering of the formulated change events.

Our work on spatiotemporal predicates [27] enables more general queries about developments [28, 24]. The idea is to specify and query changes in topological spatial relationships. Using point-set topology as a formal framework, the 9-intersection model [19] provides a canonical collection of topological predicates for each combination of spatial types. Based on the nine topologically invariant intersections of boundaries, interiors, and exteriors of the two participating objects, these predicates are mutually exclusive and cover each possible topological situation between two objects. The relevance of spatiotemporal predicates is demonstrated in [24] where we have identified an important new class of spatiotemporal queries, which is concerned with developments of spatial objects over time, which means queries ask especially about changes in spatial relationships over time. A macro mechanism is provided which allows the user to build more and more complex spatiotemporal predicates starting with a small set of elementary ones. We have demonstrated how these concepts can be realized in a relational model and how SQL can be appropriately extended to a spatiotemporal query language called *STQL* to enable the querying of developments [28].

Our work on spatiotemporal predicates has also inspired other researchers to consider the idea of spatiotemporal patterns. In [18] an approach is sketched where the developments of objects can be queried using a regular expression over one predicate. Although the approach is interesting, the expressiveness is rather limited. First, only discretely changing objects are considered. Second, since regular expressions can be formed only over one predicate at a time, queries that ask for a development described by two different predicates (such as, *Inside meet Disjoint*, see below), *cannot* be expressed. This means that the described approach is strictly less expressive than spatiotemporal predicates as defined in [27]. Finally, the approach does not support any of the additional expressiveness, such as time constraints on the duration of predicates, that is possible in the approach presented here. On the other hand the work of [18] and the described application provide a strong evidence that spatiotemporal predicates form a solid basis for the development of spatiotemporal patterns.

We have addressed the problem of user-friendly specifications of temporally changing topological situations and predicates in [25]. For this purpose, we have defined a *visual language* which extends existing concepts for visual spatial query languages, which are only capable of querying *static* topological situations. In [26] the design of the visual language is motivated in detail and a user interface is defined. The visual notation is generalized to deal with undefined parts of object developments in [29]. That paper also defines the translation into our formal model of spatiotemporal predicates. The visual notation can be used directly as a visual query interface to spatiotemporal databases, or it can provide predicate specifications that can be integrated into textual query languages leading to heterogeneous languages [23].

## 2.4 Spatiotemporal Objects and Databases

Spatial data types like *point*, *line*, and *region* have turned out to be a fundamental abstraction for modeling the two-dimensional structure of geometric entities, their properties, relationships, and operations [42, 58]. Points are elements of the Euclidean plane. Lines are two-dimensional curves. Regions describe point sets with a two-dimensional extent and are bounded by lines which in this context are called *boundaries*. To avoid anomalies and ensure well-definedness of operations formal definitions of spatial data types are often based on point-set topology [33] and on regularization functions [67].

Spatiotemporal objects are obtained by considering the changes of spatial objects over time. They can be conveniently described by *temporal functions* [30, 21], which are simply functions from time into the corresponding spatial domain. For example, the movement of a satellite (projected onto the earth's surface) can be regarded as a function $Sat : time \rightarrow point$, and the development of an oil spill can be regarded as a function $Spill : time \rightarrow region$ [20]; see Figure 2.1. We can apply functions like *Sat* or *Spill* to time values and obtain the position of the satellite or the extent of the oil spill at a particular time as spatial data.

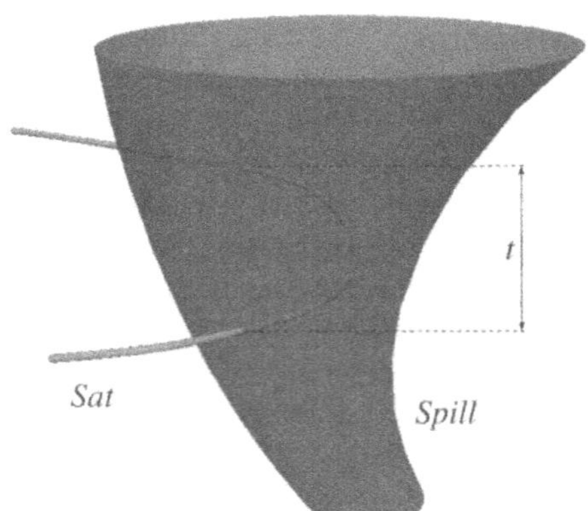


**Figure 2.1.** Satellite movement and oil spill development.

Of course, *spatial* temporal data is just a particular example of temporal data in general. For example, we can also deal with temporally changing numbers through a type like $time \rightarrow num$. Such a temporal number could give the size of the oil spill depending on the time. Using the notational convention that temporal versions of data types have their letter capitalized, we have, for example, $Sat : Point$ and $Spill : Region$. We can define functions on spatiotemporal objects, for example, a function *Area* that computes the time-dependent area of an evolving region:

$$Area : Region \rightarrow Num$$

If we apply *Area* to *Spill*, we obtain a time-dependent number reflecting the temporal development of the size of the oil spill. The design of these spatiotemporal data types has been described in [22]. A categorization and formal definition of operations has been performed in [43].

By encapsulating (spatio)temporal data in data types like *Point*, *Region*, and *Num* we can directly integrate them into data models, such as the relational data model. For example, we can define two relations containing information about satellites and pollution incidents as follows.

$$Satellites(sname : string, Pos : Point)$$
$$Pollutions(pname : string, Reg : Region)$$

The relation *Satellites* contains for each satellite one tuple consisting of the satellite's name and its movement, which is given by the attribute *Pos* that stores a moving point. Similarly, an evolving region is associated with each pollution incident through the attribute *Reg* in the relation *Pollutions.* The temporal changes are completely encapsulated by the types *Point* and *Region.* The operations offered by these data types enable us to pose spatiotemporal queries in a straightforward way. For example, we have an operation *Intersection* that computes the intersection of two spatiotemporal objects. In the case of a moving point and an evolving region this intersection is given by those parts of the moving point that lie inside the evolving region. As a query example we ask which satellites were able to observe which pollution incidents for a certain amount of time by the following query formulated in the SQL-like language STQL [20, 28].

```
SELECT sname, pname
  FROM Satellites, Pollutions
 WHERE duration(Intersection(Pos, Reg)) > 3 hours
```

The query works as follows: For all pairs of satellites and pollution incidents we consider the intersection of their spatiotemporal attributes and select those pairs for which the lifetime (indicated by $t$ in Figure 2.1) of the moving point resulting from the intersection is larger than 3 hours. In queries we use dimension conversion operators like $hours : num \rightarrow duration$ in postfix notation. The type *duration* is isomorphic to the real numbers, but it contains values describing the duration of time intervals. The *duration* function computes the total duration of the domain of a temporal function.

## 2.5 Generalizing Spatiotemporal Predicates

Although many interesting queries can be expressed by the approach described in the Section 2.4, there are also many queries that cannot be (easily) formulated. One large class of such queries are queries asking for changes in relationships between objects. For example, we might be interested in finding ships that were inside the oil spill and eventually left the spill. (This query could point to the possible polluter.) Using a spatiotemporal intersection predicate yields also all ships that entered, crossed, touched, etc. the spill and is simply not precise enough to find the desired information. Although we can express the query by adding a couple of further explicit conditions on the ship's trajectory (see for example, [28]), the query becomes incredibly complex and almost unreadable. Moreover, the situation gets much worse for more complex developments, which indicates that the available query language abstractions are not expressive enough. The same observation holds for other data models and query languages. For example, in $SQL^{ST}$ [9] queries suffer from an inflation of joins.

### 2.5.1 What are Spatiotemporal Predicates?

With the introduction of *spatiotemporal predicates* [24] we can describe precisely the developments of objects and their relationships in a relatively simple way. A (binary) spatiotemporal predicate $P$ is a function that maps a pair of spatiotemporal objects to a boolean value, that is, $P : \alpha \times \beta \to bool$ for $\alpha, \beta \in \{Point, Region\}$. An example is the predicate *Inside* (defined for the two types $Point \times Region \to bool$ and $Region \times Region \to bool$) that checks whether the first argument is always inside the evolving region. Another example is *Disjoint* that yields true for two spatiotemporal objects that have no points in common. Complex spatiotemporal predicates can be built by regular expressions over a basic set of spatial and spatiotemporal predicates. For example, the predicate describing a "leaving" development can be defined as:

$$Leaves := Inside\ meet\ Disjoint$$

The *meet* predicate is a spatial predicate that applies to two spatial objects and yields true when they touch each other. The semantics of a "string" of spatiotemporal and spatial predicates is formally defined in [27]. Intuitively, the meaning is that one predicate has to hold after the other. In other words, for a sequence of alternating spatiotemporal and spatial predicates $P_1\, p_1\, P_2\, \ldots\, p_{k-1}\, P_k$ to hold for an object (pair), there has to be a connected sequence of time intervals and instants $I_1, t_1, I_2, \ldots, t_{k-1}, I_k$ such that each spatiotemporal predicate $P_j$ holds for $I_j$ and each spatial predicate $p_j$ holds at $t_j$. Hence, *Leaves* is true for two spatiotemporal objects $P_1$ and $P_2$ if for some time $P_1$ was inside $P_2$, then touched the border, and finally was disjoint from $P_2$. In addition to predicate "concatenation" there are many other predicate compositions, such as alternation, reflection, and repetition, which have nice properties (proved in [27]) that can be used, for example, for optimization.

Assume we have the following schema for the relation *Ships*.

$$Ships(sname : string, Pos : Point)$$

With the spatiotemporal predicate *Leaves* we can formulate the query for ships leaving the oil spill as follows.

SELECT *sname*
  FROM *Ships*, *Pollutions*
WHERE *pname* = *'The Big Spill'* AND *Pos Leaves Reg*

This example demonstrates how spatiotemporal predicates offer a basic abstraction for expressing conditions on sequences of changes.

### 2.5.2 Spatiotemporal Predicates Based on Numerical Operations

The concept of spatiotemporal predicates is based on the notion of changes in *topological* relationships between spatiotemporal objects. A logical next step is the generalization to *numerical* predicates. For example, with spatiotemporal predicates we can currently not formulate a query that asks for regions (like oil spills) that after a period of growth were only shrinking (to check, for example, a successful treatment).

The modular, two-step definition of spatiotemporal predicates indicates possibilities for generalizations: A spatiotemporal predicate $P$ is obtained from a spatial predicate $p$ by (1) *lifting* it to a function $\uparrow p$ that computes for two spatiotemporal objects a temporal boolean that gives the predicate value for each time point and (2) aggregating all the boolean values using $\forall$ quantification into one boolean value. For example, consider the spatial predicate $\mathit{inside} : \mathit{point} \times \mathit{region} \rightarrow \mathit{bool}$ that tells whether or not a point is inside a region. The lifted function $\uparrow \mathit{inside} : \mathit{Point} \times \mathit{Region} \rightarrow \mathit{Bool}$ computes for a moving point and an evolving region a temporal boolean that tells for each time point whether or not the point is inside the region. The temporal aggregation of all these boolean values yields true if the point was always inside the evolving region.[1]

If we now consider numerical functions, such as *area* or *distance*, their lifted versions $\uparrow \mathit{area}$ and $\uparrow \mathit{distance}$ yield temporal numbers, which can be aggregated in two ways into booleans. First, we can apply a lifted numerical numerical predicate (such as a function checking for "$> 3$"). This results in a temporal boolean that can be aggregated further by $\forall$ as already described. With that approach, however, we are not able to check, for example, growth since boolean values are obtained for individual time values, independently from other time values. However, we have to compare the numbers of different time values in order to identify a growth trend in time. This can be performed by an operator $\Delta$ that maps any temporal object of type $\alpha$ (if a total ordering is defined on $\alpha$) to an object of type $\{\mathit{negative}, \mathit{none}, \mathit{positive}\}$ that gives for each time the change in the value at that point. Thus, $\Delta$ is basically a differential

[1] There are some subtle details regarding the time intervals over which the aggregation has to range. These considerations are not important here; a detailed analysis can be found in [27].

operator. Now, "growth" can be checked by comparing the result of $\Delta$ to *positive* and aggregating with $\forall$. We can generalize spatiotemporal predicates to work for numerical functions by parameterizing the $\forall$ quantification by a function that maps non-boolean values into booleans. This function will be lifted and applied to the temporal object just before the $\forall$ aggregation is performed. For example, to define a predicate for checking the growth of a particular *Region* value $R$, we have to map $R$ into a temporal number by applying *Area* and then determining the change at each time point by $\Delta$. Next we can define a predicate *grows* as follows.

$$grows(x) \;:\Longleftrightarrow\; x = positive$$

This predicate can be applied to all values returned by $\Delta$ and aggregated with $\forall$. We combine the last two steps in the definition of a parameterized *Always* operation that takes two parameters: (i) the predicate to be applied (here *grows*) and (ii) the spatiotemporal object to which the first parameter expression is applied (here: $\Delta(Area(R))$). Then we find out whether or not $R$ always grows through the expression:

$$Always(grows, \Delta(Area(R)))$$

The precise definition of *Always* has to take into account different lifetimes of objects, which is discussed in detail in [27]. Another issue to investigate is the result type of $\Delta$. For some applications the type $\{negative, none, positive\}$ is not general enough. It seems that we rather need $\Delta$ to be of type $\{negative, none, positive\} \times \alpha$. Then, for numerical functions $\Delta$ gives the change rate (for example, speed). However, for spatial types the situation is more involved. For example, when we apply $\Delta$ to an evolving region, we might get marked regions *and* curves as a result.

### 2.5.3 Adding Time Constraints

Another limitation of spatiotemporal predicates is demonstrated by the query Q1: While we can express the multiple-time-observation condition with the help of repetition and the definition of a *Cross* predicate, the minimum time constraint *cannot* be expressed by the current form of spatiotemporal predicates. Similarly, the query Q2 about suspicious mailbox visitors requires a constraint on the duration of the validity of a spatiotemporal predicate.

One possible way to extend spatiotemporal predicates by temporal constraints is to define an operation that can superimpose these constraints on spatiotemporal predicates. Such an operation, say *For*, would take a predicate on durations and would restrict a spatiotemporal predicate further by this condition. If *STP* denotes the type of all spatiotemporal predicates (that is, $STP = \alpha \times \beta \to bool$ for $\alpha, \beta \in \{Point, Region\}$), then *For* has the type $For : (duration \to bool) \times STP \to STP$. With this operation we can define two spatiotemporal predicates for expressing the minimum-observation-time requirement and the minimum-number-of-visits requirement, respectively.

$$\begin{array}{l} \mathit{IntersectFor}(d, P, Q) := \mathit{For}(\geq d, \mathit{Intersection})(P, Q) \\ \mathit{AtLeastFour}(P) := P\ (\neg P)\ P\ (\neg P)\ P\ (\neg P)\ P^{+} \end{array}$$

Note that we are using partial function application as a convenient means to formulate duration predicates. For example, the expression "$\geq d$" is a short form for the function $f$ that is defined by $f(x) = x \geq d$. The advantage of using expressions like "$\geq d$", in which the $\geq$ function is only partially applied to its second argument, is that we do not have to introduce a name for that function (for a detailed discussion, see [5]). Moreover, note that *For* is a *higher-order* predicate, that is, it takes two predicates as arguments and returns a predicate as a result. This fact also explains the notation $\mathit{For}(\geq d, \mathit{Intersection})(P, Q)$: $\mathit{For}(\geq d, \mathit{Intersection})$ is the application of *For* to its two argument predicates, and the result is a predicate that is applied to the pair of objects $(P, Q)$.

The alternating sequence of predicates and their negation in the definition of *AtLeastFour* is required to describe the repeated satisfaction of $P$ on arbitrary but non-overlapping time intervals. Using the two predicates we could express query Q1 as follows.

```
SELECT sname
  FROM Satellites, Pollutions
 WHERE pname = 'The Big Spill' AND
       AtLeastFour(IntersectFor(5 min, Pos, Reg))
```

With the *For* operator we can also express query Q2. Assume that we have two relations *Cars* and *Mailboxes* with the following schemas.

$$\begin{array}{l} \mathit{Cars}(\mathit{license\text{-}plate} : \mathit{string}, \mathit{Location} : \mathit{Point}) \\ \mathit{Mailboxes}(\mathit{id} : \mathit{string}, \mathit{pos} : \mathit{point}) \end{array}$$

Since mailboxes usually do not move, their position is represented by a spatial point and not a moving point. To express the spatiotemporal condition of "meeting", we have to lift the mailbox points into moving points that actually do not move. We can use the already mention lifting operation "$\uparrow$", which, when applied to objects and not functions, yields a constant temporal function that returns the argument object for each time point. For example, for a given point $p : \mathit{point}$, the expression $\uparrow p$ (of type *Point*) denotes a stationary position. The query Q2 can now be expressed as follows.

```
SELECT license-plate
  FROM Cars, Mailboxes
 WHERE For(≤ 4 hours, AtLeastTwo(
              For(≥ 2 min, Meet)(Location, ↑pos)))
```

This query works as follows. The innermost *For* refines the *Meet* predicate to hold for at least two minutes. The *AtLeastTwo* requires an occurrence of this predicate at least twice. Finally, the outermost *For* constrains this whole predicate to hold only if the total time is less than four hours. In this example, we observe the need for another operation *AtLeastTwo* or a generalization *AtLeast* that takes an integer parameter. We will discuss a generalization of these function in Section 2.6.

We can also see that the definition of *For* is not general enough. To express more complex conditions to refine spatiotemporal predicates it seems that *For* needs access to the spatiotemporal objects that are under consideration by the predicates to be refined. Then we can express constraints on speed (which is required for the query Q3) and other object properties. Moreover, we also have to consider versions of *For* that take a predicate on time instants and constrains spatial predicates (when used inside developments).

## 2.6 Spatiotemporal Patterns

The extended form of spatiotemporal predicates outlined in Section 2.5 can be used to define a language of spatiotemporal patterns.

One limitation of spatiotemporal predicates is demonstrated by the definition of the predicate *AtLeastFour*, which shows that a more detailed control over repetition is needed to express developments/queries that mention an explicit number of repetitions. This is one of many ways to compose individual predicates into *spatiotemporal patterns*.

As far as repetition is concerned, we can define in addition to the Kleene star operation a combinator *Repeat* that takes a number $k$ and a spatiotemporal predicate $P$ and returns a spatiotemporal predicate that is true if the $k$-fold composition/sequence of $P$ is true. It is interesting to observe that although we can easily express a combinator *AtLeast* as a derived predicate based on *Repeat*, namely simply by

$$AtLeast(P, k) := Repeat(P, k)\ P^*$$

it is not possible to define the combinator *AtMost* (without using recursion).

Sequential composition, repetition, and alternation (see [27]) of spatiotemporal predicates are just three simple example combinators to express structure for different relationships among spatiotemporal objects. In more advanced applications we might have to find out about more complex relationships among different parts of a development. The following query illustrates this aspect.

**Q4.** Which forests suffered from multiple fires of increasing duration?

We can express the multiple-fire constraint by an *AtLeast* operator similar to what we have done for Q1 and Q2. However, the second part of the query refers to the duration of the predicates, but not just with a condition that is evaluated for each predicate individually, but rather with a condition that relates different subpredicates. Here, it is the condition that the duration of any intersection predicate is greater than the duration of the preceding one. Since the *For* combinator is only able to constrain individual predicates, we need a more general mechanism.

On possibility is to allow aggregations over spatiotemporal predicate sequences, that is, their values and time periods. By working in a generalized model of numerical predicates, values and time periods can be exposed as part of the evaluation of a spatiotemporal predicate as a sequence of (pairs of) objects and the corresponding predicate whose evaluation on the objects has contributed to the overall result. In the example query Q4 this would be a list of forest/fire pairs plus the predicates *Intersection*, $\neg$*Intersection*, *Intersection*. (Here we assume for simplicity to having just one fire object.) Then we can imagine a function that scans this list, computes the relevant information for each element (here, the duration of the first and third object pairs), and performs an aggregation, which in this case could be simply a test whether the computed list of durations is increasing.

This approach generalizes the notion of spatiotemporal predicates from mere predicates that yield a boolean value to object filters or object partitioning functions. It is the generalization of the result type that "opens" spatiotemporal predicates for further computations. Two open questions in this context are:

- How can the notion of spatiotemporal predicates be generalized without compromising their current useful applications?
- What is a suitable language to specify aggregations over spatiotemporal predicate results?

Eventually, this generalization might lead to hierarchical sequences since multiple aggregations could follow one another offering the opportunity for another aggregation. In other words, spatiotemporal patterns provide a way to describe hierarchical or nested spatiotemporal predicates. One way of describing such hierarchies is to use a grammar formalism where terminals represent basic spatiotemporal predicates and nonterminals represent spatiotemporal patterns that can be expanded according to the grammar rules into arbitrarily nested expressions of spatiotemporal predicates. Evaluating such a general form of spatiotemporal patterns means to parse predicate sequences according to a grammar that defines a pattern. The current version of spatiotemporal predicates result as a special case by not using recursion in grammars, which leads to regular expressions.

Another issue to consider is the ability to existentially quantify objects used in development queries. Under the current interpretation, a spatiotemporal predicate is a binary predicate that can be applied to two spatiotemporal objects and evaluates to either true or false. For example, a spatiotemporal predicate that is able to express query Q4 can succeed only for a forest and *one particular* fire that repeatedly affected that forest. However, the query is rather concerned with the forest being affected repeatedly by *any* fire, that is, it should be possible to satisfy the condition by different fires. One possibility to allow for such a generalization is to introduce local quantifiers (existential and universal) into the language for specifying predicates. Nonquantified variables can be assumed to be universally quantified by default to make this a conservative extension. Another possibility is to introduce on the conceptual level "relation sequences" so that instances for a predicate can be chosen independently from different relations.

In summary, spatiotemporal patterns seem to be a logical next step following spatiotemporal predicates in the definition of more expressive query predicates. Their integration into spatiotemporal query languages, such as STQL, can extend the expressiveness of such query languages considerably.

## 2.7 Visual Query Interfaces

Spatiotemporal predicates allow the formulation of many queries in a rather direct way by expressing spatiotemporal relationships directly on attributes of the involved objects. However, as we have seen in Section 2.6, more advanced queries that require spatiotemporal patterns quickly tend to become quite complex. Moreover, the fast growing set of available spatial and spatiotemporal data and its wide dissemination through the Internet increases the class of possible users for these data. Offering ordinary users access to spatiotemporal data is therefore becoming a more and more important issue that can be addressed by developing a visual query language and a corresponding user interface. In particular, it is important to make databases accessible to users without formal training in databases and query languages, such as scientists, administrators, or other end users. In many cases, these users do not have the time or are not willing to learn a formal query language. From this point of view, even spatiotemporal predicates embedded into a query language like STQL [24, 28] is beyond what most of these users can deal with.

A visual notation for spatiotemporal predicates can help solving this problem. In particular, since spatial objects and their relationships have a natural graphical representation, a visual notation can express relationships in many cases implicitly where textual notations require the explicit application of operations and predicates. A similar situation arises in the case of spatiotemporal objects when the time dimension is visualized as one spatial dimension [25]. However, special care must be taken to keep the notation simple enough to avoid overspecifications by exposing too many details. The idea of our approach is to represent traces of object developments by two-dimensional pictures. The relationships of the drawn objects can then be interpreted as spatiotemporal predicates. An example is shown in Figure 2.2 that specifies the *Leaves* predicate. (The interpretation of pictures like these and their formal translation into spatiotemporal predicates is described in detail in [29].)

Additional improvements and simplifications can be achieved by embedding the query notation in a tailor-made user interface that supports the cre-

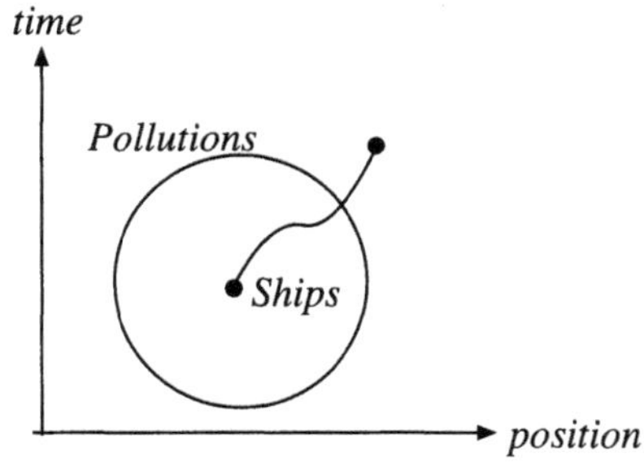


**Figure 2.2.** Visual Specification of the *Leaves* predicate.

ation and adaptation of queries by direct manipulation of the objects whose relationships represent spatiotemporal predicates and patterns [26]. Moreover, by selecting attributes from a database schema and associating them with the drawn objects, pictures can not only define spatiotemporal predicates, but also complete queries. For example, if the moving point in Figure 2.2 is associated with the *Pos* attribute of the *Ships* relation and the circle is associated with the *Reg* attribute of the *Pollutions* relation, the picture describes the following query and can be automatically translated into it.

SELECT *sname*, *pname*
FROM *Ships*, *Pollutions*
WHERE *Pos* (*Inside meet Disjoint*) *Reg*

Concreteness and directness are two strategies that have been used successfully by some end-user programming languages [8], and they should be employed in visual query interfaces for spatiotemporal databases as well. Concreteness has to do with working with values instead of with abstract variables. The strategy of directness is particularly important in our design plans. The term *directness* as used in the HCI community expands on the term *direct manipulation*, first coined by Shneiderman to describe three principles: continuous representation of the objects of interest, physical actions or presses of labeled buttons instead of complex syntax, and rapid incremental reversible operations whose effect on the object of interest is immediately visible [61]. Hutchins, Hollan, and Norman expand on these notions, suggesting that directness is inversely proportional to the cognitive effort needed to use the interface [46]. They describe directness as having two aspects. The first aspect is the distance between one's goals and the actions required by the system to achieve those goals. We argue that distance is small in our approach since the objects to be queried are displayed and directly manipulated on the screen. For example, the goal "describe a ship leaving a region" is expressed by a picture in which the trace of the ship start inside the region and ends outside of it, instead of requiring a textual expression involving predicate and operator applications. Green and Petre enumerate several examples showing the unfortunate lack of this aspect of directness (termed *closeness of mapping* in their work) in commonly used programming languages [38]. The second aspect is a feeling of direct engagement: "the feeling that one is directly manipulating the objects of interest". The notion of aiming for directness as a programming language design goal has in recent years begun to influence other kinds of end-user programming languages and domain-specific languages as well. In our approach direct engagement is supported by the user interface, which lets users place objects on a drawing region and move them around to create traces. Simultaneously to the user action, the spatiotemporal predicate that is specified through the current picture is displayed in a small area of the screen. This enables the user to move objects and observe at the same time the effect on the specified predicate. An additional benefit is that users can learn about the formal, textual notation by examples if they want to.

To extend the developed visual query notation (and the user interface) to cope with more general forms of spatiotemporal predicates and patterns, a number of issues have to be addressed.

One example is to add conditions to query pictures. Consider, for example, a refinement of the visual query in Figure 2.2 that is obtained by adding a constraint on the time the ship needed to leave the pollution region. We can express this by projecting two points onto the time axis and adding a condition to it. The resulting visual query asks only for those ships that have left the region within 30 minutes, see Figure 2.3.

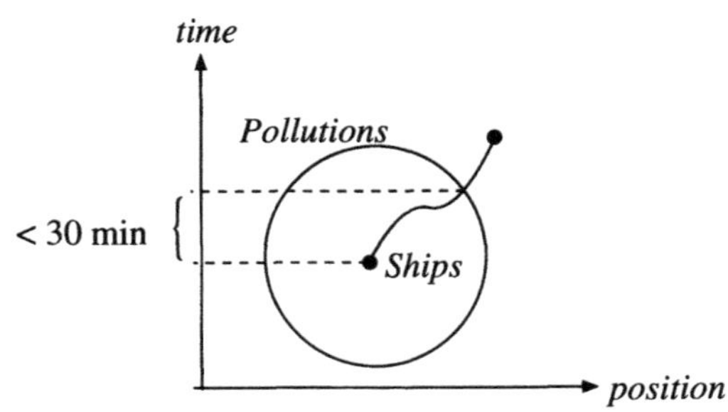


**Figure 2.3.** Refining the *Leaves* predicate with a condition.

This visual query will then be translated into the following STQL query.

SELECT *sname*, *pname*
FROM *Ships*, *Pollutions*
WHERE *Pos* (*For*($<$ 30 *min*, *Inside*) *meet Disjoint*) *Reg*

Two other useful extensions are to express numerical spatiotemporal predicates and to allow for hierarchical specifications. To use numerical spatiotemporal we need additional visual elements to refer to numerical spatiotemporal operations like *Distance* to be able to express numerical constraints on developments. In order to be able to express general spatiotemporal patterns that require nested sequence conditions we need a mechanism to refer to visual specifications from within other specifications. Nesting of visual expressions can sometimes simplify the notation considerably [13]. Coupled with zooming capabilities on the interface level, this approach can lead to convenient specification of hierarchical predicates and patterns.

## 2.8 Conclusions

We have introduced a concept of spatiotemporal patterns that can be employed in query languages to express queries about the development of spatiotemporal objects and their relationships. To help coping with the complexity of the query language we also have sketched the design of a visual query language that makes the expressiveness of spatiotemporal patterns accessible to a large group of users.

We believe that providing effective access to spatiotemporal data through expressive query languages is an important step to help scientists and ordinary users to exploit the information that is contained in fast growing spatiotemporal data sets. Moreover, we believe that the concept of spatiotemporal patterns can provide a key to solving this problem. A group of researchers at the recent NSF workshop on "Workshop on Spatio-Temporal Data Models of Biogeophysical Fields for Ecological Forecasting" has concluded the same thing [44] (emphasis added):

> [...] there is a critical need for theories and tools that will enable efficient and reliable characterization of *spatio-temporal patterns* contained in image time series.

The general goal of the described work can be summarized by the term "universal geo-data access". First, the developed query language concepts can help to improve the ability to extract information from large geo-data sets. Second, through the development of visual query interfaces, the access to geo data is provided to a large group of users.

## References

1. Agarwal P.K., Arge L., Erickson J. (2000) Indexing Moving Points. In: Proc. of the 19th ACM SIGACT-SIGMOD-SIGART Symp. on Principles of Database Systems, pp. 175–186.
2. Allen J.F. (1983) Maintaining Knowledge about Temporal Intervals. Communications of the ACM 26:832–843.
3. Allen J.F. (1984) Towards a General Theory of Action and Time. Artificial Intelligence 23:123–154.

4. Ariav G. (1986) An Overview of TQuel. ACM Transactions on Database Systems 11(4):499–527.
5. Bird R.S. (1998) Introduction to Functional Programming Using Haskell. Prentice-Hall International, London, UK.
6. Bittner T. (2001) Rough Sets in Spatio-Temporal Data Mining. In: Proc. of the Int. Workshop on Temporal, Spatial and Spatio-Temporal Data Mining, LNAI 2007, pp. 89–104.
7. Böhlen M.H., Jensen C.S., Skjellaug B. (1998) Spatio-Temporal Database Support for Legacy Applications. In: Proc. of the ACM Symp. on Applied Computing, pp. 226–234.
8. Burnett M.M. (1999) Visual Programming. In: Webster, J. G. (Ed.) Encyclopedia of Electrical and Electronics Engineering, John Wiley & Sons, pp. 215–283.
9. Chen C.X., Zaniolo C. (2000) $\text{SQL}^{ST}$: A Spatio-Temporal Data Model and Query Language. In: Proc. of the 19th Int. Conf. on Conceptual Modeling, LNCS 1920, pp. 96–111.
10. Cheng T.S., Gadia S.K. (1994) A Pattern Matching Language for Spatio-Temporal Databases. In: Proc. of the ACM Conf. on Information and Knowledge Management, pp. 288–295.
11. Chomicki J., Revesz P.Z. (1999) A Geometric Framework for Specifying Spatiotemporal Objects. In: Proc. of the 6th Int. Workshop on Temporal Representation and Reasoning, pp. 41–46.
12. Chomicki J. (2002) Querying Spatiotemporal Databases. In: Proc. of the NSF/BDEI Workshop on Spatio-Temporal Data Models of Biogeophysical Fields for Ecological Forecasting, San Diego Supercomputer Center, La Jolla, CA, USA.
13. Citrin W., Hall R., Zorn B. (1995) Programming with Visual Expressions. In: Proc. of the 11th IEEE Symp. on Visual Languages, pp. 294–301.
14. Craglia M., Haining P., Wiles P. (2000) A Comparative Evaluation of Approaches to Urban Crime Pattern Analysis. Urban Studies 37(4):711–729.
15. Cui Z., Cohn A.G., Randell D.A. (1993) Qualitative and Topological Relationships in Spatial Databases. In: Proc. of the 3rd Int. Symp. on Advances in Spatial Databases, LNCS 692, pp. 296–315.
16. Hand D.J., Mannila H., Smyth P. (2001) Principles of Data Mining. MIT Press, Cambridge, MA.
17. Walgraef D. (1997) Spatio-Temporal Pattern Formation: With Examples from Physics, Chemistry, and Material Science. Springer Verlag, New York, NY.
18. Djafri N., Fernandez A.A.A., Paton N.P., Griffiths T. (2002) Spatio-Temporal Evolution: Querying Patterns of Change in Databases. In: Proc. of the 10th ACM Int. Symp. on Advances in Geographic Information Systems.
19. Egenhofer M.J., Franzosa R.D. (1991) Point-Set Topological Spatial Relations. Int. Journal of Geographical Information Systems 5(2):161–174.
20. Erwig M. (2002) Design of Spatio-Temporal Query Languages. In: NSF/BDEI Workshop on Spatio-Temporal Data Models of Biogeophysical Fields for Ecological Forecasting, San Diego Supercomputer Center, La Jolla, CA, USA.
21. Erwig M., Güting R.H., Schneider M., Vazirgiannis M. (1998) Abstract and Discrete Modeling of Spatio-Temporal Data Types. In: Proc. of the 6th ACM Symp. on Geographic Information Systems, pp. 131–136.
22. Erwig M., Güting R.H., Schneider M., Vazirgiannis M. (1999) Spatio-Temporal Data Types: An Approach to Modeling and Querying Moving Objects in Databases. GeoInformatica 3(3):269–296.

23. Erwig M., Meyer B. (1995) Heterogeneous Visual Languages – Integrating Visual and Textual Programming. In: Proc. of the 11th IEEE Symp. on Visual Languages, pp. 318–325.
24. Erwig M., Schneider M. (1999) Developments in Spatio-Temporal Query Languages. In: Proc. of the IEEE Int. Workshop on Spatio-Temporal Data Models and Languages, pp. 441–449.
25. Erwig M., Schneider M. (1999) Visual Specifications of Spatio-Temporal Developments. In: Proc. of the 15th IEEE Symp. on Visual Languages, pp. 187–188.
26. Erwig M., Schneider M. (2000) Query-By-Trace: Visual Predicate Specification in Spatio-Temporal Databases. In: Arisawa H., Catarci T. (Eds.) Advances in Visual Information Management – Visual Database Systems, Kluwer Academic Publishers, Boston, MA, pp. 199–218.
27. Erwig M., Schneider M. (2002) Spatio-Temporal Predicates. IEEE Transactions on Knowledge and Data Engineering 14(4):881–901.
28. Erwig M., Schneider M. (2002) STQL: A Spatio-Temporal Query Language. In: Ladner R., Shaw K., Abdelguerfi M. (Eds.) Mining Spatio-Temporal Information Systems, Kluwer Academic Publishers, Norwell, MA, chapter 6, pp. 105–126.
29. Erwig M., Schneider M. (2003) A Visual Language for the Evolution of Spatial Relationships and its Translation into a Spatio-Temporal Calculus. Journal of Visual Languages and Computing 14(2):181–211.
30. Erwig M., Schneider M., Güting R.H. (1998) Temporal Objects for Spatio-Temporal Data Models and a Comparison of Their Representations. In: Proc. of the Int. Workshop on Advances in Database Technologies, LNCS 1552, pp. 454–465.
31. Kuper G.M., Libkin L., Paredaens J. (Eds.) (2000) Databases. Springer-Verlag, Berlin.
32. Piatetsky-Shapiro G., Frawley W. (1991) Knowledge Discovery in Databases. MIT Press, Cambridge, MA.
33. Gaal S. (1964) Point Set Topology. Academic Press, 1964.
34. Gadia S.K., Nair S.S. (1993) Temporal Databases: A Prelude to Parametric Data. In: Tansel A.U., Clifford J., Gadia S., Jajodia S., Segev A., Snodgrass R. (Eds.) Temporal Databases: Theory, Design, and Implementation. The Benjamin/Cummings Publishing Company, pp. 28–66.
35. Galton A. (1995) Towards a Qualitative Theory of Movement. In: Proc. of the 2nd Int. Conf. on Spatial Information Theory, LNCS 988, pp. 377–396.
36. Galton A. (2001) A Formal Theory of Objects and Fields. In: Proc. of the 5th Int. Conf. on Spatial Information Theory, LNCS 2205, pp. 458–473.
37. Geake E. (1993) How PCs Predict where Crime will Strike. New Scientist 140(1896):17.
38. Green T.R.G., Petre M. (1996) Usability Analysis of Visual Programming Environments: A 'Cognitive Dimensions' Framework. Journal of Visual Languages and Computing 7(2):131–174.
39. Grumbach S., Rigaux P., Segoufin L. (1998) Spatio-Temporal Data Handling with Constraints. In: Proc. of the 6th ACM Int. Symp. on Advances in Geographic Information Systems, pp. 106–111.
40. Grumbach S., Rigaux P., Scholl M., Segoufin L. (2000) The DEDALE Prototype. In: Kuper G.M., Libkin L., Paredaens J. (Eds.) Constraint Databases. Springer-Verlag, Berlin, chapter 17, pp. 365–382.

41. Grumbach S., Rigaux P., Segoufin L. (2001) Spatio-Temporal Data Handling with Constraints. GeoInformatica 5:95–115.
42. Güting R.H. (1988) Geo-Relational Algebra: A Model and Query Language for Geometric Database Systems. In: Proc. of the Int. Conf. on Extending Database Technology, LNCS 303, pp. 506–527.
43. Güting R.H., Böhlen M.H., Erwig M., Jensen C.S., Lorentzos N.A., Schneider M., Vazirgiannis M. (2000) A Foundation for Representing and Querying Moving Objects. ACM Transactions on Database Systems 25(1):1–42.
44. Henebry G.M., Chomicki J., Fountain T., Ranson K.J. (Eds.) (2002) Proc. of the NSF/BDEI Workshop on Spatio-Temporal Data Models of Biogeophysical Fields for Ecological Forecasting, San Diego Supercomputer Center, La Jolla, CA, USA.
45. Hoerling M.P., Kumar A. (2002) Atmospheric Response Patterns Associated with Tropical Forcing. Journal of Climate 15(16):2184–2203.
46. Hutchins E., Hollan J., Norman D. (1986) Direct Manipulation Interfaces. In: Norman D., Draper S. (Eds.) User Centered System Design: New Perspectives on Human-Computer Interaction, Lawrence Erlbaum Assoc. Hillsdale, NJ, pp. 87–124.
47. Imfeld S. (2000) Time, Points and Space – Towards a Better Analysis of Wildlife Data in GIS. Dissertation, University of Zürich.
48. Knopoff L., Gabrielov A., Ghil M. (Eds.) (2001) Proc. of the IMA Workshop on Spatio-Temporal Patterns in the Geosciences, University of Minnesota, Minneapolis, MN, USA.
49. Kollios G., Gunopulos D., Tsotras V.J. (1999) On Indexing Mobile Objects. In: Proc. of the 18th ACM SIGACT-SIGMOD-SIGART Symp. on Principles of Database Systems, pp. 261–272.
50. Mesrobian E., Muntz R., Shek E., Nittel S., La-Rouche M., Kriguer M., Mechoso C., Farrara J., Stolorz P., Nakamura H. (1996) Mining Geophysical Data for Knowledge. IEEE Expert 11(5):34–44.
51. Mielke H.W. (1989) Patterns of Life: Biogeography of a Changing World. Unwin Hyman, Inc., Winchester, MA.
52. Miller T. (1994) Computers Track the Criminal's Trail. American Demographics 16(1):13–14.
53. Naggle G. (1995) Urban Crime: A Geographic Appraisal. Geographical Magazine 67(4):56–57.
54. Peuquet D.J., Duan N. (1995) An Event-Based Spatiotemporal Data Model (ESTDM) for Temporal Analysis of Geographical Data. Int. Journal of Geographical Information Systems 9(1):7–24.
55. Peuquet D.J. (2002) Representations of Space and Time. The Guilford Press, New York, NY.
56. Renz J., Nebel B. (1999) On the Complexity of Qualitative Spatial Reasoning: A Maximal Tractable Fragment of the Region Connection Calculus. Artificial Intelligence 108(1-2):69–123.
57. Romanowicz B. (1993) Spatiotemporal Patterns in the Energy Release of Great Earthquakes. Science 260(5116):1923–1926.
58. Scholl M., Voisard A. (1989) Thematic Map Modeling. In: Proc. of the 1st Int. Symp. on Large Spatial Databases, LNCS 409, pp. 167–190.
59. Segev A., Shoshani A. (1987) Logical Modeling of Temporal Data. In: Proc. of the ACM SIGMOD Conf. on Management of Data, pp. 454–466.

60. Segev A., Shoshani A. (1993) A Temporal Data Model Based on Time Sequences. In: Tansel A.U., Clifford J., Gadia S., Jajodia S., Segev A., Snodgrass R. (Eds.) Temporal Databases: Theory, Design, and Implementation. The Benjamin/Cummings Publishing Company, pp. 248–270.
61. Shneiderman B. (1983) Direct Manipulation: A Step Beyond Programming Languages. Computer 16(8):57–69.
62. Sistla A.P., Wolfson O., Chamberlain S., Dao S. (1997) Modeling and Querying Moving Objects. In: Proc. of the Int. Conf. on Data Engineering, pp. 422–432.
63. Snodgrass R.T. (Ed.) (1995) The TSQL2 Temporal Query Language. Kluwer Academic Publishers, Boston, MA.
64. Stonebraker M. (1986) Inclusion of New Types in Relational Database Systems. In: Proc. of the Int. Conf. on Data Engineering, pp. 262–269.
65. Suzue T., Shinoda Y. (1999) Highly Reproducible Spatiotemporal Patterns of Mammalian Embryonic Movements at the Developmental Stage of the Earliest Spontaneous Motility. European Journal of Neuroscience 11(8):2697–2710.
66. Theodoridis Y., Sellis T., Papadopoulos A., Manolopoulos Y. (1998) Specifications for Efficient Indexing in Spatiotemporal Databases. In: Proc. of the 10th Int. Conf. on Scientific and Statistical Database Management, pp. 123–132.
67. Tilove R.B. (1980) Set Membership Classification: A Unified Approach to Geometric Intersection Problems. IEEE Transactions on Computers C-29:874–883.
68. Tourre Y.M., Rajagopalan B., Kushnir Y. (1999) Dominant Patterns of Climate Variability in the Atlantic Ocean during the Last 136 Years. Journal of Climate 12(8):2285–2299.
69. Tsoukatos I., Gunopulos D. (2001) Efficient Mining of Spatiotemporal Patterns. In: Proc. of the 7th Int. Symp. on Spatial and Temporal Databases, LNCS 2121, pp. 425–442.
70. Tumeo M.A., Larson M.K. (1994) Movement of Fuel Spills in the Ross Ice Shelf. Antarctic Journal of the United States 29(5):373–374.
71. Vazirgiannis M., Wolfson O. (2001) A Spatiotemporal Query Language for Moving Objects. In: Proc. of the 7th Int. Symp. on Spatial and Temporal Databases, LNCS 2121, pp. 20–35.
72. Wang F., Minor W.W. (2002) Where the Jobs Are: Employment Access and Crime Patterns in Cleveland. Annals of the Association of American Geographers 92(3):435–451.
73. Weigand T., Moloney K.A., Milton S.J. (1998) Population Dynamics, Disturbance, and Pattern Evolution: Identifying the Fundamental Scales of Organization in a Model Ecosystem. The American Naturalist 152(3):321–337.
74. Worboys M.F. (1994) A Unified Model for Spatial and Temporal Information. The Computer Journal 37(1):25–34.
75. Yeh T.S., de Cambray B. (1993) Time as a Geometric Dimension for Modeling the Evolution of Entities: A 3D Approach. In: Proc. of the Int. Conf. on Integrating GIS and Environmental Modeling.
76. Yeh T.S., de Cambray B. (1995) Modeling Highly Variable Spatio-Temporal Data. In: Proc. of the 6th AustraliAsian Database Conf., pp. 221–230.

# 3

# Toward Extensible Spatio-Temporal Databases: An Approach Based on User-Defined Aggregates

Cindy X. Chen[1], Haixun Wang[2], and Carlo Zaniolo[3]

[1] Department of Computer Science
University of Massachusetts
Lowell, MA 01854, U.S.A.
cchen@cs.uml.edu

[2] IBM T.J. Watson Research Center
P.O. Box 704, Yorktown Heights, NY 10598, U.S.A.
haixun@us.ibm.com

[3] Department of Computer Science
University of California
Los Angeles, CA 90095, U.S.A.
zaniolo@cs.ucla.edu

## 3.1 Introduction

The need for spatial, temporal, and spatio-temporal databases is ubiquitous and touches several application domains, including geographical information systems, autonomous navigation, tracking, medical imaging, and many others. To answer this need, database researchers have proposed many approaches that cover a wide spectrum of data models and query languages: a very incomplete list include [7, 15, 3, 11]. The number and variety of alternative solutions proposed reflect the complexity of technical problems and the diversity of applications and requirements encountered in the field, whereby a solution proposed for a certain spatio-temporal application might not satisfy the requirements of another. To solve this problem, we propose an approach whereby the primitives built in the system can be further customized and extended by the end-users via the query language itself —native extensibility— rather than the writing and importation of foreign functions defined in an external procedural language.

Extensibility represents a long-sought-after but hard-to-attain goal for database systems. Relational DBMSs provided no extensibility whatsoever, and even in the latest generation of O-R (Object-Relational) systems extensibility comes with many limitations and requires significant expertise and programming efforts. Indeed, the main extensibility mechanism of O-R systems is to allow SQL queries to call external functions coded in a procedural language (such as C/C++ or Java). Function libraries for specific application domains are now marketed aggressively by vendors under different brand names, such as datablades, DB extenders, cartridges, or snapins. However, they all share the same severe limitations with respect to power and flexibility, inasmuch as they can only support a predefined set of operators on single records: e.g., the functions cannot access the database tables either directly or through embedded SQL calls. Furthermore, procedural attachments to SQL, such as UDFs (User Defined Functions) or stored procedures, are notorious for being inordinately hard to develop and debug [4]. Thus current datablades are not conducive to end-user extensibility (even if their source code were available, which is seldom the case).

A breakthrough in extensibility was recently achieved by the ATLaS system [23, 24, 2]. ATLaS supports the standard SQL (on top of Berkeley DB [21]), and also supports calls to functions written in procedural languages (as O-R do). But, in addition to these, ATLaS allows end-users to define new table functions and aggregates *by programming them in SQL.*

Based on ATLaS, we pursued a new approach to spatio-temporal extensions, by

- starting with a minimal set of spatio-temporal constructs as built-in C++ functions, and
- using user defined new aggregates that implement more powerful spatio-temporal operators, designed for the specific application domain.

This chapter is organized as follows. In the next section, we describe the system ATLaS developed at UCLA which deals with UDAs. In Section 3, we introduce the data model and query language of $\mathrm{SQL}^{ST}$ [6] by sample queries. In Section 3.4, we demonstrate how to use ATLaS UDAs to support spatio-temporal queries. The performance of $\mathrm{SQL}^{ST}$ is analyzed and discussed in Section 3.5. In Section 3.6 we discuss how to support more abstract representations and alternative ways to express $\mathrm{SQL}^{ST}$ queries at that level. Section 3.7 concludes the chapter.

## 3.2 ATLaS

ATLaS (Aggregate Table Language and System) [23, 24, 2] implements the SQL language on top of Berkeley DB record manager [21]. But in addition to supporting SQL, and allowing the users to introduce new functions by programming them in C/C++ (as O-R systems do), ATLaS supports the definition of powerful user-defined aggregates (UDAs) expressed in a SQL-like language. By programming new UDAs in SQL, end users can easily extend the database system and support a variety of advanced database applications [23]. Furthermore, ATLaS is efficient. It comes with a very limited overhead over a direct implementation in a procedural language such as C/C++ (see performance results in Section 3.5).

To create an aggregate in ATLaS, a user needs to define three SQL routines, under the labels of INITIALIZE, ITERATE and TERMINATE, which, respectively, specify the computation to be performed for the first value in the stream, for each successive value, and when the end of the stream is detected. Results of the aggregation can be returned anytime during these routines by inserting tuples into an append-only RETURN stream. All the values in the RETURN stream will be returned to users at the end of the program.

For instance, to create an online average aggregate which returns results (current averages) for every 100 new values, we can define the following aggregate.

*Example 1.* MOVING_AVERAGE – a user-defined aggregate in ATLaS.

```
AGGREGATE MYAVG(next INT) : REAL
{
   TABLE state(sum INT, count INT);
   INITIALIZE : {
      INSERT INTO state VALUES(next, 1);
   }
   ITERATE : {
      UPDATE state SET sum = sum + next, count = count + 1;
      INSERT INTO return SELECT sum/count
         FROM state WHERE count % 100 = 0;
   }
   TERMINATE : {
      INSERT INTO return SELECT sum/count FROM state;
   }
}
```

The first line of this aggregate function declares a local table, STATE, to keep (in memory) the sum and count of the values processed so far. While, for this particular example, STATE contains only one tuple, it is in fact a table that can be queried and updated using SQL statements. These SQL statements are grouped into the three blocks labeled INITIALIZE, ITERATE and TERMINATE, respectively. Thus, the INITIALIZE statements insert the value taken from the input stream and sets the count to 1. The ITERATE statements update the table by adding the new input value to the sum and 1 to the count. The TERMINATE statements return the final result of computation by appending it to RETURN; for conformity with SQL, RETURN is viewed as a table which can have multiple values, and thus an INSERT INTO construct is used. We also add intermediate results from the computation to the RETURN table as part of the ITERATE statements.

Let us now consider the well known problem of coalescing after projection in a temporal table. We define the aggregate coalesce, which takes two parameters: from is the start time, to is the end time. Under the assumption that tuples are sorted by increasing start time, we can perform the task in one scan of the data. In the ITERATE routine, when the new interval overlaps the current interval stored in the state table, we coalesce the two intervals into one which ends with the later of the end time. Otherwise, the current interval is inserted into the RETURN table while the new interval becomes the current one.

*Example 2.* Coalescing

```
AGGREGATE COALESCE(from TIME, to TIME) :
            (start TIME, end TIME)
{
   TABLE state(cFrom TIME, cTo TIME);
   INITIALIZE : {
      INSERT INTO state VALUES(from, to);
   }
   ITERATE : {
      UPDATE state SET cTo = to
         WHERE cTo >= from AND cTo < to;
      INSERT INTO return SELECT cFrom, cTo
         FROM state WHERE cTo < from;
      UPDATE state SET cFrom = from, cTo = to
         WHERE cTo < from;
   }
   TERMINATE : {
      INSERT INTO return SELECT cFrom, cTo FROM state;
   }
}
```

The fact that UDAs in ATLaS are written in SQL achieves compatibility of data types and programming paradigms and inherits the well-known benefits of database query languages, such as scalability, data independence and plausibility. The ability of introducing and manipulating new tables is one of the first cornerstones of ATLaS' power. The second is the ability of one aggregate calling another, or calling itself recursively.

The ATLaS compiler translates ATLaS programs into C++ code. ATLaS adopts an open interface for its physical data model, so that the system can link with a variety of physical database implementations. The Berkeley DB library [21] is now used as ATLaS' main storage manager, but we have now added support for in-memory tables, and there is current work to support R-trees [16].

ATLaS UDAs can either be used as stand-alone programs or, imported into O-R systems such as DB2 (with limitations due to the fact that we use UDFs that return a single value for each call).

## 3.3 SQL$^{ST}$

In this section, we give a conceptual level description of SQL$^{ST}$ [6], which was designed following the general framework proposed by Worboys [26], where a spatio-temporal entity is modeled as a unified object having orthogonal spatial and temporal extents. Furthermore, Worboys' approach based on simplexes by representing 2-simplexes (triangular areas) [25] was improved by *counterclockwise directed triangles.* The virtue of counterclockwise directed triangles lies in the simple solution to the point-location problem [19], i.e., testing whether a point is *inside* a polygon, where a polygon is represented as set of triangles. The point-location problem is an important computational geometry problem and it is the basis to determine spatial relationships between spatial objects.

**Definition 1:** A triangle is a **counterclockwisely directed triangle** if its three vertexes, point$_1$=$(x_1, y_1)$, point$_2$=$(x_2, y_2)$, and point$_3$=$(x_3, y_3)$ are *counterclockwisely* orientated, i.e.,

$$\begin{vmatrix} x_1 & y_1 & 1 \\ x_2 & y_2 & 1 \\ x_3 & y_3 & 1 \end{vmatrix} > 0$$

Given a counterclockwise directed triangle T and an arbitrary point P, shown in Figure 3.1, below,

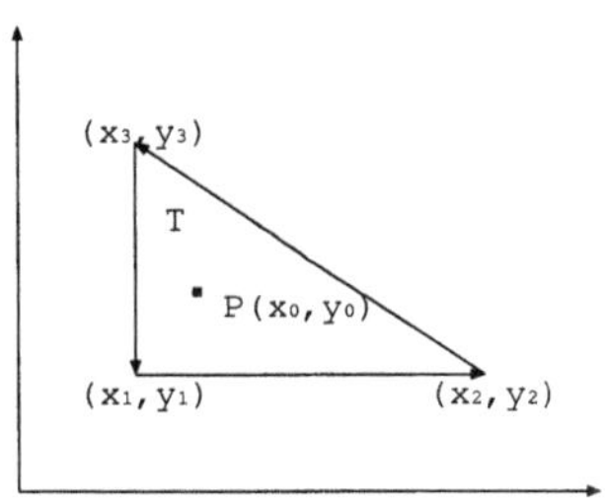


**Figure 3.1. An example of counterclockwise directed triangle**

the determination of whether P is inside the triangle is performed by the predicate function

**inside(point P, triangle T)** *which is true if the property*

$$\begin{vmatrix} x_i & y_i & 1 \\ x_j & y_j & 1 \\ x_0 & y_0 & 1 \end{vmatrix} \geq 0$$

*holds for:* $(i, j) = (1, 3), (2, 3)$, *and* $(3, 1)$.

If a point $P = (x_0, y_0)$ is on the *left* of all the directed edges of the triangle $T = ((x_1, y_1), (x_2, y_2), (x_3, y_3))^4$, then P is *inside* T.

Our basic spatial objects are *points, lines (finite line segments)*, and *counterclockwise directed triangles*; thus, a polygon is represented by a set of counterclockwise directed triangles. For example, the Hawaii island Maui is represented by more than 100 adjacent triangles with vertices specified in geographic coordinates, i.e., longitude and latitude.

By decomposing polygons into sets of triangles, hard-to-express spatial relationships between two spatial objects can be evaluated easily [9, 25]. The algorithms to decompose a polygon into triangles and to merge triangles into polygons can be found in [20, 18].

The $SQL^{ST}$ system supports the following built-in operators:

- **perimeter**
- **area**
- **inside**
- **distance**
- **intersect**
- **equal**
- **overlap**
- **contain**
- **disjoint**
- **adjacent**

The first two operators are unary operators that take as argument a polygon, and the others are binary operators. The **inside** operator takes as first operand a point and as a second operand a polygon. All the remaining operators take as operand two polygons.

We now introduce $SQL^{ST}$ through examples that illustrate its use in supporting spatio-temporal queries posed on cyclone-tracking database at a weather center. Specifically, we consider a spatio-temporal application from the NSF Arctic System Science (ARCSS) Research Program's Ocean Atmosphere Ice Interactions (OAII) Project. The marine environment is an interactive system comprising the water, ice, air, biota, dissolved chemicals and sediments. ARCSS/OAII seeks to enhance understanding of this system and its role in climate and global change. The extratropical cyclone data set [1] at ARCSS provides information about cyclones in the Northern Hemisphere. We will use this data set throughout our study. Included in this database is information about cyclone activities such as the cyclones' trajectories and the pressure along the trajectories during certain time intervals.

*Example 3.* Define the Cyclone relation.

CREATE TABLE **Cyclone (ID INT, Pressure REAL, Trajectory LINE,**
**Tstart DATE, Tend DATE)**

[4] A point is considered to be on the left of a line if it is on the (extended) line.

In this relation, the Tstart and Tend columns indicate the start and end instants of a time interval, and have a granularity of one day while the granularity could be larger or smaller, such as one month or one hour. Trajectory is a line segment whose start and end points are the positions (longitude, latitude) of the cyclone at time Tstart and Tend, respectively. Pressure is the average pressure between Tstart and Tend, and its unit is *milibar*. ID is the cyclone identifier. In this test case, the movement of cyclones are measured every day, so ID and Tstart together serve as the key of the relation.

A second type of information contained in the database is about regions of interest. For instance, all the islands in an archipelago might be described as follows.

*Example 4.* Define the Island relation.

```
CREATE TABLE Island (Name CHAR(30), Region TRIANGLE)
```

In this relation, the Region column stores the geometry of the island. Each polygon is decomposed into non-overlapping triangles. For example, our table contains about 100 rows each with Name = "Maui".

Following is a set of typical cyclone-related queries for storm tracks data set. We have also used similar queries to analyze the computer simulation data produced by a climatic weather model on another NASA-sponsored project [17].

*Example 5.* Find the cyclones which have pressure greater than 1000mb for more than three days during their life cycle.

```
SELECT ID
FROM Cyclone
WHERE Pressure > 1000
GROUP BY ID
HAVING DURATION(Tstart, Tend) > 3
```

This query will return the IDs of the cyclones whose overall high pressure stage have lasted more than three days.

As it can be easily inferred from the syntactic structure of this query, the GROUP BY and HAVING constructs, the operator DURATION is in fact a user-defined aggregate. The definition and implementation in SQL$^{ST}$ of this aggregate, as well as others that appear in this section will be discussed in detail in Section 3.4.

*Example 6.* Find the cyclones whose trajectory was enclosed by Maui.

```
SELECT ID
FROM Cyclone, Island
WHERE Name = "Maui"
GROUP BY ID
HAVING CONTAIN(Region, Trajectory)
```

This query returns the ID of the cyclones that have been completely inside the island "Maui". Aggregates in ATLaS can return any number of results. For instance, the aggregate CONTAIN(Region, Trajectory) returns a result only if the trajectory is completely inside the region. Thus CONTAIN(Region, Trajectory) in this query corresponds to the boolean condition that some result is returned.

*Example 7.* Find how long each cyclone has traveled when it was over Maui.

```
SELECT ID, SUM(length(intersect(Trajectory, Region)))
FROM Cyclone, Island
WHERE Name = "Maui"
GROUP BY ID
HAVING OVERLAP(Trajectory, Region)
   OR CONTAIN(Region, Trajectory)
```

This query returns the total distance the cyclones traveled when their trajectories overlapped with or were contained by the region of the island "Maui".

The functions intersect and length are built-in C++ functions. When a line (Trajectory) intersects a triangle (Region), the intersection is a line or a point. The length of these lines is then calculated and summed up.

Other spatial relationships similar to CONTAIN and OVERLAP include ADJACENT, DISJOINT, etc. The implementation of these operators is discussed in the next section.

*Example 8.* Identify all cyclones that have come within 50 miles of the coast of Lanai.

```
SELECT ID
FROM Cyclone, Island
WHERE Name = "Lanai"
GROUP BY ID
HAVING EDGE_DISTANCE(Trajectory, Region) <= 50
```

This query finds the cyclones that have come within 50 miles to the coast of the island "Lanai" in their trajectories .

*Example 9.* Find the cyclones that have traveled more than 300 miles continuously.

```
SELECT ID
FROM Cyclone
GROUP BY ID
HAVING MOVING_DISTANCE(Trajectory, Tstart, Tend) > 300
```

This query returns the IDs of the cyclones whose center have moved continuously over 300 miles.

In the above examples, we find SQL$^{ST}$ is a natural minimalist extension to SQL. Besides operators such as DURATION, CONTAIN, OVERLAP, EDGE_DISTANCE and MOVING_DISTANCE, users can define their own operators such as ADJACENT, DISJOINT, etc., also as user-defined aggregates. We will explain more about user-defined aggregates in the next section.

## 3.4 Spatio-Temporal Operators in ATLaS

Many abstractions have been proposed to deal with the complexity and diversity of temporal reasoning and spatial objects encountered in reality. Here we study how spatio-temporal operators used in the queries of the previous section can be expressed using (user-defined aggregates (UDAs); we first discuss temporal operators, then spatial ones, and finally the two combined.

### 3.4.1 Temporal Aggregates

We implemented our $\text{SQL}^{ST}$ system by first adding the built-in functions into ATLaS, and then using its UDA mechanism to implement the spatio-temporal aggregates needed in spatio-temporal queries. For instance, the temporal aggregate DURATION in Example 5 is implemented as follows:

*Example 10.* DURATION

```
AGGREGATE DURATION(Tstart DATE, Tend DATE) : INT
{
  TABLE state(i INT);
  INITIALIZE : {
    INSERT INTO state VALUES(Tend - Tstart + 1);
  }
  ITERATE : {
    UPDATE state SET i = i + (Tend - Tstart + 1);
  }
  TERMINATE : {
    INSERT INTO return SELECT i FROM state;
  }
}
```

This aggregate calculates the total length of the time intervals. For example, if we have a set of tuples as

```
cyclone(930001, _, _, '1993-01-01', '1993-01-05')
cyclone(930001, _, _, '1993-01-11', '1993-01-15')
cyclone(930001, _, _, '1993-01-21', '1993-01-25')
```

The result of DURATION will be 15 days.

### 3.4.2 Spatial Aggregates

UDAs can be used to implement the operators discussed in Section 3.3 on general polygons represented as sets of triangles, using the built-in operators on individual triangles. The implementation of EDGE_DISTANCE, CONTAIN and OVERLAP is discussed next.

The EDGE_DISTANCE between two polygons, each represented as a set of triangles, is the smallest distance between the vertices of triangles in either polygon and the edges of triangles in the other.

*Example 11.* EDGE_DISTANCE

```
AGGREGATE EDGE_DISTANCE(Object1 TRIANGLE,
          Object2 TRIANGLE) : REAL
{
  TABLE state(d REAL);
  INITIALIZE : ITERATE : {
    INSERT INTO state
      SELECT distance(V,E)
      FROM TABLE(vertex(Object1)) AS V,
           TABLE(edge(Object2)) AS E;
    INSERT INTO state
      SELECT distance(V,E)
      FROM TABLE vertex(Object2)) AS V,
           TABLE(edge(Object1)) AS E;
  }
  TERMINATE : {
    INSERT INTO return SELECT MIN(d) FROM state;
  }
}
```

In this example, we use the table functions vertex and edge which take as input an object of type triangle, and return its vertices and sides, respectively. Table functions are a very useful SQL:1999 [10] construct supported in most commercial DBMSs. ATLaS also supports table functions, where they can either be written in an external procedural language, or in SQL itself, as in the case of UDAs. The function distance, used in Example 11, is instead a standard scalar function that calculates the Euclidean distance between a point and a line segment.

We next discuss the implementation of CONTAIN. Here we take advantage of inheritance and overloading characteristics of O-R systems [22]. Both point and line are subtypes of triangle. A point can be considered as a triangle whose three vertices are the same and a line as a triangle whose first two vertices are the start and end points of the line and the last vertex is the center of the line.

*Example 12.* CONTAIN

```
AGGREGATE CONTAIN(Object1 TRIANGLE, Object2 TRIANGLE)
          : INT
{
  TABLE state(b INT);
  TABLE triangles(Object TRIANGLE);
  TABLE points(Vertex POINT);
  TABLE edges1(line1 LINE, count1 INT);
  TABLE edges2(line2 LINE, count2 INT);
  INITIALIZE : {
    INSERT INTO state VALUES 1;
  }
  ITERATE : {
    INSERT INTO triangle VALUES(Object1);
    INSERT INTO points VALUES(Object2.Vertex);
    INSERT INTO edges1 VALUES(Object1.Edge);
    UPDATE count1 = count1 + 1 WHERE Object1.Edge IN
        (SELECT line1 FROM edges1);
    INSERT INTO edges2 VALUES(Object2.Edge);
    UPDATE count2 = count2 + 1 WHERE Object2.Edge IN
        (SELECT line2 FROM edges2);
  }
  TERMINATE : {
    UPDATE state SET b = 0 WHERE NOT EXIST
        (SELECT Vertex FROM points, triangles
           WHERE inside(Vertex, Object) = 1);
      OR EXIST
        (SELECT intersect(line1,line2) FROM edges1, edges2
           WHERE count1 = 1 AND count2 = 1);
    INSERT INTO return SELECT b FROM state WHERE b = 1;
  }
}
```

The function inside is a built-in function that will evaluate to 1 if a point is *inside* a triangle; 0 otherwise; and the function intersect is a built-in function that will return the common part of two intersect lines. It does not return anything if the lines do not overlap or cross.

The spatial aggregate CONTAIN($Object_1$, $Object_2$) then uses inside and intersect to test if $Object_1$ contains $Object_2$. The aggregate first iterate through all the records and insert the triangles belonging to $Object_1$ into one table, the vertices of $Object_2$ into another table, and their edges into two other different tables. The edges are associated with a count. If the count is 1, then the edge only occurs once in the set of triangles; thus the edge is an outer edge, i.e., an edge of the original polygon.

Initially, b in the STATE is set to 1. If there exists a vertex of $Object_2$ that is not contained in any triangle of $Object_1$, or if their outer edges cross, then b will be set to 0, thus not inserted into the RETURN table. As a result, the aggregate CONTAIN will not return anything; otherwise, it will return 1, the value of b.

Another example is the aggregate OVERLAP, which can be defined as follows.

*Example 13.* OVERLAP

```
AGGREGATE OVERLAP(Object1 TRIANGLE, Object2 TRIANGLE)
            : INT
{
   TABLE state(b INT);
   TABLE edges1(line1 LINE);
   TABLE edges2(line2 LINE);
   INITIALIZE : {
      INSERT INTO state VALUES 0;
   }
   ITERATE : {
      INSERT INTO edges1 VALUES(Object1.Edge);
      INSERT INTO edges2 VALUES(Object2.Edge);
   }
   TERMINATE : {
      UPDATE state SET b = 1 WHERE EXIST
          (SELECT intersect(line1,line2) FROM edges1, edges2 ;
      INSERT INTO return SELECT b FROM state WHERE b = 1;
   }
}
```

Similar to CONTAIN, the spatial aggregate OVERLAP uses intersect to test if an edge of a triangle belonging to $\text{Object}_1$ crosses an edge of a triangle belonging to $\text{Object}_2$. Therefore, the aggregate first iterates through all the records and inserts the edges of $\text{Object}_1$ and $\text{Object}_2$ into two different tables. If there exists two crossing edges, then the value of b will be set to 1, which will then be returned by the aggregate OVERLAP($\text{Object}_1$, $\text{Object}_2$); otherwise, OVERLAP($\text{Object}_1$, $\text{Object}_2$) will not return anything.

In a similar fashion, end-users can easily define additional operators such as ADJACENT, DISJOINT, thus producing a spatio-temporal database system that has powerful primitives and is customized for their application.

### 3.4.3 Spatio-Temporal Aggregates

Some operators may require dealing with both spatial and temporal information at the same time, such as MOVING_DISTANCE.

*Example 14.* MOVING_DISTANCE

```
AGGREGATE MOVING_DISTANCE(Object, Tstart DATE,
            Tend DATE) : REAL
{
   TABLE state(d REAL, x REAL, y REAL, time DATE);
```

```
    INITIALIZE : {
       INSERT INTO state
           VALUES(0, center_x(Object), center_y(Object), Tend);
    }
    ITERATE : {
       UPDATE state SET
           d = d + sqrt((center_x(Object)-x)^2 + (center_y(Object)-y)^2),
            x = center_x(Object), y = center_y(Object), time = Tend
           WHERE Tstart = time + 1);
       INSERT INTO state
           VALUES(0, center_x(Object), center_y(Object), Tend);
           WHERE Tstart ≠ time + 1);
    }
    TERMINATE : {
       INSERT INTO return SELECT d FROM state;
    }
}
```

This aggregate calculates the distance an object travels in each continuous time period. We calculate and sum up the distance between two positions of the center of an object at two consecutive time intervals. When the start of a new time interval is not consecutive to the ends of any existing time intervals in the state table, we start a new round of calculation. Upon termination, all the d values in the state table, i.e., the distance traveled for each continuous time period, are returned. The built-in functions $\mathsf{center}_x$ and $\mathsf{center}_y$ calculates the $x$ and $y$ coordinates of the center of an object, respectively.

## 3.5 Performance

A key question to evaluate the effectiveness of the $\mathrm{SQL}^{ST}$ approach is how much overhead there is to have our extensions programmed in SQL rather than in C or C++. Therefore we performed the following experiment on the cyclone data set obtained from [1]. The data set contains a 28-year record (1 May 1966 through 31 December 1993) of daily cyclone statistics for the Northern Hemisphere. The data set includes the position and pressure of each cyclone, together with the ID of the cyclone and a date field. The data set has about 200,000 records and is over 14MB. Also, we used an island relation that contains 1000 tuples. Then, we compared the performance of four queries written in C++ (accessing the data through the Berkeley-DB API), against those written in ATLaS. An index has been created on the ID, Tstart columns of the Cyclone table and on the Name column of the Island table. We compare the results of the queries in four different cases: (1) ATLaS using indexes; (2) ATLaS not using indexes; (3) C++ using indexes; and (4) C++ not using indexes.

We tested the performance of the following four queries on a Pentium III with a single 500HZ processor and 160MB memory, running LINUX.

*Example 15.* Find the duration of the cyclones occurred in June, 1966.

```
SELECT ID, DURATION(Tstart, Tend)
FROM Cyclone
WHERE '1966-06-01' <= Tstart AND '1966-07-01' > Tstart
GROUP BY ID
```

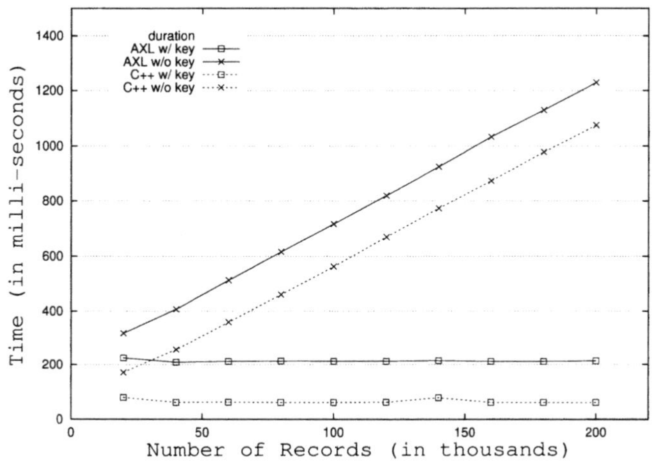


**Figure 3.2. Performance result of Query 15**

*Example 16.* Find the distance traveled by the cyclones occurred in June, 1966.

```
SELECT ID, MOVING_DISTANCE(position, Tstart, Tend)
FROM Cyclone
WHERE '1966-06-01' <= Tstart AND '1966-07-01' > Tstart
GROUP BY ID
```

Figures 3.2 and 3.3 are the results of the aggregates DURATION and MOVING_DISTANCE. Since the number of records that meet the selection criterion is fixed, so when a key is used, we only need to retrieve the qualified records, thus the result of performance of ATLaS and C++ are constant. When no key is used, we need to search the entire database, so the result of the performance of both ATLaS and C++ increase linearly. In both cases, the program written directly in C++ against the Berkeley DB API outperforms ATLaS by a constant.

*Example 17.* Find the cyclones which occurred in June, 1966 and landed on the island Oahu.

```
SELECT ID
```

```
FROM Cyclone, Island
WHERE '1966-06-01' <= Tstart AND '1966-07-01' > Tstart
      AND Name = "Oahu"
GROUP BY ID
HAVING CONTAIN(Region, Trajectory)
```

*Example 18.* Find the distance between cyclones which occurred in June, 1966 and the coast of island Oahu.

```
SELECT ID, EDGE_DISTANCE(Trajectory, Region)
FROM Cyclone, Island
WHERE '1966-06-01' <= Tstart AND '1966-07-01' > Tstart
      AND Name = "Oahu"
GROUP BY ID
```

Figure 3.4 and 3.5 show, in logarithmic scale, the results of the aggregates CONTAIN and EDGE_DISTANCE. Still, when a key is used, the results of performance of ATLaS and C++ are constant. When no key is used, the results of the performance of both ATLaS and C++ increase. Again, In both cases, C++ API of Berkeley-DB is slightly faster than ATLaS. Unlike the aggregates DURATION and MOVING_DISTANCE which handle the incoming data as soon as it comes in without storing it, the aggregates CONTAIN and EDGE_DISTANCE need to insert the incoming data into temporary tables first, and then upon termination, query these tables to get the desired results. Thus the performance curve of these two aggregates is different from that of DURATION and MOVING_DISTANCE.

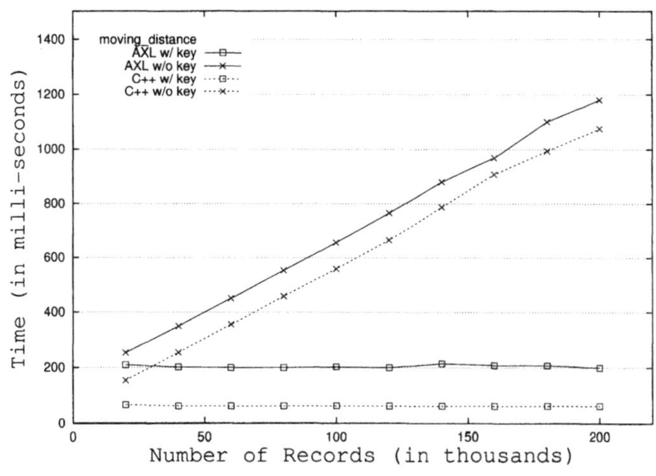


**Figure 3.3. Performance result of Query 16**

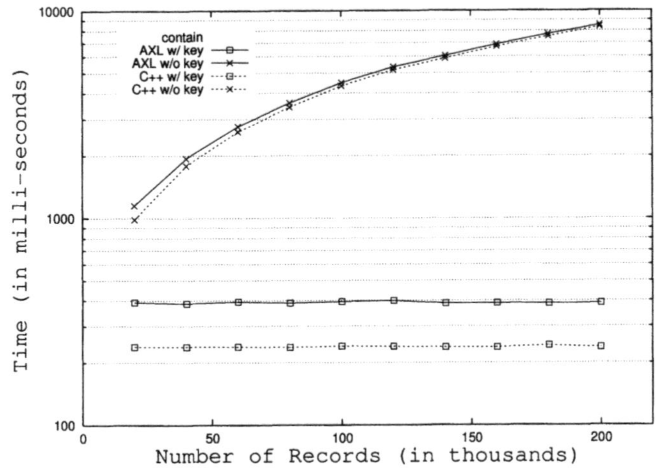


**Figure 3.4. Performance result of Query 17**

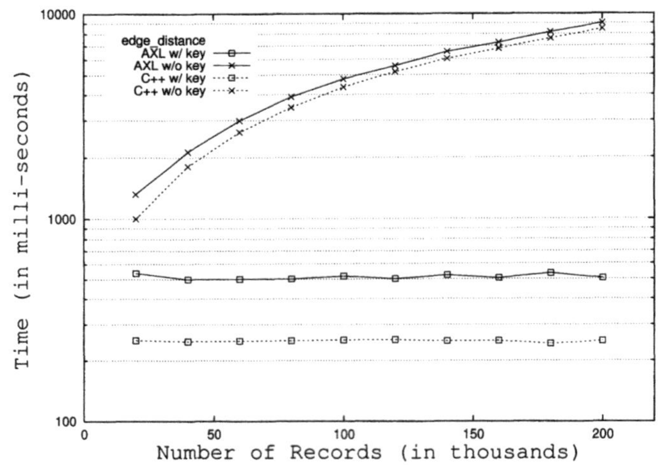


**Figure 3.5. Performance result of Query 18**

The performance results show that the user-defined aggregates of SQL$^{ST}$ only incur in a modest performance penalty for the ease of use and extensibility they provide.

## 3.6 Toward More Abstract Representations

The extensibility mechanisms of SQL$^{ST}$ can also be used to elevate the representation to richer and more abstract levels. For instance, at the conceptual level, polygons provide a powerful representation for spatial objects. In terms of temporal representation, users may want to see the movement of the cyclones as a series of snapshot of their positions. In this section, we show how $\text{SQL}^{ST}$ can be further extended to support these more abstract representations. In fact, we will now express the queries in Section 3.3 using *points* and *polygons* as spatial data types and *time instants* as temporal data type.

### 3.6.1 Schema Definition

The schema of the databases in Section 3.3 have been changed as follows.

*Example 19.* Define the Cyclone relation. (Example 3)

```
CREATE TABLE
      Cyclone (ID INT, Pressure REAL, Position POINT, Time DATE)
```

In the Cyclone relation, the spatial data column is Position which has a data type point and specifies the $x$ and $y$ coordinates of the center of a cyclone. The temporal data column is Time which has a granularity of one day and captures the time instants of cyclones' movements.

*Example 20.* Define the Island relation. (Example 4)

```
CREATE TABLE
      Island (Name CHAR(30), Extent POLYGON)
```

In the Island relation, the Extent column has a spatial data type as polygon and specifies the geometry of the island.

### 3.6.2 Spatio-Temporal Queries

Since the spatio-temporal operators are supported at the internal level, so table expressions in O-R systems [4] are used to transform the queries.

For instance, Example 5 in Section 3.3 is expressed as follows at the conceptual level.

*Example 21.* Find the cyclones which have pressure greater than 1000mb for more than three days during their life cycle (Example 5).

```
SELECT New.ID
FROM (SELECT ID, MAPPING(Position, Time)
        FROM Cyclone
        WHERE C.Pressure > 1000 GROUP BY ID)
        AS New(ID, Trajectory, Tstart, Tend)
GROUP BY New.ID
HAVING DURATION(New.Tstart, New.Tend) > 3
```

In this query, we use table expressions in the FROM clause to accomplish the transformation between different levels of abstraction.

To map a pairs of points and time instants into a line and a time interval, we define the UDA MAPPING:

*Example 22.* MAPPING

```
AGGREGATE MAPPING(position POINT, time DATE) :
            (LINE, DATE, DATE)
{
   TABLE state(l LINE, d1 DATE, d2 DATE);
   TABLE tmp(p POINT, t DATE);
   INITIALIZE : {
      INSERT INTO tmp VALUES(position, time);
   }
   ITERATE : {
      INSERT INTO state VALUES((p, position), t, time)
         WHERE time = t + 1;
      UPDATE tmp SET p = position, t = time;
   }
   TERMINATE : {
      INSERT INTO return SELECT l, d1, d2 FROM state;
   }
}
```

Similarly, Example 6 can be expressed as follows.

*Example 23.* Find the cyclones whose trajectory was enclosed by Maui (Example 6).

```
SELECT New.ID
FROM (SELECT ID, MAPPING(Position, Time), T.Region
         FROM Cyclone, Island, TABLE(decompose(Extent)) AS T
         WHERE Name = "Maui")
         AS New (ID, Trajectory, Tstart, Tend, Region)
GROUP BY New.ID
HAVING CONTAIN(New.Region, New.Trajectory)
```

Decompose is a built-in function which triangulates a polygon (Extent in this query) into a set of triangles (Region). The algorithm of decompose can be found in [20, 18].

## 3.7 Conclusions

In this chapter, we described a novel approach to extensible data models and query languages to manage spatio-temporal information, and demonstrated its benefits in implementing the $\text{SQL}^{ST}$ system. By building on the native extensibility mechanisms of ATLaS, we minimized the external procedural extensions required, since only user-defined aggregates and table functions defined in SQL are now needed to support powerful spatio-temporal constructs.

In $\text{SQL}^{ST}$, end-users can easily introduce their own spatio-temporal operators, and retain the flexibility of working at different levels of abstraction as best fit their needs. Current work focuses on adding new spatial and temporal indices, including R-trees [16], and the optimization of complex queries with temporal joins and spatial joins.

## References


1. ARCSS Data and Information Archive Cyclone Track Data Set. *http://arcss.colorado.edu/Catalog/arcss003.html*
2. ATLaS User Manual. *http://wis.cs.ucla.edu/atlas*
3. Cai M., Keshwani D., Revesz P.Z. (2000) Parametric Rectangles: A Model for Querying and Animation of Spatiotemporal Databases. In: Proc. of the 7th International Conference on Extending Database Technology, pp. 430–444.
4. Chamberlin D. (1998) A Complete Guide to DB2 Universal Database. Morgan Kaufmann.
5. Chen C.X., Zaniolo C. (1999) Universal Temporal Extensionsfor Data Languages. In: Proc. of the 15th International Conference on Data Engineering, pp. 428–437.
6. Chen C.X., Zaniolo C. (2000) $SQL^{ST}$: A Spatio-Temporal Data Model and Query Language. In: Proc. of the 19th International Conference on Conceptual Modeling, pp. 96–111.
7. Chomicki J., Revesz P.Z. (1997) Constraint-Based Interoperability of Spatiotemporal Databases. In: Advances in Spatial Databases, LNCS 1262, Springer, pp. 142–161.
8. Clarkon K.L., Tarjan R.E., Van Wyk C.J. (1989) A Fast Las Vegas Algorithm for Triangulating a Simple Polygon. Discrete & Computational Geometry 4(5):423–432.
9. Egenhofer M.J., Frank A.U., Jackson J.P. (1989) A Topological Data Model for Spatial Databases. In: Proc. of the Symposium on the Design and Implementation of Large Spatial Databases, pp. 271–286.
10. Eisenberg A., Melton J. (1999) SQL: 1999, formerly known as SQL 3. SIGMOD Record 28(1):131–138.
11. Forlizzi L., Güting R.H., Nardelli E., Schneider M. (2000) A Data Model and Data Structures for Moving Objects Databases. In: Proc. of ACM-SIGMOD International Conference on Management of Data, pp. 319–330.
12. Fournier A., Montuno D.Y. (1984) Triangulating simple polygons and equivalent problems. ACM Transactions on Graphics 3(2):153–174.
13. Güting R.H., Böhlen M.H., Erwig M., Jensen C.S., Lorentzos N.A., Schneider M., Vazirgiannis M. (2000) A Foundation for Representing and Querying Moving Objects. ACM Transactions on Database Systems 25(1):1–42.
14. Garey M.R., Johnson D.S., Preparata F.P., Tarjan R.E. (1978) Triangulating a Simple Polygon. Information Processing Letters 7(4):175–179.
15. Grumbach S., Rigaux P., Segoufin L. (1998) The DEDALE System for Complex Spatial Queries. In: Proc. of ACM-SIGMOD International Conference on Management of Data, pp. 213–224.
16. Guttman A. (1984) R-Trees: A Dynamic Index Structure for Spatial Searching. In: Proc. of ACM-SIGMOD International Conference on Management of Data, pp. 47–57.

17. Muntz R., Shek E., Zaniolo C. (1995) Using $LDL++$ For Spatio-Temporal Reasoning in Atmospheric Science Databases. In: Ramakrishnan R. (Eds.) Applications of Logic Databases, pp. 101–119.
18. O'Rourke J. (1998) Computational Geometry in C. Cambridge University Press.
19. Preparata F., Shamos M. (1988) Computational Geometry: An Introduction. Springer.
20. Seidel R. (1991) A Simple and Fast Incremental Randomized Algorithm for Computing Trapezoidal Decompositions and for Triangulating Polygons. Computational Geometry: Theory and Applications 1(1):51–64.
21. Sleepycat Software. The Berkeley Database (Berkeley DB). *http://www.sleepycat.com*
22. Stonebraker M., Brown P., Moore D. (1999) Object-relational DBMSs: tracking the next great wave. Morgan Kaufmann.
23. Wang H., Zaniolo C. (2000) Using SQL to Build New Aggregates and Extenders for Object-Relational Systems. In: Proc. of the 26th International Conference on Very Large Databases, pp. 166–175.
24. Wang H., Zaniolo C. (2003) ATLaS: A Native Extension of SQL for Data Mining. In: Proc. of the 2nd SIAM International Conference on Data Mining.
25. Worboys M.F. (1992) A Generic Model for Planar Geographical Objects. International Journal of Geographic Information Systems 6(5):353–372.
26. Worboys M.F. (1994) A Unified Model for Spatial and Temporal Information. Computer Journal 37(1):26–34.

# 4

# A Declarative Framework for Reasoning on Spatio-Temporal Data

Mirco Nanni[1], Alessandra Raffaetà[2], Chiara Renso[1], and Franco Turini[3]

[1] ISTI CNR - Pisa
{nanni, renso}@isti.cnr.it
[2] Dipartimento di Informatica - Università Ca' Foscari Venezia
raffaeta@dsi.unive.it
[3] Dipartimento di Informatica - Università di Pisa
turini@di.unipi.it

## 4.1 Introduction

New technologies in the field of mobile computing and communication can provide a wealth of spatio-temporal information. Collected data are useful as far as they can be used to analyse phenomena and to take informed decisions. The first step for allowing one to make a profitable use of data is to provide a query language for them, and we believe that the query language, even more in the case of spatio-temporal data, must be able to handle not only data but also rules, and exhibit both deductive and inductive capabilities. Rules can be used to represent general knowledge about the collected data, and deductive capabilities can provide answers to queries that require some inference besides the crude manipulation of the data. Induction can help extracting implicit knowledge from data and, according to the impressive success in the knowledge discovery in database field, is a powerful support to decision making.

In order to support both deductive and inductive reasoning on spatio-temporal data we propose the formalism STACLP (Spatio-Temporal Annotated Constraint Logic Programming) [20], based on constraint logic programming extended with annotations. The basic constraint logic programming provides the deductive capabilities, and annotations allow a neat representation of temporal, spatial and spatio-temporal knowledge. On this ground we can construct the inductive capabilities of the language by implementing knowledge extraction methods. The computation of clusters of trajectories is the case we discuss in detail in the chapter. However we believe that it is paradigmatic for the general approach to the definition of knowledge discovery methods in STACLP.

Our proposal fits in a new and promising research field, that is the integration of declarative paradigms and systems for dealing with spatial and/or temporal information, such as spatial databases and Geographical Information Systems (GISs). In the literature we can find other attempts to exploit the deductive capabilities of logics to *reason* on geographic data [25, 23]. For instance [1] expresses spatial data in an object-oriented paradigm and builds a deductive part to infer knowledge from the spatial objects. The system DEDALE [12, 13], instead, relies on a logical model based on linear constraints, which generalises the constraint database model of [16]. In DEDALE both time and space are modelled in a uniform way by using constraints (for more details see Sect. 4.4). Finally, we recall MuTACLP [21, 18], a constraint logic based knowledge representation language that offers facilities for handling spatio-temporal information and some basic operators for combining different spatio-temporal knowledge bases. In MuTACLP temporal information is represented by annotations whereas spatial information is encoded into the formulae, by using constraints. This leads to a mismatch of conceptual levels and a loss of simplicity. In STACLP we overcome this mismatch, by defining a uniform setting where spatial information is represented by means of annotations, so that the advantages of using annotations apply to the spatial dimension as well.

Information induction over spatio-temporal data is mainly an unexplored field, yet. In this context, [2] suggests two main kinds of information to induce: *meta-rules*, i.e., regularities shown along time by the rules obtained in each snapshot, and *evolution rules*, i.e., rules computed over pre-computed spatio-temporal features of entities. So far, very little attention has been paid to the mining of spatio-temporal classifiers, association rules and time sequences, while a few proposals exist for spatio-temporal clustering. In addition to the trivial extensions of spatial clustering methods, which treat time as just another generic dimension, three main approaches can be found in literature, focusing on the clustering of moving objects: [17] considers generic sequences together with a conceptual hierarchy over the sequence elements, used to compute both the cluster representatives and the distance between two sequences. In [11], a model-based clustering method for continuous trajectories is proposed, which puts together objects which can be obtained from a common core trajectory by adding noise with normal distribution. Finally, [9] proposes a general mapping from any data space to an Euclidean space, where any standard clustering algorithm can be applied.

In Sects. 4.2 and 4.3 we introduce syntax and semantics of STACLP along with some simple examples, that illustrate its knowledge representation capabilities. In Sect. 4.4 we deal with the representation of one of the most fundamental spatio-temporal objects, i.e. the trajectory. In Sect. 4.5 we set the stage for the rest of the chapter, by introducing the requirements of a complex application dealing with the analysis of the behaviour of a nice group of animals, crested porcupines, living in a natural park of Tuscany (Italy). In Sect. 4.6 we present how STACLP allows us to solve part of the above problems by using the deductive capabilities of the language, while Sect. 4.7 deals with implementing a clustering method in STACLP, how to specialise it for trajectories, and how to solve the problems of the application that require such a kind of inductive reasoning. Finally, Sect. 4.8 draws some conclusions and outlines future research directions.

## 4.2 STACLP: a Spatio-Temporal Language

In this section we present the language STACLP [20], which extends Temporal Annotated Constraint Logic Programming [10], a constraint logic programming language with temporal annotations, by adding spatial annotations. The pieces of spatio-temporal information are given by pairs of annotations which specify the spatial extent of an object at a certain time period. The use of annotations makes time and space explicit but avoids the proliferation of spatial and temporal variables and quantifiers. Moreover, it supports both definite and indefinite spatial and temporal information, and it allows one to establish a dependency between space and time, thus permitting to model continuously moving points and regions.

### 4.2.1 Time and Space

Time can be discrete or dense. Time points are totally ordered by the relation $\leq$. We denote by $\mathcal{T}$ the set of time points and we suppose to have a set of operations (e.g., the binary operations $+$, $-$) to manage such points. The time-line is left-bounded by 0 and open to the future, with the symbol $\infty$ used to denote a time point that is later than any other. A *time period* is an interval $[r, s]$ with $r, s \in \mathcal{T}$ and $0 \leq r \leq s \leq \infty$, which represents the convex, non-empty set of time points $\{t \mid r \leq t \leq s\}$. Thus the interval $[0, \infty]$ denotes the whole time line.

Analogously space can be discrete or dense and we consider as spatial regions *rectangles* represented as $[(x_1, x_2), (y_1, y_2)]$, where $(x_1, y_1)$ and $(x_2, y_2)$ denote the lower-left and upper-right vertex of the rectangle. Precisely, $[(x_1, x_2), (y_1, y_2)]$ models the region $\{(x, y) \mid x_1 \leq x \leq x_2, y_1 \leq y \leq y_2\}$. Rectangles are the two-dimensional counterpart of convex sets of time points.

### 4.2.2 Annotations and Annotated Formulae

An *annotated formula* is of the form $A\,\alpha$ where $A$ is an atomic formula and $\alpha$ an annotation. We define three kinds of temporal and spatial annotations inspired by similar principles:

- `at` $T$ and `atp` $(X, Y)$ are used to express that a formula holds in a time or spatial point.
- `th` $I$, `thr` $R$ are used to express that a formula holds *throughout*, i.e., at *every* point, in the temporal interval or the spatial region, respectively.
- `in` $I$, `inr` $R$ are used to express that a formula holds at *some* point(s), in the interval or the region, respectively. They account for indefinite information.

The set of annotations is endowed with a partial order relation $\sqsubseteq$. Given two annotations $\alpha$ and $\beta$, the intuition is that $\alpha \sqsubseteq \beta$ if $\alpha$ is "less informative" than $\beta$ in the sense that for all formulae $A$, $A\,\beta \Rightarrow A\,\alpha$. This partial order is used in the definition of new inference rules. In addition to Modus Ponens, STACLP has the two inference rules below:

$$\frac{A\,\alpha \qquad \gamma \sqsubseteq \alpha}{A\,\gamma}\ \textit{rule}\ (\sqsubseteq) \qquad\qquad \frac{A\,\alpha \qquad A\,\beta \qquad \gamma = \alpha \sqcup \beta}{A\,\gamma}\ \textit{rule}\ (\sqcup)$$

The rule ($\sqsubseteq$) states that if a formula holds with some annotation, then it also holds with all annotations that are smaller according to the partial ordering. The rule ($\sqcup$) says that if a formula holds with some annotation $\alpha$ and the same formula holds with another annotation $\beta$ then it holds with the least upper bound $\alpha \sqcup \beta$ of the two annotations.

Next, we introduce the *constraint theory for temporal and spatial annotations.* A constraint theory is a non-empty, consistent first order theory that axiomatises the meaning of the constraints. Besides an axiomatisation of the total order relation $\leq$ on the set of points, the constraint theory includes the axioms in Table 4.1 defining the partial order on temporal and spatial annotations. The first two axioms state that $\mathtt{th}\,I$ and $\mathtt{in}\,I$ are equivalent to $\mathtt{at}\,t$ when the time period $I$ consists of a single time point $t$. Next, if a formula holds at every point of a time period, then it holds at every point in all sub-periods of that period (($\mathtt{th}\ \sqsubseteq$) axiom). On the other hand, if a formula holds at some points of a time period then it holds at some points in all periods that include this period (($\mathtt{in}\ \sqsubseteq$) axiom). The axioms for spatial annotations are analogously defined.

| | |
|---|---|
| ($\mathtt{at\,th}$) | $\mathtt{at}\,t = \mathtt{th}\,[t,t]$ |
| ($\mathtt{at\,in}$) | $\mathtt{at}\,t = \mathtt{in}\,[t,t]$ |
| ($\mathtt{th}\ \sqsubseteq$) | $\mathtt{th}\,[s_1,s_2] \sqsubseteq \mathtt{th}\,[r_1,r_2] \Leftrightarrow r_1 \leq s_1,\ s_2 \leq r_2$ |
| ($\mathtt{in}\ \sqsubseteq$) | $\mathtt{in}\,[r_1,r_2] \sqsubseteq \mathtt{in}\,[s_1,s_2] \Leftrightarrow r_1 \leq s_1,\ s_2 \leq r_2$ |
| ($\mathtt{atp\,thr}$) | $\mathtt{atp}\,(x,y) = \mathtt{thr}\,[(x,x),(y,y)]$ |
| ($\mathtt{atp\,inr}$) | $\mathtt{atp}\,(x,y) = \mathtt{inr}\,[(x,x),(y,y)]$ |
| ($\mathtt{thr}\ \sqsubseteq$) | $\mathtt{thr}\,[(x_1,x_2),(y_1,y_2)] \sqsubseteq \mathtt{thr}\,[(x_1',x_2'),(y_1',y_2')] \Leftrightarrow$ $x_1' \leq x_1,\ x_2 \leq x_2', y_1' \leq y_1,\ y_2 \leq y_2'$ |
| ($\mathtt{inr}\ \sqsubseteq$) | $\mathtt{inr}\,[(x_1',x_2'),(y_1',y_2')] \sqsubseteq \mathtt{inr}\,[(x_1,x_2),(y_1,y_2)] \Leftrightarrow$ $x_1' \leq x_1,\ x_2 \leq x_2', y_1' \leq y_1,\ y_2 \leq y_2'$ |

**Table 4.1.** Axioms for the partial order on annotations

### 4.2.3 Combining Spatial and Temporal Annotations

In order to obtain spatio-temporal annotations the spatial and temporal annotations are combined by considering pairs of annotations as a new class of annotations. Let us first introduce the general idea of pairing of annotations.

**Definition 1.** *Let* $(A, \sqsubseteq_A)$ *and* $(B, \sqsubseteq_B)$ *be two disjoint classes of annotations with their partial order. Their* pairing *is the class of annotations* $(A{*}B, \sqsubseteq_{A*B})$ *defined as* $A * B = \{\alpha\beta, \beta\alpha \mid \alpha \in A,\ \beta \in B\}$ *and* $\gamma_1 \sqsubseteq_{A*B} \gamma_2$ *whenever*

$$((\gamma_1 = \alpha_1\beta_1 \wedge \gamma_2 = \alpha_2\beta_2) \vee (\gamma_1 = \beta_1\alpha_1 \wedge \gamma_2 = \beta_2\alpha_2)) \wedge (\alpha_1 \sqsubseteq_A \alpha_2 \wedge \beta_1 \sqsubseteq_B \beta_2)$$

In our case the spatio-temporal annotations are obtained by considering the pairing of spatial and temporal annotations.

**Definition 2.** *The class of spatio-temporal annotations is the pairing of the spatial annotations* Spat *built from* `atp`, `thr` *and* `inr` *and of the temporal annotations* Temp, *built from* `at`, `th` *and* `in`, *i.e.* Spat$*$Temp.

To clarify the meaning of our spatio-temporal annotations, we present some examples of their formal definition in terms of `at` and `atp`. Let $t$ be a time point, $J = [t_1, t_2]$ be a time period, $s = (x, y)$ be a spatial point and $R = [(x_1, x_2), (y_1, y_2)]$ be a rectangle.

- The equivalent annotated formulae $A\,\mathtt{atp}\,s\,\mathtt{at}\,t$ and $A\,\mathtt{at}\,t\,\mathtt{atp}\,s$ mean that $A$ holds at time point $t$ in the spatial point $s$.
- The annotated formula $A\,\mathtt{thr}\,R\,\mathtt{th}\,J$ means that $A$ holds *throughout* the time period $J$ and at every spatial point in $R$. The definition of such a formula in terms of `atp` and `at` is:

$$A\,\mathtt{thr}\,R\,\mathtt{th}\,J \;\Leftrightarrow\; \forall t \in J.\ \forall s \in R.\ A\,\mathtt{atp}\,s\,\mathtt{at}\,t.$$

  The formula $A\,\mathtt{th}\,J\,\mathtt{thr}\,R$ is equivalent to the formula above because one can be obtained from the other just by swapping the universal quantifiers.
- The annotated formula $A\,\mathtt{thr}\,R\,\mathtt{in}\,J$ means that there exist(s) *some* time point(s) in the time period $J$ in which $A$ holds throughout the region $R$. The definition of such a formula in terms of `atp` and `at` is:

$$A\,\mathtt{thr}\,R\,\mathtt{in}\,J \;\Leftrightarrow\; \exists t \in J.\ \forall s \in R.\ A\,\mathtt{atp}\,s\,\mathtt{at}\,t.$$

  In this case swapping the annotations swaps the universal and existential quantifiers and hence results into a different annotated formula $A\,\mathtt{in}\,J\,\mathtt{thr}\,R$, meaning that for every spatial point in the region $R$, $A$ holds at some time point(s) in $J$. Thus we can state $\mathit{snow}\,\mathtt{thr}\,R\,\mathtt{in}\,[\mathit{jan}, \mathit{mar}]$ in order to express that there exists a time period between January and March in which the whole region $R$ is completely covered by the snow. On the other hand $\mathit{snow}\,\mathtt{in}\,[\mathit{jan}, \mathit{mar}]\,\mathtt{thr}\,R$ expresses that from January to March each point of the region $R$ will be covered by the snow, but different points can be covered in different time instants.

### 4.2.4 Least Upper Bound and its Constraint Theory

For technical reasons related to the properties of annotations (see [10, 20]), we restrict the rule ($\sqcup$) to least upper bounds that produce valid, new annotations, i.e., rectangular regions and temporal components which are time

(1) $\mathtt{thr}\,[(x_1,x_2),(y_1,y_2)]\mathtt{th}\,[t_1,t_2] \sqcup \mathtt{thr}\,[(x_1,x_2),(z_1,z_2)]\mathtt{th}\,[t_1,t_2] = \mathtt{thr}\,[(x_1,x_2),(y_1,z_2)]\mathtt{th}\,[t_1,t_2] \Leftrightarrow y_1 \le z_1, z_1 \le y_2, y_2 \le z_2$

(1′) axiom obtained by swapping the annotations in (1).

(2) $\mathtt{thr}\,[(x_1,x_2),(y_1,y_2)]\mathtt{th}\,[t_1,t_2] \sqcup \mathtt{thr}\,[(z_1,z_2),(y_1,y_2)]\mathtt{th}\,[t_1,t_2] = \mathtt{thr}\,[(x_1,z_2),(y_1,y_2)]\mathtt{th}\,[t_1,t_2] \Leftrightarrow x_1 \le z_1, z_1 \le x_2, x_2 \le z_2$

(2′) axiom obtained by swapping the annotations in (2).

(3) $\mathtt{thr}\,[(x_1,x_2),(y_1,y_2)]\mathtt{th}\,[s_1,s_2] \sqcup \mathtt{thr}\,[(x_1,x_2),(y_1,y_2)]\mathtt{th}\,[r_1,r_2] = \mathtt{thr}\,[(x_1,x_2),(y_1,y_2)]\mathtt{th}\,[s_1,r_2] \Leftrightarrow s_1 \le r_1, r_1 \le s_2, s_2 \le r_2$

(3′) axiom obtained by swapping the annotations in (3).

(4) $\mathtt{inr}\,[(x_1,x_2),(y_1,y_2)]\mathtt{th}\,[s_1,s_2] \sqcup \mathtt{inr}\,[(x_1,x_2),(y_1,y_2)]\mathtt{th}\,[r_1,r_2] = \mathtt{inr}\,[(x_1,x_2),(y_1,y_2)]\mathtt{th}\,[s_1,r_2] \Leftrightarrow s_1 \le r_1, r_1 \le s_2, s_2 \le r_2$

(5) $\mathtt{in}\,[t_1,t_2]\mathtt{thr}\,[(x_1,x_2),(y_1,y_2)] \sqcup \mathtt{in}\,[t_1,t_2]\mathtt{thr}\,[(x_1,x_2),(z_1,z_2)] = \mathtt{in}\,[t_1,t_2]\mathtt{thr}\,[(x_1,x_2),(y_1,z_2)] \Leftrightarrow y_1 \le z_1, z_1 \le y_2, y_2 \le z_2$

(6) $\mathtt{in}\,[t_1,t_2]\mathtt{thr}\,[(x_1,x_2),(y_1,y_2)] \sqcup \mathtt{in}\,[t_1,t_2]\mathtt{thr}\,[(z_1,z_2),(y_1,y_2)] = \mathtt{in}\,[t_1,t_2]\mathtt{thr}\,[(x_1,z_2),(y_1,y_2)] \Leftrightarrow x_1 \le z_1, z_1 \le x_2, x_2 \le z_2$

**Table 4.2.** Axioms for least upper bound of annotations

periods. Thus we consider the least upper bound in the cases illustrated in Table 4.2.

Axioms (1), (1′), (2) and (2′) allow one to enlarge the region in which a property holds in a certain interval. If a property $A$ holds both throughout a region $R_1$ and throughout a region $R_2$ in every point of the time period $I$ then it holds throughout the region which is the union of $R_1$ and $R_2$, throughout $I$. Notice that the constraints on the spatial variables ensure that the resulting region is still a rectangle. Axioms (3) and (3′) concern the temporal dimension: if a property $A$ holds throughout a region $R$ and in every point of the time periods $I_1$ and $I_2$ then $A$ holds throughout the region $R$ in the time period which is the union of $I_1$ and $I_2$, provided that $I_1$ and $I_2$ are overlapping. By using axiom (4) we can prove that if a property $A$ holds in some point(s) of region $R$ throughout the time periods $I_1$ and $I_2$ then $A$ holds in some point(s) of region $R$ throughout the union of $I_1$ and $I_2$, provided that such intervals are overlapping. Finally, the last two axioms allow to enlarge the region $R$ in which a property holds in the presence of an `in` temporal annotation.

### 4.2.5 Clauses

The clausal fragment of STACLP, which can be used as an efficient spatio-temporal programming language, consists of clauses of the following form:

$$A\,\alpha\beta \leftarrow C_1,\ldots,C_n, B_1\,\alpha_1\beta_1,\ldots,B_m\,\alpha_m\beta_m \quad (n,m \ge 0)$$

where $A$ is an atom, $\alpha$, $\alpha_i$, $\beta$, $\beta_i$ are (optional) temporal and spatial annotations, the $C_j$'s are constraints and the $B_i$'s are atomic formulae. Constraints $C_j$ cannot be annotated. A *STACLP program* is a finite set of STACLP clauses.

*Example 1.* Assume that a person is described by his/her name, the activity and the spatial position(s) in a *certain time interval.* For instance, from 1am to 10am John sleeps, from 11am to 12am he has breakfast and then in the afternoon he goes skiing up to 4pm, while Monica skies from noon to 4pm. This can be expressed by means of the following clauses.

```
does(john,sleep) atp (2,12) th [1am,10am].
does(john,eat) atp (2,6) th [11am,12am].
does(john,ski) inr [(500,2000),(1000,2000)] th [12am,4pm].
does(monica,ski) inr [(500,2000),(1000,1500)] th [12am,4pm].
```

The temporal information is represented by a `th` annotation because the property holds throughout the time period. Instead the spatial location is expressed by using an `atp` annotation when the exact position is known, or by an `inr` annotation if we can only delimit the area where the person can be found.

Furthermore, a place can be described by its name and its area represented by a `thr` annotation.

```
place(refuge) thr [(700,710),(1200,1205)].
place(ski) thr [(50,2000),(1000,2000)].
```

Below we show how some queries involving the spatial and/or temporal knowledge can be formulated in our language.

1. Where is John between 12am and 2pm?
   `does(john,_) inr R in [12am,2pm]`
   The answer to this query consists of (possibly different) regions where John stays during that time period. We use the `in` annotation because we want to know all the different positions of John between 12am and 2pm while the `inr` annotation allows one to know the region John is in during that time period, even if his exact position is unknown.
   If we asked for `does(john,_) atp R th [12am,2pm]` then we would have constrained John to stay in only one place for the whole time period.
   The query `does(john,_) atp R in [12am,2pm]` asks for definite positions of John sometime in $[12am, 2pm]$.
2. Where is John while Monica is in the refuge?
   `does(john,_) inr R th I, does(monica,_) inr R1 th I, place(refuge) thr R1`
   This query is a composition of a spatial join and a temporal join.

## 4.3 Semantics of STACLP

In the definition of the semantics, without loss of generality, we assume all atoms to be annotated with `th`, `in`, `thr` or `inr` labels. In fact, `at` $t$ and `atp` $(x, y)$ annotations can be replaced with `th` $[t, t]$ and `thr` $[(x, x), (y, y)]$ respectively by exploiting the (`at th`) and (`atp thr`) axioms. Moreover, each atom in the object level program which is not two-annotated, i.e., which is labelled by at most one kind of annotation, is intended to be true throughout the whole lacking dimension(s). For instance an atom $A\,\mathtt{thr}\,R$ is transformed into the two-annotated atom $A\,\mathtt{thr}\,R\,\mathtt{th}\,[0,\infty]$. Constraints remain unchanged.

The meta-interpreter for STACLP is defined by the following clauses:

$$demo(empty). \tag{4.1}$$

$$demo((B_1, B_2)) \leftarrow demo(B_1), demo(B_2) \tag{4.2}$$

$$demo(A\ \alpha\beta) \leftarrow \alpha \sqsubseteq \delta, \beta \sqsubseteq \gamma, clause(A\ \delta\gamma, B), demo(B) \tag{4.3}$$

$$\begin{aligned} demo(A\ \alpha'\beta') \leftarrow\ & \alpha_1\beta_1 \sqcup \alpha_2\beta_2 = \alpha\beta, \alpha' \sqsubseteq \alpha, \beta' \sqsubseteq \beta, \\ & clause(A\ \alpha_1\beta_1, B), demo(B), demo(A\ \alpha_2\beta_2) \end{aligned} \tag{4.4}$$

$$demo(C) \leftarrow constraint(C), C \tag{4.5}$$

A clause $A\,\alpha\beta \leftarrow B$ is represented at the meta-level by

$$clause(A\,\alpha\beta, B) \leftarrow valid(\alpha), valid(\beta) \tag{4.6}$$

where *valid* is a predicate that checks whether the interval or the region in the annotation is not empty.

The first two clauses are the ordinary ones to solve the empty goal and a conjunction of goals. The resolution rule (clause (4.3)) implements both the Modus Ponens rule and the rule ($\sqsubseteq$). It states that given a clause $A\,\delta\gamma \leftarrow B$ whose body $B$ is solvable, we can derive the atom $A$ annotated with any annotation $\alpha\beta$ such that $\alpha \sqsubseteq \delta$ and $\beta \sqsubseteq \gamma$. Such constraints are processed by the constraint solver using the constraint theory for temporal and spatial annotations shown in Sect. 4.2.2. Clause (4.4) implements the rule ($\sqcup$) (combined with Modus Ponens and rule ($\sqsubseteq$)). It states that if we can find a clause $A\,\alpha_1\beta_1 \leftarrow B$ such that the body $B$ is solvable, and if the atom $A$ can be proved with annotation $\alpha_2\beta_2$, then we can derive the atom $A$ labelled with any annotation less or equal than the least upper bound of $\alpha_1\beta_1$ and $\alpha_2\beta_2$. The constraint $\alpha_1\beta_1 \sqcup \alpha_2\beta_2 = \alpha\beta$ is solved by means of the axioms defining the least upper bound introduced in Sect. 4.2.4. Clause (4.5) manages constraints by passing them directly to the constraint solver.

## 4.4 Modelling and Representing Trajectories

One of the basic forms of spatio-temporal information is given by spatio-temporal objects, namely objects which move along time within a spatial environment. Therefore, in order to properly approach several spatio-temporal analysis problems, it is essential to model and represent such objects and their movement in a suitable way.

In this section, we will briefly summarise some of the most common spatio-temporal data models presented in literature, and thus we will introduce the model adopted in this work to represent the movement of spatio-temporal objects, together with its formalisation within the STACLP language.

## Spatio-Temporal Data Models

Although a quite large number of ideas were proposed in literature for modelling spatio-temporal data (see for example [3]), in recent years only a few proposals have been widely discussed. Moreover, no general purpose database system does exist which integrates spatial, temporal and thematic information in a satisfactory way.

In what follows, we provide a brief overview of four of the main data models recently discussed in literature.

**Parametric 2-spaghetti:** Chomicki and Revesz [6] have introduced a new model, called Parametric 2-spaghetti model, that generalises the 2-spaghetti model (able to represent points, lines and polygons) by allowing an interaction between spatial and temporal attributes. Vertex coordinates can now be *linear functions of time*, thus permitting to model continuous movements along time. As an example, a dynamic region $r_1$, modelled as a dynamic triangle, could be represented by a set of generalised tuples of the following kind:

| $ID$ | $x$ | $y$ | $x'$ | $y'$ | $x''$ | $y''$ | $From$ | $To$ |
|---|---|---|---|---|---|---|---|---|
| $r_1$ | 3 | 3 | 10 | 10 | 20 | $11-t$ | 8 | 10 |
| $r_1$ | $13-t$ | 3 | 10 | $t$ | $2t$ | $t-9$ | 10 | $\infty$ |

Notice that the values in the relation are parametric in the time $t$, whose value ranges, in each tuple, between *From* and *To*. Query languages for the Parametric 2-spaghetti data model have not been defined yet. This model has been used in [5] to animate spatio-temporal objects.

**Type lifting:** In this approach, proposed by Erwig, Güting, Schneider and Vazirgiannis [7, 14], collections of *abstract data types* for spatial values changing over time are defined. Essentially, the authors introduce data types for moving points and moving regions together with a set of operations on such entities. The design of the model for spatio-temporal data is based on a type constructor $\tau$ which transforms any given atomic data type $\alpha$ into a type $\tau(\alpha) = time \rightarrow \alpha$ where $time = \mathbb{R}$, that is a continuous model of time is supported. In particular this constructor is applied to two spatial data types, *point* and *region*, allowing one to represent two fundamental abstractions *moving point* (*mpoint*) and *moving region* (*mregion*). Then a set of operations are defined on these types, such as the distance $mdistance : mpoint \times mpoint \rightarrow \tau(real)$, which returns the distance between two moving points at all times. The presented data types can be embedded into any DBMS data model as attributes data types, and the operations be used in queries.

**ST-simplexes:** In this model, introduced by Worboys [24], each elemental spatial object – called *simplex* and composed by either a single point, finite straight line segment or triangular area – is assigned with a bitemporal element, which is a finite set of pairs of intervals, representing respectively

the database and the event times of the object. A complex spatial object, temporally referenced, is called *ST-complex* and it is modelled by a finite set of ST-simplexes, subject to some constraints that we do not report. The query language provides operators to make the union, difference and intersection of two ST-complexes, and the selection on the temporal and spatial component. Moreover, it is also given a topological operator to return the boundary of an ST-complex.
In the Worboys' model, the temporal and the spatial dimensions are independent. This prevents from describing spatial relations which may be parametric w.r.t. time, i.e. w.r.t. their evolution, for instance the relationship that exists between time and the area covered by an incoming tide.

**Constraints:** The idea behind Constraints Databases is to finitely represent infinite relations by means of (finite) sets of constraints over a given domain (the *linear constraints over rationals* or *real polynomial constraints* is often used as logical theory of constraints). Among the approaches presented in literature dealing with spatial information, Constraints Databases is the only one to provide a uniform data model for both time and space. In fact, if we consider as class of constraints linear polynomials, we can model both temporal and spatial data with efficient implementations.
In particular, in order to allow an efficient evaluation, Grumbach et al. [13] require that the $d$-dimensional databases to be considered consist only of relations with a bounded *orthographic dimension* $l \leq d$, informally defined as the maximal number of dependent components in the relation. A typical setting consists of relations over $\mathbb{Q}^2 \times \mathbb{Q}$, the first component for space and the second one for time, with no interaction among the two, obtaining an orthographic dimension equal to 2. An example is:

$$object(S, x, y, t) :- y \leq 6, 1 \leq y, 5x - y - 34 = 0, t \leq 8.$$
$$object(S, x, y, t) :- 7 \leq x, x \leq 11, x - 4y - 3 = 0, t > 8.$$

Notice the absence of interaction between time and the two spatial coordinates.

## A Trajectory Data Model

From an abstract point of view, the movement of a spatio-temporal object $o$ – i.e., its *trajectory* – can be represented by a continuous function of time which, given a time instant $t$, returns the position at time $t$ of the object in a $d$-dimensional space (typically $d \in \{2, 3\}$). Formally $o : \mathbb{R}^+ \to \mathbb{R}^d$.

In a real-world application, however, object movements are given by means of a finite set of *observations* – or *control points* –, i.e. a finite subset of points taken from the actual continuous trajectory. Moreover, it is reasonable to expect that observations are taken at irregular rates within each object, and that there is not any temporal alignment between the observations of different objects. As a result, it is possible to have couples of objects for which at all time points the observation of at least one of the objects is missing, as in the simple examples depicted in Fig. 4.1, where the empty and filled circles represent the observations of two different objects. A very basic operation

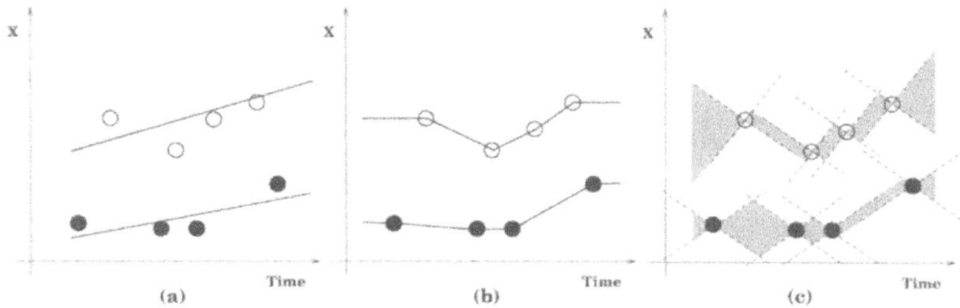


**Figure 4.1.** Raw control points and possible reconstruction of trajectories

such as the comparison between objects, then, cannot be performed by simply comparing their raw observations. To allow the comparison between objects, an (approximate) reconstruction of their full trajectory is needed. Among the several possible solutions, we can distinguish three simple yet interesting categories:

- *Global regression*: if the trajectories are known to follow a simple regular behaviour which can be represented by some analytical time-dependent expression, the movement of an object $o$ can be described by a single regression function. Fig. 4.1(a) shows an example where a linear regression function is used. Notice that the resulting curve does not always touch all the given control points.
- *Local interpolation*: although there is not a global function describing the whole trajectory, objects are assumed to move between the observed points following some rule. For instance, a linear interpolation function models a straight movement with constant speed, while other polynomial interpolations can represent smooth changes of direction. The above mentioned linear (local) interpolation, in particular, seems to be a quite standard approach to the problem, essentially adhering to Chomicki and Revesz's parametric 2-spaghetti model. Fig. 4.1(b) depicts a simple example.
- *Domain knowledge-based reconstruction*: in principle, when no interpolation function (global or local) is known, the position of objects out of the control points can be any point in $\mathbb{R}^d$. However, sometimes a "domain knowledge" of a different kind can help to understand how objects behave between observed points, exploiting information about either the objects or the space they move in. Typically, such knowledge is expressed as constraints on the position of objects, which might strongly narrow the range of choices of their movements. A simple example, graphically shown

in Fig. 4.1(c), is given by information on the maximum speed that objects can reach: in this case all positions of an object in space-time must fall within (hyper-)cones having the edge in a control point, axes parallel to the time axes, and aperture representing the maximum speed of the object. Thus, the set of all possible positions of such object between two control points can be obtained as the intersection of the two corresponding (hyper-)cones – corresponding to the gray areas of the picture.

In this work, we will focus on the second proposal, following the parametric 2-spaghetti approach, modelling the movement of objects through linear interpolation between control points and assuming stationarity beyond the extreme control points. As mentioned, this solution yields a good tradeoff between flexibility and simplicity.

### Trajectory Representation

In this section we show how trajectories can be modelled in STACLP. Representing trajectories in STACLP has several benefits. First of all, the language allows for a high level representation and manipulation of time as well as space thus providing primitive support for reasoning on spatio-temporal data. Secondly, it allows us to mix an inductive and deductive steps to perform complex kinds of analysis on the behaviour of moving objects as we will show in Sect. 4.6.

Given an object $o$ the *observations* which describe the trajectory of the object is a finite set of triplets $(x, y, t)$, where $x$ and $y$ are the coordinates of the object tracked at time $t$. As mentioned above, the method adopted in this work to reconstruct the full trajectory from the observations computes the linear interpolation between consecutive localisation points. This kind of interpolation is particularly easy to represent since given two localisation points $(x_1, y_1, t_1)$ and $(x_2, y_2, t_2)$ with no other localisation in the time interval $[t_1, t_2]$ (i.e., the two observations are consecutive), we have that the estimated coordinate at time $t$, with $t_1 \leq t \leq t_2$, is $(x, y)$ where:

$$\begin{aligned}(x - x_1)(t_2 - t_1) \; - \; (x_2 - x_1)(t - t_1) \; &= \; 0, \; \textit{and} \\ (y - y_1)(t_2 - t_1) \; - \; (y_2 - y_1)(t - t_1) \; &= \; 0\end{aligned}$$

This approximation of trajectories can be easily modelled in STACLP: the localisation points are represented by using the `atp/at` annotation and then the straight line between the two end points is expressed as a constraint.

Specifically, the $N$ localisations of each object $o$, $(x_i, y_i, t_i)$ for $i = 1, \ldots, N$, can be represented by the following $N$ STACLP facts:

```
fix(o) atp (x1, y1) at t1
fix(o) atp (x2, y2) at t2
  ⋮
fix(o) atp (xN, yN) at tN
```

Such localisations will define the *core* of the trajectory of object $o$, which is then *completed* by defining all the intermediate points through linear interpolation using the following STACLP rules:

```
traj(O) atp (X, X) at T :- fix(O) atp (X, Y) at T.
traj(O) atp (X, Y) at T :- fix(O) atp (X1, Y1) at T1,
                           fix(O) atp (X2, Y2) at T2,
                           succ(T1,T2), T1 < T < T2,
                           X=(X1(T2-T)+X2(T-T1))/(T2-T1),
                           Y=(Y1(T2-T)+Y2(T-T1))/(T2-T1).
```

In the body of the second rule, approximated points $(x, y)$ are computed by using the equation for the line passing through two given points, shown above. The presence of the (standard) *successor* predicate `succ`, defined as true for all and only the couples of (strictly) consecutive localisation points, ensures that no other observation exists between times $t_1$ and $t_2$, i.e., the interpolation is performed only between consecutive localisation points.

## 4.5 An Application Example

Behavioural ecology is the science which studies animal behaviour with special interest in the relation to the environment where animal lives. This can be an interesting application domain for our framework since the problems coped with require the analysis of large spatio-temporal datasets and would definitely benefit from the availability of high level reasoning capabilities.

A typical technique used by biologists to collect information on the studied animals consists of tracking their movements by means of special collars which the animals are equipped with. In this way, large datasets containing spatio-temporal localisations, called *fixes*, of tracked animals are built. Each fix includes the identifier of the animal, the position expressed by the spatial coordinates $X$, $Y$ and the time $T$ of the localisation.

The set of fixes allows to view animals as spatio-temporal objects. The study of their trajectories can be quite helpful in determining the *home range* of animals, i.e. their life area, varying along the time, and it can be useful to establish relationships with other animals in order to understand the modalities of social and spatial aggregation.

The geographical features of the area of interest, such as vegetation, rivers, buildings, roads, which play a basic role in the analysis, are usually stored in a GIS, which thus represents the natural tool support for the biologists. Several commercial GISs offer a support for spatial analyses. For instance ArcView 3.2 provides a number of extensions, such as *Spatial Analyst* [8] by ESRI and *Movement* [15] by USGS Alaska, that implement a set of biological/statistical functions for home range calculations and other spatial analyses applied in animal ecology. However, the lack of tools to analyse in a correlated way space and time makes such systems not completely satisfactory when dealing with these applications.

To exemplify the usefulness of spatio-temporal analysis, we next list a number of relevant problems in behavioural ecology specifically involving spatio-temporal aspects. Such questions emerge from a research leaded by biologists from the University of Siena about the behavioural ecology of crested porcupines in the Maremma Regional Park (Tuscany, Italy) [4]. It is worth recalling that the crested porcupine is mainly nocturnal, lives in natural or artificial burrows and there is very few information available on the behaviour of such species. For this reason there is currently much interest, in the animal ecology field, in studying its habits.

### Den Localisation

Crested porcupines are nocturnal: they spend the day inside dens, typically burrows on the ground, whereas during the night they leave their dens looking for food. They usually change dens over time, and it is particularly relevant for researchers to determine when this happens and where the new den is located in order to formulate some hypotheses on the reasons of the displacement as well as to discover unknown behavioural patterns. A technique used by biologists to detect den locations is called *homing-in*. Essentially the researcher physically follows the signal of the radio collar of an animal by means of a portable radio receiver and tries to reach the animal when it is inside the den.

Unfortunately, homing-in is a quite expensive procedure and thus it cannot be applied extensively. Usually it is done on a bi-monthly basis thus creating a two months gap where the location of dens is unknown. Therefore a strong concern consists of understanding where dens are located in the period of time in which there is no homing-in localisation. This can be done exploiting the animals fixes and, in particular, analysing those fixes that have been taken between dawn and sunset, when, according to the expert knowledge the crested porcupine stays inside or close to the den. Given the set of dawn-sunset fixes a probable den can be inferred as a spatial location that has in its neighbourhood a high number of such fixes.

Observe that this kind of analysis requires a correlation of spatial and temporal information since relevant fixes have a specific temporal property (they have been taken between dawn to sunset) and, by using such data, a location which represents a possible den, is determined considering a spatial neighbourhood property such as a high density of fixes.

### Relations Among Animals

For understanding the habits and the social behaviour of animals it is extremely relevant to discover the relationships existing among individuals. Notice that this research includes a wide range of cases such as finding pairs of individuals of different sex that move together and possibly share the same den (couples), or groups of individuals that move together (herds) or couple/groups of individuals that avoid each other (territoriality).

To assess the degree of association among individuals, a possibility consists of comparing the overlap of their home ranges, estimated at given time intervals (typically one month or, more generally, $k$ months). However, such method is quite raw because it can happen that animals stay in common areas but in different periods of time or the overlap might be so large that the simultaneous presence of the animals in the area does not ensure that they are really close to each other.

A more precise procedure requires the calculation of the inter-individual distance between animals localised at the same time. We say that two fixes are *contemporary* if they refer to localisations of animals in the same place and at the same time, i.e., we consider a kind of spatio-temporal closeness among individuals. Since the tracking technique usually presents several sources of error, in the analysis two fixes are assumed to be *contemporary* if they fall within a given time interval and the corresponding localisations are within a certain distance. The effective values for the temporal and spatial thresholds are established by biologists.

Analysing this kind of inter-individual distance between animals, it is possible to make hypotheses about which animals can be considered a couple, form a herd, or avoid other individuals. For instance, two animals of different sex are likely to be a couple in a given period of time if they have a high quantity of contemporary fixes in the considered interval. On the other hand, an animal avoids another one if their inter-individual distance is always greater than a minimal fixed value.

### Changes in the Home Ranges

One of the basic objectives in the analysis of animal behaviour is a better understanding of how the animals change their *home ranges* along time. This objective requires, among other tasks, the detection of seasonal variations of the home ranges, both in location and size, and also, to infer the factors determining the dimensions of home ranges for these species.

A simple approach to face this problem consists of partitioning the time period covered by the analysis, $[t_{start}, t_{end}]$, into consecutive time sub-intervals of proper duration $\Delta T$, thus evaluating the home ranges within each subinterval and analysing the sequence of results obtained. However, in general, determining a suitable value for $\Delta T$ results to be not easy. On one hand, it has to be large enough to allow the computation of a significant home range, but, on the other one, it has to be small enough to catch swift changes – *swiftness* being a concept which is highly relative and dependent on the animal species under analysis. In several contexts, such as the study of the crested porcupines, analysts have not enough a priori information on the animals under study to deduce a right time interval. In some cases, moreover, it is even possible that no $\Delta T$ value satisfying the two constraints mentioned above exists at all. The definition of $\Delta T$ is often subjective and conditioned by our perception of the temporal units (weeks, months, seasons, years), leading to choose temporal intervals which are shifted w.r.t. the observed phenomena. Therefore, there is a clear need for a technique which provides an estimate of home ranges, regardless of the a-priori definition of time intervals.

A simple solution to the above mentioned problem, is to calculate home ranges in a "continuous" way within a given time interval (e.g., a season). The "continuity" can be obtained by replacing the time partitioning approach, described above, with a temporal *sliding window*: now, two parameters $\delta_t$ and $W_t$ are defined, and home ranges are computed on overlapping time intervals, each of duration $W_t$ and each shifted of $\delta_t$ w.r.t. the previous one, starting from interval $[t_{start}, t_{start} + W_t]$. Notice that the partitioning approach described first, is just a particular case where $W_t = \delta_t = \Delta T$. For example, with $t_{start}$ = "January 1st" and $\Delta T$ = "30 days", the partitioning approach would result in home ranges computed separately (approximatively) on each month of the year. In the case of animals which change their life areas at a quicker rate (e.g., once a week), the resulting home ranges would mix together several distinct movements, and thus would not provide a sound description of the animal activity. With sliding windows, on the contrary, setting $W_t$ ="30 days" and $\delta_t$ = "1 day", we would obtain the home ranges for the intervals from January 1st to January 31st, from January 2nd to February 1st, and so on. This way, even animal movements at a scale *finer* than the window size produce a detectable change in the home range – the larger and the longer is the animal movement, the larger is the change in the home range, in a continuous way –, and thus distinct movements can be distinguished between each other.

We notice that computing home ranges over a sliding window produces a *smoothly-evolving* home range, which is perfectly analogous to the smooth curve obtained by applying any (of the several well known) sliding window-based smoothing operator on a time series. Such operators have a wide application in several fields, such as noise-reduction filters in signal processing tools. In this sense, the sequence of home ranges obtained for an object $o$ can be viewed as an alternative, improved representation of its trajectory. Therefore, depending on how home ranges are computed and the complexity of their representation, they can be used in some analysis tasks in place of the original trajectory, in order to improve the quality of the results (e.g., in any analysis where the global *trend* of $o$'s movement is relevant, while its single movements are uninteresting and potentially misleading).

### Spatio-Temporal Relations Between Localisations and Events

The spatio-temporal relation between an animal and a particular event, defined in time and space, is one of the problems which are often addressed by animal ecology researchers. Indeed very frequently, there is a need for assessing whether and to what extent an event (e.g. a change in a crop cover, a hunting chase, a meteorological or geo-morphological occurrence) defined in time and space, has caused variations in the movement of the monitored animals.

Since the problem requires a co-location analysis task (between animals and events), it presents difficulties which are similar to those mentioned for the localisation of dens. However, since general events have more complex relations with moving animals than simple dens, several additional issues arise.

The first is that spatial and temporal limitations of the effect of the event on animals, must be defined. For instance, we could exclude all events taking place within a spatial distance from individual greater than a fixed threshold or those events occurring later than the occurrence of the localisation itself. Alternatively one could exclude the events occurring before a given period of time.

Therefore, we should expect that if an event concretely influences the animals, it would then "attract" or "reject" them, or in general modify in some way the movement they performed *before* the event, to an extent which is proportional to their spatio-temporal distance.

Moreover this technique should allow us to estimate the time interval between the occurrence of the event and its effects (as well as their end) on the animal.

These general problems allow several different concrete formulations, which can require simple or complex analysis tasks, mainly depending on the assumptions that can be made.

In a simple setting, we can assume that the event has visible effects on the level of single animals (thus not requiring the concept of herd or group), and that such effects can only be of the attraction and rejection types. In this case, the analysis task requires to compute spatio-temporal distances between animals and events, in order to find out which animals might potentially be influenced by which events. Therefore, for each pair *(event, animal)*, the existence of a rejection or attraction movement should be checked, by analysing how the spatial distance between the pair changes after the event.

In a more complex situation, we could be required to view animal groups (e.g., herds) as a whole. This is particularly true when either (i) the behaviour of a single animal is not significant, and thus it is potentially misleading (e.g., it could have an anomalous reaction to the event, not representative of its species or its group); or (ii) we are simply interested in comparing the behaviour of different herds; or (iii) the event influences the animals by altering some properties of the group which are not evident by looking at its single members (e.g., the event could induce the group to compact itself, as a self-defensive reaction, even though its members movement is not sensibly altered). In this case, the notion of animal group has to be defined and computed, prior to evaluate the spatio-temporal nearness between animals (considering each group as a whole) and events. Then, the candidate groups, i.e. those which were close enough to an event to be potentially influenced, are analysed in order to assess which ones show a behavioural alteration w.r.t.

their corresponding events. The latter step will require a suitable formalisation of "behaviour change", whose complexity can greatly vary depending on the properties it analyses.

## 4.6 Support for Spatio-Temporal Analysis in STACLP

In the previous section we have described some examples of analysis of spatio-temporal data which are relevant in a specific application domain. Generally speaking a framework for dealing with spatio-temporal (but also general data) should offer the ability of querying such data and of reasoning over them.

Current systems allow the user to exploit some consolidated mechanisms such as SQL-like queries, statistical functions and some spatial operations. Despite the fact that these systems provide efficient and mature technology to query standard relational or spatial data, they do not offer the user high level operations to handle the special time and space related dimensions. Indeed, nowadays the development of a spatio-temporal application with commercial systems requires the programming of some ad-hoc components by means of procedural languages provided by the system itself for customisation purposes. However, the absence of specific spatio-temporal operations makes the customisation of applications extremely complex, especially for non-expert users.

In our opinion, more sophisticated mechanisms are needed to provide the user with high-level and, at the same time, more general support for reasoning. We believe that the language for "programming" the extensions should be a real knowledge representation language, or, better to say, a very high level query language as much as the last extensions to SQL are a declarative programming language for extending relational databases. This view is shared by a number of other researchers, who all work on the integration of declarative paradigms and GISs [13, 25, 23].

Our proposal is mainly based on the use of computational logics as knowledge representation formalisms and as inference engine supporting reasoning.

In particular, in our framework both *deductive* and *inductive* inferences are conciliated as well as enriched with spatio-temporal primitive operations. Rules can be used to represent general knowledge about the collected data, and deductive capabilities can provide answers to queries that require some inference besides the crude manipulation of the data. On the other hand, induction is in some sense the inverse of deduction, that is deriving general principles from concrete data, therefore synthesising new knowledge from data or experience.

Next subsection focuses on deduction, whereas Sect. 4.7 is devoted to introduce mechanisms to support inductive analysis and it describes how these tools can be successfully used in our case study.

### 4.6.1 Deductive Analysis in STACLP

In this section we will show how the deductive capability of our language allows us to solve several problems presented in Sect. 4.5, namely, finding out the estimated position of the den whenever its real location is unknown, discovering the animals which are likely to be a couple and determining how the home range of a given animal changes.

As described in Sect. 4.4, we model the spatio-temporal localisations of each crested porcupine by a collection of facts of the kind:

```
fix(id) atp (x,y) at t.
```

specifying the position `x,y`, and the time `t` (expressed in seconds) of a localisation for the animal `id`.

Below we will show the STACLP code that implements the expert criteria by which we successfully solve some of the questions proposed by the biologists. The rules are slightly simplified by removing some implementation details, like the use of the predicate *constraint* which indicates that the atom is a constraint. Such a presentation allows us to focus on the knowledge representation ability of the language. Furthermore the code can be made executable by a simple precompilation step. The rules extensively use the Prolog meta-predicate `findall(X,G,L)` which computes the list `L` of elements `X` that satisfy the goal `G`.

Let us now formalise the questions we want to cope with. From the expert knowledge we know that the crested porcupine stays in its den during the day whereas it usually spends the night far from it. Hence, in order to determine the position of the den we collect all the fixes of the animal ranging between an hour before dawn and an hour after sunset, the idea being that in this time period the animal is probably close to its den. The rules that implement our analysis criteria are the following:

```
possible_loc(Id,Lloc) at T :-
   findall(loc(X,Y), (fix(Id) atp(X,Y) at T1, dawn_sunset(T1) at T),
          Lloc).

dawn_sunset(T1) at T :- light(D,S) at T, D-3600 <= T1, T1 <= S+3600.

prob_den(Id,Rad,Prob,L) in [T1,T2] :- possible_loc(Id,Lloc) in [T1,T2],
                                      neighbour_list(Lloc,Rad,Prob,L).
```

The first clause returns the list of positions `Lloc` of the animal `Id` between dawn and sunset in a certain day. The predicate `dawn_sunset`, occurring in this rule, checks whether the time `T1` of the localisation falls between one hour before dawn and one hour after sunset of the day `T` (i.e., the day time T belongs to). Notice that, since the dawn and sunset times vary along the year, the predicate `light` is used to record a monthly estimate of such times.

The third clause extracts from the list of the localisations computed by the predicate `possible_loc` the positions which are likely to be dens. The predicate `neighbour_list` (not defined for sake of brevity) selects those positions whose neighbourhood (with radius `Rad`) includes a great quantity of fixes between dawn and sunset (precisely greater than the value `Prob`). This quantity is estimated by considering the ratio between the number of fixes in the neighbourhood and the total number of fixes for the given animal in the period of time of interest.

Consider now the second problem illustrated in Sect. 4.5, namely providing a high level mechanism to determine relations among individuals. We focus here on the specific problem of finding possible couples, that is animals of different sex that move together. In order to find out the likely couples we exploit the notion of contemporary fixes: two fixes are contemporary if they are within a certain distance and in a certain time interval. We recall that the effective values for the threshold have to be chosen by the domain experts.

At this point we compute the number of contemporary fixes for a pair of animals and according to this number the expert can decide whether the pair is a couple or not.

```
couple_in_day(Id1,Id2,R,S,N) at T :-
    findall(c(Id1,Id2), (fix(Id1) atp(X1,Y1) at T1,
                         fix(Id2) atp(X2,Y2) at T2,
                         sex(Id1, S1), sex(Id2, S2), S1 != S2,
                         contem(X1,Y1,X2,Y2,R,S,T1,T2) at T),
            L),
    length(L,N).
contem(X1,Y1,X2,Y2,Rad,Sec,T1,T2) at T:- in_day(T1,T2) at T,
                dist(X1,Y1,X2,Y2,D), D<Rad, abs(T2-T1)<Sec.
couple(Id1,Id2,R,S,Ratio) in [T1,T2] :-
                    couple_in_day(Id1,Id2,R,S,N) in [T1,T2],
                    couple_in_day(Id1,Id2,1000000,S,M) in [T1,T2],
                    Ratio is (N/M).
```

The predicate `couple_in_day` returns the number `N` of contemporary fixes in a day for the pair of crested porcupines `Id1,Id2`. Two fixes are considered contemporary if their spatial and temporal distance is bounded by `R` and `S` respectively, as encoded by the predicate `contem`. The atom `in_day(T1,T2) at T` checks whether either `T1` or `T2` is within the day `T`. Finally, the predicate `couple` returns the ratio between the number of contemporary fixes of the crested porcupines `Id1,Id2` and the number of observations of `Id1,Id2` within `S` seconds at arbitrary distance (concretely this is obtained by setting a very large bound for the distance parameter) in a certain time period.

Finally, we show the STACLP code to compute a representation of animal home ranges which can enable an effective detection of their changes, following the ideas presented in Sect. 4.5.

We recall that we want to compute the home range of an animal within a given time interval `[t_start,t_end]` in a "continuous" way. Concretely, home ranges are computed on overlapping time intervals, each of duration `Wt` and each shifted of $\delta$`t` w.r.t. the previous one. We assume that `Wt <= t_end - t_start`, i.e., at least one of such intervals can fit in `[t_start,t_end]`.

```
homerange(Fix_list, Home) :- <ad hoc query/external call>
home(Id, Home) at t_start:- findall((T,X,Y),
    (fix(Id) atp (X,Y) at T, T>= t_start, T<t_start+Wt), Fix_list),
    homerange(Fix_list, Home).
home(Id, Home) at T:- home(Id, _) at T_prev,
        T=T_prev+δt, T + Wt <= t_end,
        findall((T_fix,X,Y),
                (fix(Id) atp (X,Y) at T_fix, T_fix >= T, T_fix < T+Wt),
                Fix_list),
        homerange(Fix_list, Home).
```

Given a list of fixes, the predicate `homerange` returns the corresponding home range `Home` by simply calling a routine provided by an external application. More details on the use of built-in predicates to directly invoke external functions can be found in [18, 21].

The predicate `home` returns the home range `Home` for an animal `Id` at regular time points, i.e. at `t_start` and at time points shifted from `t_start` of a multiple of `δt`. The first rule states that the home range for animal `Id` at the time instant `t_start` is obtained by finding all the fixes included in the interval `[t_start, t_start+Wt]` and then applying the home range routine to these fixes. The second rule manages the remaining time points and is defined in a similar way: it computes the home range using the fixes within the interval `[T, T+Wt]`, provided that `T` is shifted from `t_start` by a multiple of `δt` and `T+Wt <= t_end`.
It is worth noticing how the use of a declarative language in general, and STACLP in particular, which supports spatio-temporal reasoning, allows for a simple, compact and readable formalisation of the expert knowledge as well as it provides a very expressive query language. Moreover, STACLP offers the ability of performing external calls to ad-hoc functions provided by independent applications. These features make the STACLP code a kind of abstraction level over the "low level" tools used by domain experts. Actually, we do not intend STACLP as a framework to replace existing specialised systems, on the contrary, it should be integrated with these systems, thus it can be considered as a "middleware" in which very sophisticated spatio-temporal reasoning can be performed.

## 4.7 Inductive Analysis in STACLP

As seen in the previous section, deductive reasoning can be useful to solve several analysis problems which essentially require to find entities and values having some, possibly complex, properties. The STACLP language already showed to have the expressive power needed to fully formalise – and thus to solve – such problems. Moreover, thanks to its high-level nature, in most cases the formalisation step is simple and intuitive, yielding a very user-readable STACLP program.

However, dealing with more complex analysis tasks, it is quite common to meet concepts and abstract entities whose definition through deductive rules can be extremely difficult. It is especially the case when only vague, ambiguous or incomplete formalisations of the concept are available. A simple example, already introduced in this chapter, is the notion of animal group, or herd. While its meaning is intuitive and clear from a human view-point, formalising it can be difficult and highly dependent on the specific application – that is to say, a general definition of herd, suitable for any context, does not exist. In many cases, a suitable solution to this problem requires the *extrapolation* of new information from those already available. In other words, knowledge induction capabilities can be needed to properly tackle some difficult problems.

For this reason, the STACLP language can be fruitfully extended with induction capabilities, such as data mining algorithms. In this section we show (i) how a basic data mining tool, the k-means clustering algorithm, specifically tailored around trajectories, can be defined as STACLP rules, and (ii) how it can be used to provide a solution for some of the more complex problems introduced in Sect. 4.5.

### 4.7.1 Clustering

The clustering task is aimed at identifying clusters embedded in the data, i.e. to partition (although not necessarily in a crisp way) the dataset into collections of data objects, such that within each partition the objects are "similar" to one another, while they are "different" from the objects contained in other partitions. The greater the similarity within the group, and the greater the difference between groups, the better or more distinct the clustering. In our context, the aim is to partition the moving objects which populate our database, especially animals, into groups of individuals following similar trajectories.

Among the classical clustering algorithms, K-means is one of the best known and widely used, for its simplicity and its low computational complexity. It is a centre-based algorithm, meaning that clusters are represented by means of artificial objects (the *centres* or *representatives*) which summarise the properties of all the objects in their cluster. The k-means algorithm, described by the following pseudo-code, is essentially an iterative convergence process which tries to find "stable" centres:

***K-means***

1. Select k random points as initial centres;
2. *REPEAT*
3. for all points $p$
4. Assign $p$ to its closest centre;
5. *for all clusters C*
6. *Re-compute the centre of C;*

7. UNTIL no centre changed in last iteration or max n. iterations reached;

The general k-means clustering schema can be instantiated to a specific k-means algorithm by specifying the two key operations used in the schema: (i) computing the distance between two objects, and (ii) computing the representative of a set of objects (i.e., the centre of a cluster). Different definitions for these two steps can yield completely different notions of clustering.

The most straightforward and most common way of computing cluster centres in standard contexts where objects are described by vectors of basic attributes (e.g., the coordinates of points in a 2-dimensional space) is to define the attributes of the centre as the average of the values taken by the objects in the cluster. The extension of this idea to the spatio-temporal context is quite simple, since trajectories are essentially moving points, thus providing a very natural and intuitively sound instantiation of the centre computation: a centre will be an object whose position at time $t$ is the average of the positions of all objects in the cluster at time $t$.

On the contrary, it is more difficult to find a single most natural instantiation of the distance operation between objects, although some definitions exist which can cover some common concrete problems. For this reason, we first provide the STACLP rules which define a k-means algorithm for trajectories, independently of the distance definition. Later on, a set of possible choices for the distance function is given, together with the implementation of one of them.

For ease of presentation, we assume that `k` is the (fixed) number of clusters to find, and all objects to be clustered have an Id of the form "`objs(name_of object)`".

```
objs_to_cluster(O_list) :-
                findall(X, (fix(X) atp (_,_) at _, X=objs(_)), O_list).
assign(Iter, [], []).
assign(0, [A1|A],[Obj1|Objs]) :- K=random(k),
           A1=cluster(Obj1,K), assign(0, A, Objs).
assign(Iter, [A1|A],[Obj1|Objs]) :- Iter>0,
              closest(Iter-1, Obj1, Cluster),
              A1=cluster(Obj1,Cluster), assign(Iter, A, Objs).
```

The `objs_to_cluster` predicate defines the set of objects to be clustered, which can be easily customised by rewriting the body of its rule with any suitable criterion (here we selected all `objs(X)` objects having some fixes defined). For any iteration `Iter`, the `assign` predicate associates every object with its closest cluster centre, based on the results recursively obtained at the previous iteration, representing this information as terms of the form `cluster(object_ID, cluster_number)`. At iteration zero, the assignment object-cluster is random.

```
closest(Iter, Obj, Cluster) :- best_dist(Iter, Obj, k, Cluster, D).
best_dist(Iter, Obj, 1, 1, D) :- distance(centre(Iter, 1), Obj, D).
```

```
best_dist(Iter, Obj, K, Cluster, D) :- K>1,
              distance(centre(Iter, K), Obj, D1),
              best_dist(Iter, Obj, K-1, Cluster2, D2),
              if D1 < D2 then Cluster=K, D=D1
                         else Cluster=Cluster2, D=D2.
```

Here the selection of the closest cluster centre is implemented, simply scanning all the k centres obtained for the previous iteration, searching for the minimum value of the distance w.r.t. the object to assign. Here it appears the distance predicate, defined later in this section. In the following last set of rules, the centre of each cluster for any iteration is defined by setting its coordinates to the average values taken by all objects in the cluster (notice that a predicate sum_pairs is used to sum the single components of couples: since it is quite trivial to implement, its definition is omitted).

```
fix(centre(Iter,K)) atp (X,Y) at T :-
    objs_to_cluster(O_list), assign(Iter, A, O_list),
    member(cluster(Obj,K), A), fix(Obj) atp (_, _) at T,
    compute_avg_position(A, K, T, X, Y).
compute_avg_position(A, K, T, X, Y) :-
      findall((X1, Y1),
              (member(cluster(O,K),A), traj(O) atp (X1, Y1) at T), L),
      sum_pairs(L, (Xsum, Ysum)), length(L,N), N>0,
      X=Xsum/N, Y=Ysum/N.
fix(centre(K)) atp (X,Y) at T :-
              fix(centre(max_n_iters, K)) atp (X,Y) at T.
assignments(A) :- objs_to_cluster(O_list),
                  assign(max_n_iters, A, O_list).
```

Notice that in the definition of compute_avg_position the traj predicate is used to interpolate the position of objects, since the fixes of an object could be not aligned to the fixes of the others. The final result of the clustering process is represented by the cluster assignments and the means of the centres obtained when the maximum number of iterations has been reached – in particular, such centres will coincide with a local minimum of the clustering process if the algorithm converges in less than max_n_iters iterations.

The distance between objects can be defined in several ways, depending, e.g., on the meaning given to clusters or the coarseness allowed in the computation. One example of coarse but simple distance has been implicitly given in the previous section, where the similarity between two animals were defined as the percentage of mutually contemporary fixes (see the couple predicate in Sect. 4.6.1). In that case, only the explicit information on fixes has been exploited, not considering the whole trajectory followed by objects. A different and more precise solution, then, should take into account the position of objects for each time instant. Following this idea, a simple general approach to compute the distance $D(o_1, o_2)$ between two objects $o_1$ and $o_2$, whose positions along time $o_1(t)$ and $o_2(t)$ are defined over a time interval $T$, can be described by the following expression:

$$D(o_1, o_2) = \Phi(d_{o_1,o_2})\big|_T$$

where the first parameter of the schema, $d_{o_1,o_2}(t)$, is a distance measure between $o_1(t)$ and $o_2(t)$, and the second one, $\Phi(f)|_T$, is a functional computed over function $f$ and domain $T$ and returns a real value. In the STACLP rules given below, $d()$ is instantiated as the Euclidean distance on $\mathbf{R}^2$, and $\Phi()$ is the *average* functional, thus modelling $D(o_1, o_2)$ as the average Euclidean distance between $o_1$ and $o_2$. However, such parameters are modular components of the clustering algorithm, and therefore can be easily instantiated with other functions, as those described in [19] (e.g., other Minkowski's metrics for $d()$, and *min* or *max* functionals for $\Phi()$), thus defining new distance notions for $D(o_1, o_2)$.

Computing the average Euclidean distance between moving objects requires to calculate an integral of the Euclidean distance formula over a given time interval $[t_{start}, t_{end}]$. Thanks to the linear interpolation model adopted, such computation can be realised in linear time w.r.t. the number of fixes of each object [19]. This is due to the fact that the integration interval can be broken down to subintervals and in each of them the integral can be symbolically solved and thus computed in constant time. The following rules essentially find such subintervals, use a predicate `compute_sub` (not described here, as well as `sort_without_duplicates`, for sake of brevity) to compute local integrals, and aggregate them.

```
distance(O1,O2,D) :- collect_fixes(O1,O2,Fixes),
                     integral(O1,O2,Fixes,Int), D=I/(t_end-t_start).
collect_fixes(O1,O2,Fixes) :-
                   findall(T, fix(O1) atp (_,_) at T, L1),
                   findall(T, fix(O2) atp (_,_) at T, L2),
                   append(L1,L2,L).
                   sort_without_duplicates(L, Fixes).
integral(O1,O2,[_],0).
integral(O1,O2,[T1|[T2|T]], Int) :-
      traj(O1) atp (X11,Y11) at T1, traj(O1) atp (X12,Y12) at T2,
      traj(O2) atp (X21,Y21) at T1, traj(O2) atp (X22,Y22) at T2,
      compute(X11, Y11, X12, Y12, X21, Y21, X22, Y22, T1, T2, Int1),
      integral(O1,O2,[T2|T], Int2), Int = Int1 + Int2.
```

### 4.7.2 Knowledge Discovery on Trajectories

Some of the problems listed in Sect. 4.5 are difficult to solve with the only mean of deduction tools, while others could benefit from other capabilities, such as induction, to yield a more satisfactory solution. In this section we provide two examples where using the induction tool introduced above helps to find a simple and compact solution to two of the above mentioned problems:

detecting the relations between animals and the relations between localisations and events.

In the first case, a fully deductive approach has already been presented in Sect. 4.6.1, where a simple criterion was adopted, based on contemporary fixes, to discover animal couples. A more precise and general solution to the problem can be achieved by noticing that animal couples and animal herds are groups of animal which, in general, move together. This can be straightforwardly rephrased saying that animal herds are clusters of animal individuals whose mutual distance is, on average, small. This leads to the following STACLP formalisation, where trajectories of animals are clustered using the k-means algorithm, and focusing on a time interval $[t_{start}, t_{end}]$:

```
objs_to_cluster(O_list) :-
    findall(X, (fix(X) atp (_,_) at _, X=cut_obj(_)), O_list).
fix(cut_obj(X)) atp (X,Y) at t_start :-
                traj(obj(X)) atp (X,Y) at t_start.
fix(cut_obj(X)) atp (X,Y) at t_end :-
                traj(obj(X)) atp (X,Y) at t_end.
fix(cut_obj(X)) atp (X,Y) at T :- T > t_start, T<t_end,
                fix(obj(X)) atp (X,Y) at T.
cluster(K,O_list) :-
    findall(Obj, (assignments(A), member(cluster(Obj,K), A)), O_list).
couples([Obj1, Obj2]) :- cluster(_, [Obj1, Obj2]),
                         sex(Obj1, S1), sex(Obj2, S2), S1 != S2.
herds(O_list) :- cluster(_, O_list), length(O_list, N),
                 N>=min_herd_size.
```

The first rule redefines the `objs_to_cluster` predicate in order to cluster the new `cut_obj(X)` objects, obtained by clipping the trajectories of the original `obj(X)` objects on the $[t_{start}, t_{end}]$ time interval (see the above definitions for `fix(cut_obj(X))`). In the rules defining couples and herds, we assumed (i) to be interested in couples of different sex, and that (ii) a necessary (and sufficient) condition for a group of animals to be an herd is that its size is not smaller that a given threshold. Of course, it is easy to insert more complex conditions on the properties of the group and of the animals it contains (e.g., checking the respect of given proportions in the number of male and female individuals).

The problem of studying the relations between localisations and events requires to discover the effects produced on the behaviour of animals by an event. A natural way to face the problem is to compare the behaviour of the animals *before* the event with their behaviour *after* the event. As mentioned in Sect. 4.5, several approaches can be followed, analysing either the behaviour of single individuals, or the behaviour of herds. The individual-based class of solutions simply requires to specify a suitable notion of "behaviour", in order to be able to quantify behaviour changes. In what follows, we provide two examples of solutions falling in the (more complex) herd-based class, which exploits both the deductive engine of STACLP and the clustering tool defined at the beginning of this section.

A first formalisation of the event-effects problem can be given in terms of herd dispersion/compactness, saying that a herd reacts to the event if and only if its dispersion, measured as the sum of distances between individuals (or, equivalently, as the average of distances), grows beyond a given threshold factor `max_disperse` or shrinks beyond a factor `max_compact`. This can be implemented as a STACLP program performing a clustering of the animals before the event and checking growth and shrink factor after the event.

As a first step, for each object `objs(X)` two new objects are created: `before(X)` and `after(X)`, obtained by cutting the trajectory of `obj(X)` respectively on the time intervals $[t_{start}, t_{event}]$ and $[t_{event}, t_{end}]$, where $t_{event}$ is the instant when the event begins. This operation can be obtained by means of STACLP rules similar to those provided in the previous example for the `cut_obj(X)` objects.

Next, a clustering on the `before(X)` objects has to be performed, which can be achieved, as in the previous example, by simply rewriting the definition of the `objs_to_cluster` predicate in the following way:

```
objs_to_cluster(O_list) :-
                findall(X, (fix(X) atp (_,_) at _, X=before(_)), O_list).
```

Finally, for each cluster the cluster dispersion – i.e., the sum of distances between the objects in the cluster – is computed before and after the event, tagging each cluster as "dispersed" or "compacted" if the dispersion respectively grows or shrinks beyond the thresholds `max_disperse` and `max_compact`.

```
cluster_size(K, Before, After):- cluster(K,O_list),
                findall( (D_before, D_after),
                         ( member(before(X1), O_list),
                           member(before(X2), O_list),
                           X1 != X2,
                           distance(before(X1), before(X2), D_before),
                           distance(after(X1), after(X2), D_after),
                         D_list),
                sum_pairs(D_list, (Before, After)).
dispersed_cluster(K) :- cluster_size(K, Before, After),
                        After > Before * max_disperse.
compacted_cluster(K) :- cluster_size(K, Before, After),
                        Before > After * max_compact.
```

Another aspect of interest for the analyst is the existence of animals that, in reaction to an event, leave the herd. This can be seen as an alternative measure of dispersion of the herd, based on the behaviour of single animals w.r.t. the herd they belong to instead of considering only the overall behaviour of the heard.

Animals which abandon their herd can be defined as individuals that before the event belong to a given cluster, but, after the event, they *move* closer to other clusters. We therefore need to cluster animals w.r.t. their trajectory before the event, and then check if the cluster assignments are still valid also after event. The first step has already been performed in the previous example. To implement the second one, we compute first the centre of each cluster after the event (essentially with the same rule used to define the clustering engine), then for each object we find the closest centre to verify if it corresponds to the cluster the object belongs to:

```
fix(centre(after_event,K)) atp (X,Y) at T :-
    cluster(K, O_list), rename_objs(K, O_list, A_list),
    member(Obj, O_list), fix(Obj) atp (_,_) at T,
    compute_avg_position(A_list, K, T, X, Y).
rename_objs(K, [], []).
rename_objs(K, [before(X) | Objs], [A1|As]) :- A1=cluster(after(X),K),
            rename_objs(K, Objs, As).
run_away(obj(X)) :- cluster(K_before, O_list),
            member(before(X), O_list),
            closest(after_event, after(X), K_after),
            K_after != K_before.
```

The centre of each cluster $k$ is named `centre(after_event, k)`, in order to obtain object names syntactically compatible with the rules which define the clustering engine, and in particular those which define the `closest`, used to define the `run_away` predicate shown above. Finally, the `rename_objs` predicate has the simple purpose of converting the cluster assignment of each `before(X)` object into the equivalent assignment for the corresponding `after(X)` object (i.e., the second half of each trajectory is labelled according to the clustering obtained on the first half).

## 4.8 Conclusions

The framework we presented provides the user with high level mechanisms to represent and reason on spatio-temporal data. The peculiarity of this approach is that it exhibits both deductive and inductive capabilities, thus offering the possibility to make analysis driven by domain expert rules (deduction) and driven by observations (induction). Moreover we showed how STACLP can be successfully applied to a concrete case study concerning behavioural ecology.

Currently we are improving the implementation of STACLP, which is at a prototype stage, and we intend to study how to introduce other knowledge discovery techniques, such as classification and frequent patterns. When dealing with either spatial objects, or temporal objects, or spatio-temporal objects in order to apply the above data mining techniques, the most crucial and difficult problem is to find a clear definition of the basic concept of *item*, which is the elementary unit of data to analyse, and the derived concept of *transaction*, which is a collection of items. So far, it seems to us that the problem is quite depending on the kind of application at hand, but one of our future research goals is to establish definitions abstract enough for these concepts in order to capture most of the possible applications.

Another promising direction we addressed concerns qualitative spatial reasoning. Some preliminary results are presented in [22] and we would like to analyse forms of qualitative reasoning on trajectories.

*Acknowledgements.* We thank T. Ceccarelli and A. Massolo for providing us the case study and F. Fornasari and B. Furletti who collaborated in the implementation of the system. This work has been partially supported by European Project IST-1999-14189 - Rev!gis.

## References

1. Abdelmoty A.I., Paton N.W., Williams M.H., Fernandes A.A.A., Barja M.L., Dinn A. (1994) Geographic Data Handling in a Deductive Object-Oriented Database. In: Proc. of the DEXA Conference, LNCS 856, Springer, pp. 445–454.
2. Abraham T. (1999) Knowledge Discovery in Spatio-Temporal Databases. PhD thesis, School of Computer and Information Science, Faculty of Information Technology, University of South Australia.
3. Abraham T., Roddick J.F. (1999) Survey of Spatio-Temporal Databases. GeoInformatica 3(1):61–99.
4. Ceccarelli T., Centeno D., Giannotti F., Massolo A., Parent C., Raffaetà A., Renso C., Spaccapietra S., Turini F. (2001) The behaviour of the Crested Porcupine: the complete case study. Technical report, DeduGIS - EU WG.
5. Chomicki J., Liu Y., Revesz P.Z. (1999) Animating Spatiotemporal Constraint Databases. In: Spatio-Temporal Database Management, LNCS 1678, Springer, pp. 224–241.
6. Chomicki J., Revesz P.Z. (1999) Constraint-Based Interoperability of Spatiotemporal Databases. GeoInformatica 3(3):211–243.
7. Erwig M., Güting R.H., Schneider M., Vazirgiannis M. (1999) Spatio-Temporal Data Types: An Approach to Modeling and Querying Moving Objects in Databases. GeoInformatica 3(3):269–296.
8. ESRI. Spatial Analyst. http://www.esri.com.
9. Faloutsos C., Lin K.-I. (1995) Fastmap: a fast algorithm for indexing of traditional and multimedia databases. In: Proc. of SIGMOD Conference, ACM, pp. 163–174.
10. Frühwirth T. (1996) Temporal Annotated Constraint Logic Programming. Journal of Symbolic Computation 22:555–583.
11. Gaffney S., Smyth P. (1999) Trajectory clustering with mixture of regression models. In: Proc. of the KDD Conference, ACM, pp. 63–72.
12. Grumbach S., Rigaux P., Segoufin L. (1998) The DEDALE System for Complex Spatial Queries. In: Proc. of the SIGMOD Conference, ACM, pp. 213–224.
13. Grumbach S., Rigaux P., Segoufin L. (2001) Spatio-Temporal Data Handling with Constraints. GeoInformatica 5(1):95–115.
14. Güting R.M., Böhlen M.H., Erwig M., Jensen C.S., Lorentzos N., Schneider M., Vazirgiannis M. (2000) A Foundation for Representing and Querying Moving Objects. TODS 25(1):1–42.

15. Hooge P.N., Eichenlaub W.M., Solomon E.K. (1997) Animal Movement Extension to ArcView. Ver. 2.0., 1997. Alaska Biological Science Center, U.S. Geological Survey, Anchorage, AK, USA.
16. Kanellakis P.C., Kuper G.M., Revesz P.Z. (1995) Constraint query languages. Journal of Computer and System Sciences 51(1):26–52.
17. Ketterlin A. (1997) Clustering sequences of complex objects. In: Proc. of the KDD Conference, ACM, pp. 215–218.
18. Mancarella P., Raffaetà A., Renso C., Turini F. Integrating Knowledge Representation and Reasoning in Geographical Information Systems. International Journal of GIS. To appear.
19. Nanni M. (2002) Clustering Methods for Spatio-Temporal Data. PhD thesis, Dipartimento di Informatica, Università di Pisa.
20. Raffaetà A., Frühwirth T. (2001) Spatio-Temporal Annotated Constraint Logic Programming. In: Proc. of PADL'01, LNCS 1990, Springer, pp. 259–273.
21. Raffaetà A., Renso C., Turini F. (2002) Enhancing GISs for Spatio-Temporal Reasoning. In: Proc. of GIS'02, ACM, pp. 35–41.
22. Raffaetà A., Renso C., Turini F. (2003) Qualitative Spatial Reasoning in a Logical Framework. In: Proc. of the AI*IA Conference, LNAI 2829, Springer, pp. 78–90.
23. Spaccapietra S. (Ed.) (1999) Spatio-Temporal Data Models & Languages (DEXA Workshop). IEEE Computer Society Press.
24. Worboys M.F. (1994) A unified model for spatial and temporal information. The Computer Journal 37(1):26–34.
25. Worboys M.F. (1995) GIS – A Computing Perspective. Taylor & Francis.

# 5

# Representing and Querying Imperfect Spatio-Temporal Information in an Object-based Framework

Gloria Bordogna and Sergio Chiesa

Institute for Environmental Process Dynamic (IDPA)
Dep. of Milan – Lab. of Georesources, National Research Council (CNR)
via Pasubio 5, 24044 – Dalmine (BG), Italy
gloria.bordogna@idpa.cnr.it

## 5.1 Introduction

Storing and managing large collections of spatio-temporal information is necessary in several applications to infer hidden knowledge from the study of the evolution of natural phenomena. For example, for the study of the variation of temperatures in a region, of the evolution of the soil coverage over a period of time, of the recurrence of natural hazards in critical areas, of the spreading of pollution or epidemics.

Often, decision strategies are based on the evaluation of spatio-temporal analysis operations, which require new capabilities for current Geographic Information Systems [3, 10]. Specifically, the ability to manage imperfect spatio-temporal information is considered very important because any observation of geographic phenomena is affected by some kind of imperfection, and decisions can be taken more accurately when such imperfection is accounted for during the decision process [35].

The two dual conceptual data models for spatial information, i.e., the exact object-based model and the field-based model, do not allow representing imperfect spatial information [11]. This is because the former only represents objects with well-defined boundaries, whereas the latter describes properties that extend smoothly over the whole spatial domain. Spatial entities characterized by either unsharp boundaries or sharp boundaries whose position is ill-known are then forced into a too rigid representation that either introduces artificial boundaries or disregards relevant irregularities. To overcome these limitations, in the literature several proposals of spatial data models managing geographic entities characterized by spatial uncertainty have been proposed [6, 11, 15, 16, 18, 27, 29].

On the other side GISs, designed to focus on management of space-related information, have traditionally ignored the complementary need for time-related information which is essential for allowing the analysis of the evolution of geographic entities over some period of time. Temporal data modelling in GISs has been approached by defining snapshot models, by introducing time-space compound attributes and spatial objects belonging to time sequences [1, 2].

More recently, unified spatio-temporal models have been proposed in which entities may have a spatial as well as a temporal extent [39, 46], the two extents being dealt with as orthogonal dimensions. The temporal extent denotes the period of time in which the observation of a dynamic phenomenon is valid, i.e., the period of time in which the spatial reference of an entity can be considered constant (stable).

However, these more powerful approaches did not address the need for representing the uncertainty that may affect spatio-temporal information. Yet temporal uncertainty arises in many real applications where temporal knowledge cannot be precisely captured. This can happen either because of the scarcity of the observations (so that it is impossible to state when exactly the change occurred), or because the timing of the observation is not exactly known (e.g., a human may observe an event without exactly recording the instant it happened, for instance, the accident occurred *around 2 pm*), or because the continuously changing nature of the observed phenomenon makes it impossible to associate a precise time to a precise value (e.g., while observing the temperature in a furnace it may not be possible to state exactly at what time the temperature has reached 350° Celsius). In the literature, approaches addressing the modelling of uncertain temporal information may be found in [20, 22, 48]. Only a few proposals aim at managing uncertainty in spatio-temporal information [21, 41, 49].

In this chapter we propose an object-based data model to represent and manage spatio-temporal phenomena that is founded on a fuzzy dual representation of uncertain spatio-temporal information dealt with as orthogonal concepts [4]. The idea at the basis of the model is that dynamic phenomena are generally represented by a sequence of observations (snapshots), each one describing the system status at a precisely known time instant. This time instant is assumed as indicating the time in which the observation can be considered a valid description of the phenomenon. However, since the transition from a given status to another one is generally smooth, and occurs with a set of micro variations over a period of time, we try to represent this transition by associating with each snapshot a fuzzy interval of validity [20, 48, 52]. This fuzzy validity interval is identified by linguistic granules of temporal information such as *around* t, *just before* t, *just after* t etc., defined as fuzzy subsets of the temporal domain. In this way we can have several time instants associated with a single snapshot, each one having its distinct degree of validity.

An alternative view of a spatio-temporal phenomenon is to account for the lack of knowledge on the spatial reference by associating with each precise time instant an indeterminate snapshot. This snapshot can be defined so as to synthesize the multitudes of the phenomenon's spatial changes occurring during the precise period of time, or to identify the possible position of the phenomenon at the given time instant [6]. To this aim linguistic granules of spatial information such as *around point* a, *just within the border* B, etc., are defined with a semantic specified by fuzzy subsets of the spatial domain. So, the elements of the spatial reference that undergo continuous changes are associated with a smaller degree. These alternative views of an uncertain spatio-temporal phenomenon can be seen as special cases of a more general representation in which both the spatial and the temporal information are represented by means of continuous functions identified by linguistic granules.

In section 5.2 of this chapter, an ontology for imperfect spatio-temporal information is introduced. In section 5.3, a dual representation within the framework of possibility theory is proposed to represent dynamic spatial phenomena whose information is affected by some kind of imperfection. In section 5.4, the evaluation of spatio-temporal constraints against the fuzzy dual representation is introduced, and, in section 5.5, Abstract Spatio Temporal Data Types (ASTDT) are defined for an object-based data model to represent imperfect information relative to dynamic spatial phenomena [6]. In particular, spatio-temporal operations between fuzzy and indeterminate spatio-temporal objects are introduced by discussing their dependence on the type of imperfection.

## 5.2 An Ontology for Imperfect Spatio-Temporal Information

Spatio-temporal information can be affected by imperfection due to several reasons. In this section we analyze some characteristics of this imperfection and highlight some relations existing between imperfection in spatial information and in temporal information. This is important to design models to represent effectively dynamic spatial entities.

Our assumption is that there is an application with its own objectives, an expert in the field that observes the spatial phenomenon and represents it on a map [17]. The imperfection may derive by one or all of the following variables: the observer or its means of observation. Figure 5.1 shows our proposal of ontology of terms used to identify distinct characteristics of imperfect spatio-temporal information. It organizes the terms into a hierarchy with increasingly specialized terms appearing low down the hierarchy. Further refinements of this ontology are conceivable.

Imperfection may affect either the time dimension, or the spatial dimension, or both. In both cases, we can distinguish two main branches: *uncertainty*, and *vagueness*. Although spatial imperfection has been analyzed by many researchers [11, 14, 17, 24, 43, 45, 47] currently there is not a unique and unambiguous vocabulary to identify the types of imperfection.

*Spatial Uncertainty* derives either from vague and incomplete knowledge of the phenomenon or errors and depends on the characteristics of the observation. A particular kind of uncertainty is *indeterminacy* that occurs when the available knowledge is insufficient to precisely delimit a phenomenon that by its nature may be crisp. *Indeterminacy* is strictly related to the granularity of the information, where for granularity we refer to the existence of grains in information, in the sense that individual elements in the grain cannot be distinguished or discerned apart. Land-cover maps derived from classification of satellite images represent a common example of indeterminacy. The features (e.g., the road network) distinguishable within such maps are obviously related to the spatial resolution of the images. Two particular kinds of indeterminacy may be distinguished: *incompleteness* and *imprecision*.

The former is the lack of information relative to a sub-region of the spatial domain. This can be due either to the fact that the observation does not cover that specific region, or to the presence of obstacles between the point of observation and the observed phenomenon (e.g., clouds in satellite images that obscure the land below).

*Imprecision* occurs when the available information refers to spatial grains having a coarser granularity with respect to the granularity of the map on which the spatial domain is represented, so that we cannot select a sharp boundary delimiting a phenomenon on the map, but only a thick boundary.

*Errors* depend on the inaccuracy of the observation. A particular kind of error is the *inconsistency* of the information. *Inconsistency* generally derives from the availability of multiple observations of the same phenomenon and may lead to contradictions and conflicts. It may occur when distinct thematic layers carry different information with respect to the same spatial unit. So, for instance, a forest inventory may indicate that a given land unit is covered by woods, whereas an agricultural map depicts the same unit as a vineyard. This is typically due to the fact that different agencies or departments carry out the forest and agriculture surveys, each one generating its own data set.

The concept of *spatial vagueness* is related to the nature of the observed spatial phenomenon characterized by some gradual physical, social, or cultural property. Many natural features are fuzzy by definition. A typical example is represented by natural ecosystems, whose boundaries can often be set only on an arbitrary basis. Different forest types, indeed, are not sharply separated, but they gradually merge into each other. It is worth noting that some phenomena may appear vague when observed at a given resolution, but by increasing the resolution, their boundary becomes sharper and sharper until it appears crisp at a limit resolution. On the other side, there are phenomena that appear invariably affected by *vagueness*, such as the soil type.
Of course there can be situations in which both *spatial uncertainty* and *spatial vagueness* coexist, such as in the case of intrinsically vague phenomena observed through a sparse network of observation points. For example the observation of a meteorological phenomenon such as the position of a high-pressure zone based on a sparse network of points of measurement.

The introduction of temporal information as a new dimension in the representation of a spatial phenomenon may lead to a further level of imperfection, depending on the semantics of the temporal information. Basically, there are

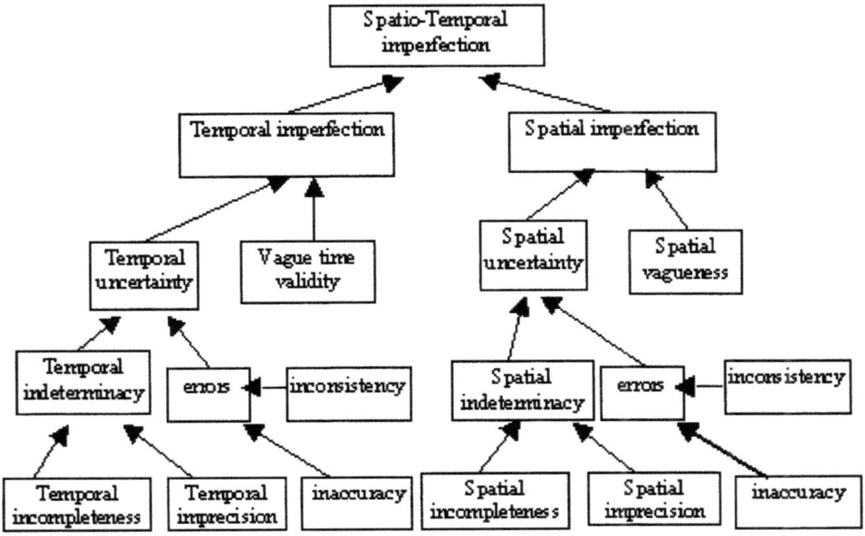


**Figure 5.1.** The ontology of imperfect spatio-temporal Information

two semantics for the time variable associated with spatial or non-spatial information [1, 2]. The most common semantic is to consider that the temporal specifications associated to the observation of a phenomenon record the evolution of the phenomenon over a period of time (by time-stamping snapshots of the phenomenon) as observed in the real world from the application perspective. This semantic interpretation is known as *valid time*, as opposed to *transaction time*, which is used to record when an observation is entered into or deleted from the database. Values for valid time are determined by the application. Values for transaction time are given by the computer system. Attribute values, relationships and objects in a database can have a period of validity associated with them. An additional semantic option is to allow representation of *time relationships* between phenomena. For instance, the temporal information associated with an object or attribute in the database may be a relative information that acquires meaning in relation with the time information of some other elements in the database. It can be an ordinal information interpreted as causality, e.g. an object has created another object, or an object has lived before another object, and so on [44].

*Temporal uncertainty* relative to the time validity of a phenomenon may occur in the case in which there is not a precise knowledge of the instant of observation of the phenomenon, and we can just specify it approximately (*temporal indeterminacy*). This may be caused by the granularity of the time information. A way to represent this information is by means of a fuzzy subset on the time reference domain interpreted as a possibility distribution [20, 48, 52]. Another cause of temporal uncertainty is due to errors of measurements.

A situation of *temporal vagueness* occurs whenever the definition of the period of validity for the observation of a phenomenon is hampered by heuristic knowledge of the phenomenon dynamics [48]. An example can be the validity period of satellite images of glaciers to identify their minimum extent, when there is not an image available taken at exactly the time the extent was minimal.

Temporal uncertainty relative to the *time relationships* between phenomena may occur when one is unable to specify a total order between events on an ordinal scale or in the case in which one is only able to imprecisely qualify the extent of the time relationship between events. An example of the latter is the specification that an event occurred long before, very long before, etc., another event.

In this chapter we consider the representation and management of the temporal imperfection relative to the concept of temporal validity of a spatial phenomenon. An object-based approach is adopted to represent spatio-temporal phenomena, as it is done in [46]. In the present approach the modelling of imperfection relies on the definition of a set of Abstract Spatio-Temporal Data Types (ASTDTs) for representing phenomena affected by distinct combinations of *vagueness* and/or *uncertainty* on the spatio-temporal reference. Nevertheless, for many real applications, we argue that some scenarios can be disregarded based on the assumption that, in some applications, imperfection of the spatial information is inversely related to the imperfection of the temporal information. We can then specify either an exact spatial information and be *vague* or *indeterminate* on its *time validity*, or we can have *indeterminacy* or *vagueness* on the *spatial information* and be precise on the time validity. For example, I can say that I know precisely the position of a taxicab on a map but I am *indeterminate* on its *time validity* at that position:

"the taxi is at the central station *around* 12.00".

Alternatively, I can be precise on the time information but *vague* on the *spatial reference* and say:

"the taxi is *close* to the Central Station at 12.00".

This inverse relationship has suggested a dual representation for imperfect spatio-temporal phenomena [4]; on the one side a phenomenon can be represented by precise spatial reference information associated with an *indeterminate or vague validity interval.* On the other side, it can be represented through an *indeterminate or vague spatial reference* and precise time reference [6]. In [4] it was shown that these alternative representations are subsumed by a more general one in which imperfection is modelled on both the temporal and the spatial reference: for example to represent the information:

"the taxi is *close* to the Central Station *around* 12".

In the next section we recall this fuzzy dual representation for imperfect dynamic spatial information.

## 5.3 The fuzzy dual representation of imperfect spatio-temporal information

In this section we define a dual representation for dynamic spatial phenomena that is suited to manage the imperfection relative to either the temporal or the spatial reference of a phenomenon.

We first introduce the representation in which we have precise spatial reference and *indeterminate or vague time reference.* A common way to represent dynamic spatial phenomena is based on sequences of snapshots, maps of the spatial reference of the phenomenon taken at specific instants of time: $\mathtt{s}_1$ at time $\mathtt{t}_1$, $\mathtt{s}_2$ at time $\mathtt{t}_2$, ..., $\mathtt{s}_n$ at time $\mathtt{t}_n$ in which $\mathtt{t}_1 < \mathtt{t}_2 < \ldots < \mathtt{t}_n$ are defined on a temporal domain T, and $\mathtt{s}_i$ for $i$=1,...$n$ are maps defined on a spatial domain X. $\mathtt{s}_i$ can identify either a grid of pixels (the raster representation of spatial information), or a graphic meta-file specifying points, lines, polygons, etc. (the vector representation). So, for example, the dynamic phenomenon "*variation of the temperatures on the Alps*" can be represented by a sequence of snapshots mapping the daily temperatures measured in observation points located on the Alps. ¿From this information we can guess that the dynamic spatial phenomenon changes its spatial reference as follows:

from $\mathtt{s}_1$ to $\mathtt{s}_2$ within the time interval $[\mathtt{t}_1,\mathtt{t}_2]$,
...,
from $\mathtt{s}_{n-1}$ to $\mathtt{s}_n$ within the time interval $[\mathtt{t}_{n-1},\mathtt{t}_n]$.

At each instant of time $\mathtt{t}\in[\mathtt{t}_{i-1},\mathtt{t}_i]$ between two consecutive snapshots, the spatial reference of the phenomenon is *in between* the ones defined by the snapshots $\mathtt{s}_{i-1}$ and $\mathtt{s}_i$. The two maps can be both possible or valid to a distinct extent for representing the phenomenon at time t. In fact, when no further information on the dynamics is available, one can guess that if t is closer to $\mathtt{t}_{i-1}$ then $\mathtt{s}_{i-1}$ would be the most likely spatial manifestation of the phenomenon, while if t is closer to $\mathtt{t}_i$, then $\mathtt{s}_i$ would be the most preferred one. Formally, we can represent a dynamic spatial phenomenon $\mathtt{o}_d$ as a set of pairs $(\tau_i, \mathtt{s}_i)$:

$$\mathtt{o}_d := \{(\tau_1,\mathtt{s}_1),(\tau_2,\mathtt{s}_2),...,(\tau_n,\mathtt{s}_n)\} \tag{5.1}$$

where $\tau_i$ is a *linguistic term* identifying the *indeterminate time* instant associated with the snapshot $\mathtt{s}_i$. For example, $\tau_i$ = "*around* 17.00". We can consider that $\mathtt{s}_i$ is the spatial manifestation of the phenomenon $\mathtt{o}_d$ taken at the indeterminate time instant $\tau_i$. The semantic of $\tau_i$ is defined by a possibility distribution $\mu_i$ on the temporal domain T:

$$\tau_i := \int_{\mathtt{t}_k \in \mathtt{T}} \mu_i(\mathtt{t}_k)/\mathtt{t}_k$$

in which $\mu_i(\mathtt{t}_k) \in [0,1]$ is the possibility degree that $\mathtt{s}_i$ represents the spatial reference of the phenomenon at time instant $\mathtt{t}_k$. In this way, the time reference is identified by linguistic granules that are dealt with by continuous functions on the temporal domain T.

In the case in which we do not have any knowledge on the dynamic laws governing the evolution of the phenomenon, we can define the semantics of $\tau_i$ by a triangular membership function centered in $\mathtt{t}_i$, the precise instant of observation of $\mathtt{s}_i$, with left and right displacements $|\mathtt{t}_i\text{-}\mathtt{t}_{i-1}|$, $|\mathtt{t}_{i+1}\text{-}\mathtt{t}_i|$ respectively (see figure 5.2). This is the usual linear interpolation. More usefully, we can derive the fuzzy validity range of a snapshot based on the application of the dynamic laws governing the evolution of the phenomenon. Since often the dynamic laws are not completely known or are too complex to be described in the classical way, some fuzzy rule-based mechanism can be defined to derive an approximate description of the dynamics of the phenomenon. For example, if the snapshot is a satellite image of a glacier taken at a given date t in which the temperature was below zero °C, we can assume that the snapshot remains fully valid until the first snowing event or increase of temperature above zero °C [4].

On the other side, there are cases in which we have information on the phenomenon dynamics detailed enough to be able to specify the degree of validity of the available snapshots in each instant within a range $[\mathtt{t}_1, \mathtt{t}_2]$. An example of this situation is the case of the trajectory of a weapon. In these situations $\tau_i$ is defined by a fuzzy subset of the time domain T and is interpreted as the fuzzy time validity range of the snapshot $\mathtt{s}_i$; in this case $\mu_i(\mathtt{t}_k)$ is the time validity degree of $\mathtt{s}_i$ at time $\mathtt{t}_k$.

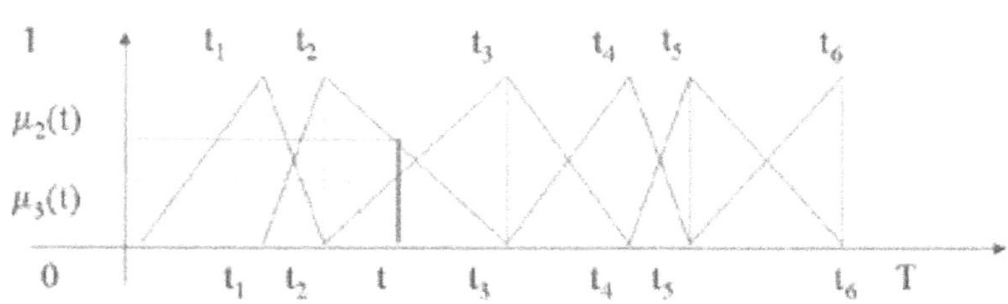


**Figure 5.2.** Semantics of the time references of the snapshots

Figure 5.3 depicts the representation of the path of a taxi cab on a map based on formula (5.1), i.e., "Taxi cab's path" := {(*Just before* 17.00, in a) , (*Around* 17, in b)}.

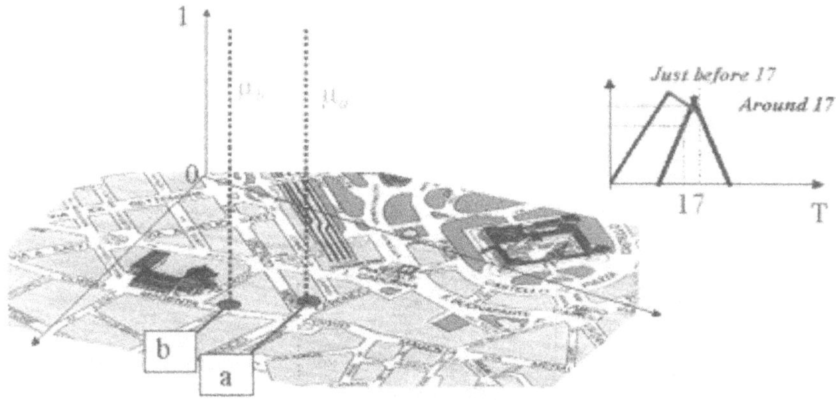


**Figure 5.3.** Representation of the path of a taxi cab based on formula (5.1)

The dual representation of (5.1) is defined with a precise time reference, and an *indeterminate* or *vague spatial reference* by a set of pairs ($\mathtt{t}_i$, $\sigma_i$):

$$\underline{\mathtt{o}}_d := \{(\mathtt{t}_1, \sigma_1), (\mathtt{t}_2, \sigma_2), \ldots, (\mathtt{t}_n, \sigma_n)\} \tag{5.2}$$

in which $\sigma_i$ represents the *possible* spatial reference (or the *vague* spatial reference) of the phenomenon at the precise time $\mathtt{t}_i$.

$\sigma_i$ is a linguistic label such as "*just within the border of region* A" whose semantic is defined by either a possibility distribution or a fuzzy subset on the spatial domain X:

$$\sigma_i := \int_{\mathtt{s}_k \in \mathtt{X}} \pi_i(\mathtt{s}_k)/\mathtt{s}_k$$

in which the value $\pi_i(\mathtt{s}_k) \in [0,1]$ can be interpreted as the possibility degree (or validity degree) of $\mathtt{s}_k$ in representing the spatial reference of the phenomenon at time instant $\mathtt{t}_i$. In other words, each snapshot $\mathtt{s}_k$ in $\sigma_i$ depicts the possible or valid position (to the extent $\pi_i(\mathtt{s}_k)$) of the phenomenon at time $\mathtt{t}_i$. Thus, by formula (5.2), we can represent either *spatial indeterminacy* or *spatial vagueness*, where the gradual property is the time validity of the spatial reference.

Figure 5.4 depicts the representation of the path of a taxi cab on a map based on formula (5.2), i.e., "Taxi cab's path" := {(at 16.50, *close to* a) , (at 17.00, *close to* b)}, in which "*close to* a" and "*close to* b" identify rectangle-shaped membership functions centered in points a and b.

To convert the representation formalized in (5.1) into its dual representation defined in (5.2), a mapping **Dual**:$\wp(\mathtt{T})\times\mathtt{T}\rightarrow \wp(\mathtt{X})$, in which $\wp(A)$ denotes the powerset of fuzzy set on A, can be defined i.e., **Dual**($\mathtt{o}_d$) $= \sum_{\mathtt{t}_j\in\mathtt{T}}(\mathtt{t}_j, \sigma_j) = \underline{\mathtt{o}}_d$. This is obtained by computing $\sigma_j = \mathtt{o}_d(\mathtt{t}_j)\ \forall \mathtt{t}_j \in \mathtt{T}$ as defined in formula (5.3) in the next section. It has been proved that **Dual**($\mathtt{o}_d$) can

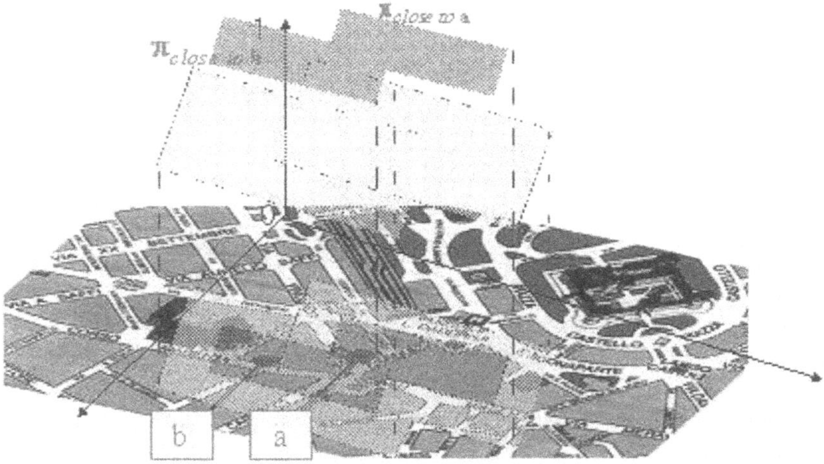


**Figure 5.4.** Representation of the path of a taxi cab based on formula (2)

be transformed back into representation (5.1) without loss of information by applying the inverse function of **Dual**. It can be noticed that the dual representations defined by (5.1) and (5.2) are subsumed by a more general one in which a dynamic spatial phenomenon is represented by a set of pairs $(\tau_i, \sigma_i)$ in which both the temporal reference $\tau_i$ and the spatial reference $\sigma_i$ are linguistic granules identifying fuzzy subsets of the temporal and spatial domain respectively.

Two linguistic variables **instant** and **interval** can be defined on the basic domain of time T such as: $\tau_i \in$ *Term set* (**instant**) $\cup$ *Term set*(**interval**) := {*around* t, *just before* t, *just after* t, ***approximately during*** [t$_1$, t$_2$],...}. Analogously, two linguistic variables **position** and **spatial extent** can be defined on the basic domain of space X such as: $\sigma_i \in$ *Term set* (**position**) $\cup$ *Term set*( **spatial extent**) := {*close to* x, *south-close to* x, *just within the border of region* A, *just outside the border of region* A, ...}

## 5.4 Evaluation of spatio-temporal constraints

In this section we introduce some basic spatio-temporal operations that can be defined and evaluated based on the fuzzy dual representation introduced in the previous section. In particular, it is discussed how the definition of such operations depends on the type of imperfection that is modelled.

Based on the representation of a dynamic phenomenon $o_d$ defined by formula (5.1) we can easily answer queries imposing temporal constraints [48]. The phenomenon is represented with a temporal reference described by either a possibility distribution or a fuzzy subset on T. We can apply one of the several approaches defined in the literature for the evaluation of crisp conditions against fuzzy attribute values [9, 40]. For example, to answer the query:

*"What is the possible (valid) position of the phenomenon* $o_d$ *at the precise instant* t*?"*

we compute the virtual snapshot $o_d(t)$ defined by a fuzzy subset of the spatial domain X as follows:

$$o_d(t) := \{(\tau_1(t), s_1), (\tau_2(t), s_2), \ldots, (\tau_n(t), s_n)\} = \sum_{i=1,\ldots,n} \mu_i(t)/s_i = \sigma \tag{5.3}$$

in which $\tau_i(t) = \mu_i(t)$ is the possibility degree (or validity degree) of the spatial reference $s_i$ at time instant t.We interpret the fuzzy set identified by $\sigma$ as the possible (or valid) spatial reference of the phenomenon at time t.

Based on representation (5.1) also soft temporal constraints can be evaluated, i.e., constraint defined by means of fuzzy subset of the time domain T. For example, one could ask for the location of a phenomenon *just before*, or *just after* a given instant or *during* a period [48]. For example, to answer a query such as:

*"What is the possible (valid) position of the phenomenon* $o_d$ *more or less during the period* $[t_A, t_B]$*?"*

we assume that *more or less* $[t_A, t_B]$ is defined with semantic specified by the trapezoidal membership function $(t_A - \varepsilon, t_A, t_B, t_B + \varepsilon)$.
$\mu_{more\ or\ less[t_A,t_B]}$ imposes the fuzzy inclusion $\subset_f$ of the possible (or valid) temporal reference of $o_d$ in the fuzzy interval $[t_A, t_B]$. The answer is then provided by the following fuzzy subset of the spatial domain:

$$o_d(more\ or\ less[t_A, t_B]) := \{(\Delta_1/s_1), (\Delta_2/s_2), \ldots (\Delta_n/s_n)\} \tag{5.4}$$

in which $\Delta_i \in [0,1]$ is the degree of a fuzzy inclusion, i.e., $\Delta_i =$ degree($\tau_i \subset_f more$ *or less*$[t_A, t_B]$).

Depending on the semantics of the temporal imperfection (either *indeterminacy* or *vagueness*) we choose a distinct definition for the fuzzy inclusion (inclusion of a possibility distribution into a fuzzy set or inclusion between fuzzy sets) [7]. $\Delta_i$ is interpreted as the possibility (or validity) of the position of the phenomenon to be $s_i$ during the fuzzy time interval $[t_A, t_B]$.

Now, let us analyze the evaluation of spatial constraints against the representation formalized in (5.1). To answer queries specifying spatial constraints such as:

*"At what time the spatio-temporal object* $\texttt{o}_d$ *overlaps the spatial entities in* $\texttt{s}$*?"*

we have to evaluate the spatial constraint $\texttt{C} :=$ "*overlap* $\texttt{s}$" against the spatial reference $\texttt{s}_1,\ldots,\texttt{s}_n$ of $\texttt{o}_d$, that is dealt with as sparse information in representation (5.1). To this aim we have to rely on the spatial operators defined in the data model adopted to represent the snapshots $\texttt{s}_i$. In current GISs two alternative spatial data models are commonly adopted to describe spatial entities, named field based model and exact object based model [12]. The first model is generally adopted to represent physical or more generally gradual properties of entities extending smoothly over the entire spatial domain and corresponds with the raster representation, while the second model is used to represent features with well-defined boundaries and corresponds with the vector representation. In our context, we are not going into the details of the spatial data model adopted to represent $\texttt{s}_i$ that can identify either a grid, i.e., a matrix, or a meta-file of vectors on the spatial domain. For the definition of the spatial operators that can be adopted to evaluate metric or topological relationships between spatial entities we rely on the literature [13, 25]. We only recall that the classical spatial operators used by current GISs are defined as binary operators admitting only two degrees of satisfaction (*False* and *True*). In general, if we adopt a classic binary definition of the spatial operator $\texttt{C}$: $\texttt{X}\times\texttt{X}\rightarrow \{0, 1\}$ with $\texttt{X}$ spatial domain, we can define as a result of its evaluation on the two snapshots ($\texttt{s}_k$ taken at time $\tau_k$, and $\texttt{s}$) the set $\{\tau_k | \; \texttt{C}(\texttt{s}_k, \, \texttt{s})=1\}$.

$\tau_k$ can be interpreted as the possible time instant (or the fuzzy period) in which the phenomenon satisfies the constraint $\texttt{C}$, for instance, the possible instants in which the regions in $\texttt{s}_k$ *overlap* those in $\texttt{s}$.

If we extend the definition of the spatial operators in order to make them soft constraints $\mathbf{C}$: $\texttt{X}\times\texttt{X}\rightarrow[0,1]$ [5] we can compute the fuzzy set $\sum_{k=1,\ldots,n}$ $\mathbf{C}(\texttt{s}_k, \, \texttt{s})/\tau_k$. In this case $\tau_k$ with $k=1,\ldots,n$, are the possible time instants (or the validity period) in which the phenomenon satisfies the constraint $\mathbf{C}$ to the extent $\mathbf{C}(\texttt{s}_k,\texttt{s}) \in[0,1]$. We can consider the fuzzy or indeterminate instant $\tau_k$ corresponding with the best matching degree $\mathbf{C}(\texttt{s}_k, \, \texttt{s})$ as the time in which the widest overlapping between the spatial reference of the phenomenon and the spatial object $\texttt{s}$ occurs.

A spatio-temporal constraint is evaluated as a compound constraint. For example, given the query:

*"Does the spatio-temporal object* $\texttt{o}_d$ *overlaps the spatial entities in* $\texttt{s}$ *during the period of time*$[\texttt{t}_A,\texttt{t}_B]$?"

it can be reformulated in the following explicit form:

*"Does the possible (valid) position of the phenomenon* $\texttt{o}_d$ *during the period of time*$[\texttt{t}_A,\texttt{t}_B]$*overlaps the spatial entities in* $\texttt{s}$ *?"*

Its evaluation is composed of two subsequent steps. First the possible (or valid)

position of the phenomenon during the interval of time $[\mathtt{t}_A,\mathtt{t}_B]$ is computed, i.e., the fuzzy set $\mathtt{o}_d([\mathtt{t}_A,\mathtt{t}_B]) = \{(\Delta_1/\mathtt{s}_1),(\Delta_2/\mathtt{s}_2),\ldots(\Delta_n/\mathtt{s}_n)\}$ as defined in formula (5.4). Then, the spatial constraint $\mathbf{C}$ := "*overlap* s" is evaluated against $\mathtt{o}_d([\mathtt{t}_A,\mathtt{t}_B])$ as in the previous sample query, thus computing the fuzzy set $\sum_{k=1,\ldots,n}\mathbf{C}(\mathtt{s}_k,\mathtt{s})/\Delta_k$.
$\Delta_k$ is the matching degree of the temporal constraint by the dynamic phenomenon $\mathtt{o}_d$, while $\mathbf{C}(\mathtt{s}_k,\mathtt{s})$ is the matching degree of the spatial constraint by $\mathtt{o}_d([\mathtt{t}_A,\mathtt{t}_B])$, i.e., by the phenomenon during the period of time $[\mathtt{t}_A,\mathtt{t}_B]$.

Based on the fuzzy dual representation of a dynamic spatial phenomenon $\underline{o}_d$ defined by formula (5.2) with a precise time reference and a *fuzzy* or *indeterminate spatial reference*, we can answer queries imposing spatial constraints. In this case, since $\sigma_i$ can be interpreted as either a possibility distribution or a fuzzy subset of the spatial domain, we have to define proper spatial operators. In this case we can rely on the operators defined for spatial entities with unsharp boundaries [13]. For example, to answer the query:

"*At what time the spatio-temporal phenomenon* $\underline{o}_d$ *is (possibly) located inside the regions depicted in* s?"

we compute the following fuzzy set identified by $\tau$ defined on the temporal domain T:

$$\underline{o}_d(\mathtt{s}) := \{(\Delta_1/\mathtt{t}_1),(\Delta_2/\mathtt{t}_2),\ldots(\Delta_n/\mathtt{t}_n)\} = \sum_{i=1,\ldots,n}(\Delta_i/\mathtt{t}_i) = \tau \qquad (5.5)$$

in which $\tau$ is interpreted as the possible (or valid) temporal reference of the phenomenon when its position is included in s, and $\Delta_i \in[0,1]$ is the degree of a fuzzy inclusion between fuzzy or indeterminate spatial objects, i.e., $\Delta_i = \mathit{degree}(\mathtt{s}_i \subset_f \mathtt{s})$ [5]. In this case, we have to adopt a definition of the fuzzy inclusion depending on the semantics of the spatial imperfection that is represented [6]. Several proposals have been defined in the literature [13, 16, 38, 53].

In the case of *spatial vagueness*, generally, in the determination of the result of a spatial operation, the contribution of each spatial grain is weighted based on the value of its membership degree to the vague region. In this way, a grain belonging to the fuzzy boundary of a spatial entity contributes to a lesser extent to determine the result of the geometric, metric or topological operation with respect to the grains that belong to the core of the spatial entity. So, for instance, the area of a vague region is computed as the cardinality of the fuzzy subset representing the spatial reference. The degree of the fuzzy inclusion between vague regions can be computed as follows:

$$\mathit{degree}(\sigma_i \subset_f \mathtt{s}) = |\sigma_i \cap \mathtt{s}|/|\sigma_i| = AREA(\sigma_i \cap \mathtt{s})/AREA(\sigma_i) \qquad (5.6)$$

in which $AREA(\mathtt{s})$ denotes the area of the regions (or fuzzy regions) in $\mathtt{s}$ defined by the membership function $\pi_{\mathtt{s}}(\mathtt{x})$, and $AREA(\mathrm{A})$ is computed as the cardinality of the fuzzy set A, i.e.:

$$AREA(\sigma_i \cap \mathtt{s}) = \sum_{\mathtt{x}\in\mathtt{X}} \min_{k=1,\dots,n} (\pi_i(\mathtt{s}_k), \pi_{\mathtt{s}}(\mathtt{x}))$$

In this way, we get values that increase as more and more portions of $\sigma_i$ are included in $\mathtt{s}$. The maximum value of 1 is obtained when $\sigma_i$ is completely inside $\mathtt{s}$.

On the other hand, when representing *spatial indeterminacy*, the evaluation of the spatial operations cannot be based on the same methods used in the case of spatial vagueness. This is basically due to the fact that the spatial reference is a possibility distribution, and then the evaluation of the spatial operations must keep into account the worst possible scenario in which the actual spatial reference satisfies the spatial constraint to the lowest degree. To reflect this desired behavior we can define the fuzzy inclusion based on a fuzzy implication $\mathtt{I}$ as follows:

$$\text{DEGREE}(\sigma_i \subset_f \mathtt{s}) = \min_{\mathtt{x}\in\mathtt{X}} \mathtt{I}_{k=1,\dots,n} \; (\pi_i(\mathtt{s}_k), \pi_{\mathtt{S}}(\mathtt{x}))$$

$\mathtt{I}$ can be the Rescher-Gaines implication defined as $\mathtt{I}_{Rescher-Gaines}(\mathrm{A,B})=1$ if $\mathrm{A}\leq\mathrm{B}$, and 0 otherwise, or the Gödel implication if $\mathtt{I}_{Gödel}(\mathrm{A,B})=1$ if $\mathrm{A}\leq\mathrm{B}$, and B otherwise, or the Goguen implication, $\mathtt{I}_{Goguen}(\mathrm{A,B})=1$ if $\mathrm{A}\leq\mathrm{B}$, and $\mathrm{B}/\mathrm{A}$ otherwise.

On the basis of representation in formula (5.2), other kinds of soft spatial constraints can be evaluated. For example, one could ask when the phenomenon is *north* of a region of interest, or *close* to a given point on the map.

On the other side, to answer queries specifying temporal constraints such as in the query:

*"Where is $\underline{o}_d$ at (after, before, during etc.) time $\mathtt{t}$?"*

we have to evaluate the temporal constraint $\chi$ against the temporal reference of the phenomenon, i.e., $\mathtt{t}_1, \mathtt{t}_2, \dots, \mathtt{t}_n$ that is dealt with as sparse information in this representation. If we define $\chi : \mathtt{T} \times \mathtt{T} \rightarrow [0,1]$ as a soft constraint, we can define as a result the fuzzy set $\sum_{k=1,\dots,n} \chi(\mathtt{t}_k,\mathtt{t})/\sigma_k$ in which $\sigma_k$ can be interpreted as the possible location of the phenomenon when $\chi$ is satisfied to the extent $\chi(\mathtt{t}_k, \mathtt{t}) \in[0,1]$.

A spatio-temporal constraint is evaluated as a compound constraint. For example, given the query:

*"Does object $\underline{o}_d$ is included in* $\mathtt{s}$ *during the period of time* $[\mathtt{t}_A,\mathtt{t}_B]$*?"*

it can be reformulated in the following explicit form:

*"Does the possible (valid) time reference of the phenomenon $\underline{o}_d$ while included in the regions depicted in* $\mathtt{s}$ *overlaps the period* $[\mathtt{t}_A,\mathtt{t}_B]$*?"*

Its evaluation is composed of two subsequent steps. First the possible (or valid) temporal reference $\underline{o}_d(\mathtt{s}) = \sum_{=1,\ldots,n}(\Delta_i/\mathtt{t}_i) = \tau$ of the phenomenon while $\underline{o}_d$ is included in the regions in $\mathtt{s}$ is computed based on formula (5.5); then the overlapping constraint $\omega$ between the fuzzy time reference $\tau$ and the interval of time $[\mathtt{t}_A,\mathtt{t}_B]$ is computed $\sum_{k=1,\ldots,n} \omega(\mathtt{t}_k,[\mathtt{t}_A,\mathtt{t}_B])/\Delta_k$.
$\Delta_k$ is the matching degree of the spatial constraint, while $\omega(\mathtt{t}_k,[\mathtt{t}_A,\mathtt{t}_B])$ is the matching degree of the temporal constraint.

## 5.5 Object based representation of imperfect spatio-temporal information

In the previous section, we have seen that imperfect spatio-temporal information can be represented in the unifying framework of fuzzy set and possibility theory through a dual model in which the time and space dimensions are dealt with as orthogonal concepts [23, 52]. In [6], an object based data model for imperfect spatial information has been proposed that integrates the two alternative views of spatial information based on either exact objects or fields. In this section, we extend this proposal to incorporate the imperfection of the temporal dimension. Specifically we define some Abstract Spatio-Temporal Data Types for representing the different kinds of imperfect information identified in the previous sections. In figure 5.5, the graphic elements used to identify the object classes, the attributes and the relationships in the object-based spatio-temporal schema are depicted. The oval shaped nodes identify basic domains of the primitive attributes, i.e., those domains that are not defined in this context but are assumed as already defined. A spatio-temporal phenomenon is then defined as an instance of one of the four classes (see figure 5.6):

*Indeterminate time-stamped phenomenon*: this class is used in the case in which we have snapshots of the phenomenon whose date of acquisition is not precisely known, or is defined on a coarser scale with respect to the one requested by the application;

*Phenomenon with fuzzy temporal validity*: this class is used in the case in which one wants to represent the temporal validity of the snapshots, that may vary depending on some dynamic laws governing the phenomenon;

*Phenomenon with indeterminate spatial validity*: this class makes it possible to describe phenomena with precise information on the time of acquisition of the spatial reference, but affected by spatial indeterminacy due to incompleteness or vagueness of the information, or to limitations of the means of representation (ex: we know the position of a taxi cab by a verbal description during a telephone call occurring in a precisely known time instant);

*Phenomenon with fuzzy spatial validity*: this class is useful in the case in which we have precise information on the time of acquisition of the spatial reference of a phenomenon and we want to represent the gradual validity of the spatial reference at that specific instant of time (ex: we want to represent the spatial variability that an ecosystem may have at a specific date).

These four main classes, besides having the primitive attributes *name* and *ID* univocally identifying their instances, have a temporal and a spatial reference. The *indeterminate* and *fuzzy spatial* reference and the *indeterminate* and *fuzzy temporal* reference attributes are defined with structured domains, i.e., the *Indeterminate time instant* class, the *Indeterminate snapshot* class, the *fuzzy validity interval* class and the *fuzzy snapshot* class. The precise classes are considered as primitive attributes having in this context a primitive domain, i.e., the temporal domain $\mathtt{T}$ and the spatial domain $\mathbb{R} \times \mathbb{R}$ respectively. Thus, in this context, we are not going to specify further the definition of the basic values of $\mathtt{T}$ and $\mathbb{R} \times \mathbb{R}$. The *Indeterminate time instant* class and the *Indeterminate snapshot* class are defined as subclasses of the *Possibility distribution* class defined on a temporal and on a spatial domain respectively, while the *Fuzzy validity interval* class and the *Fuzzy snapshot* class are defined as subclasses of the *Fuzzy set* class on a temporal and spatial domain respectively. These are the basic bricks for the definition of the object-based spatio-temporal database. The *Possibility distribution class* and the *Fuzzy set* class are defined as in figure 5.7. In this way to deal with spatio-temporal instances affected either by indeterminacy or vagueness proper methods can be defined specifically for these classes. In this schema the imperfection of spatio-temporal information is represented at distinct levels of details: in the outmost level, the one that is more visible to the user, linguistic qualifiers are used either to identify the possible/vague spatial reference or the indeterminate/fuzzy time reference of the phenomena. These qualifiers can be drawn from the *term set* defined for the linguistic variables **instant** and **interval** for the indeterminate and fuzzy temporal reference [20], and **position** and **spatial extent** for the indeterminate and vague spatial reference respectively [6]. In the intermediate level, we have the explicit representation of the semantics (meanings) of the linguistic qualifiers defined on their basic domains $\mathtt{T}$, and $\mathbb{R} \times \mathbb{R}$. The inmost level corresponds with the physical implementation of the data model.

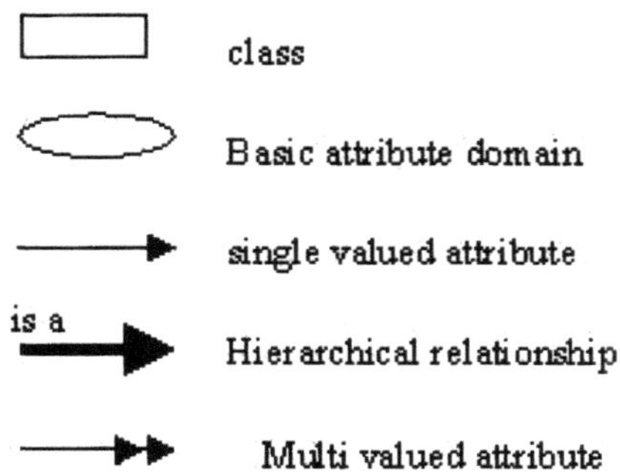


**Figure 5.5.** Graphic elements in the schema of the Object-based spatio-temporal data model

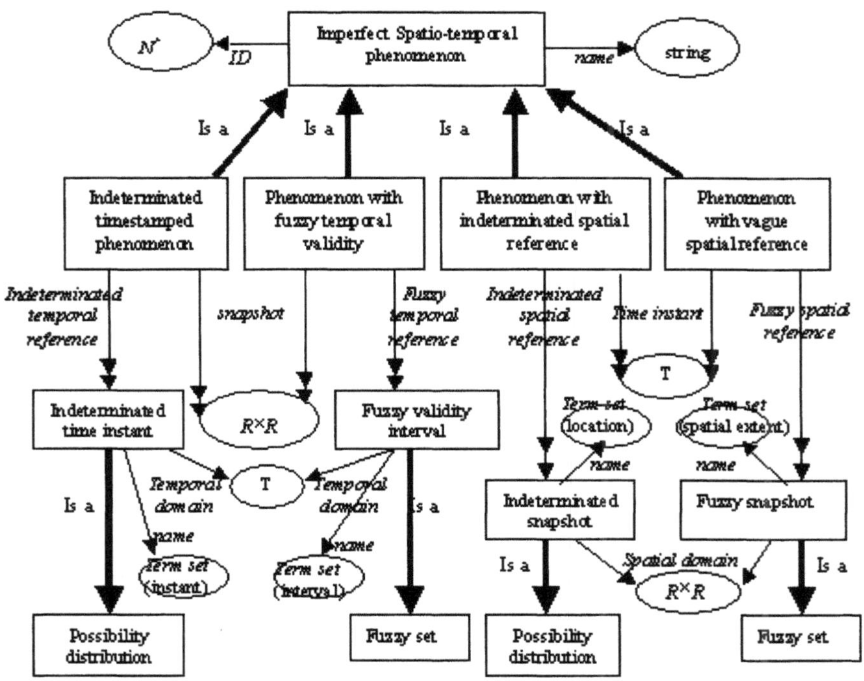


**Figure 5.6.** Schema of the spatio-temporal data model for imperfect phenomena

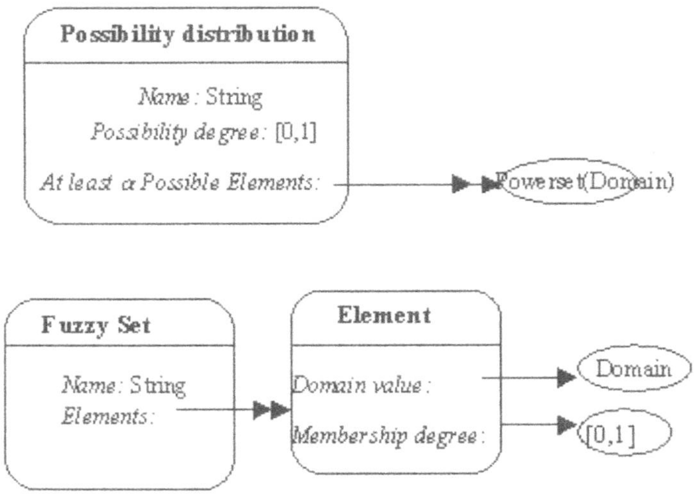


**Figure 5.7.** Schema of the possibility distribution and fuzzy set classes

## 5.6 Conclusion

The representation of dynamic spatial phenomena has to account for the uncertainty that generally characterizes observations. In this contribution, an ontology for imperfect spatio-temporal information is proposed within fuzzy set and possibility theory, and a fuzzy dual representation for such imperfect information is proposed. The time and space dimensions are seen as orthogonal dimensions and the imperfection may be represented as affecting the spatial or the temporal components of the information. In this chapter it is discussed how the definition of the spatio-temporal operations on imperfect information strongly depends on the kind of the imperfection modelled in the spatial and temporal context, and some guidelines are indicated to define methods for evaluating spatio-temporal constraints. Furthermore, Abstract Spatio-Temporal Data Types for representing imperfect dynamic spatial phenomena are proposed for an object-based data model. These special views may be used in most practical situations to index imperfect spatio-temporal data for efficient retrieval.

## References

1. Abraham T., Roddick J.F. (1999) Survey of Spatio-Temporal Databases. GeoInformatica 3(1):61–99.
2. Al-Taha K.K., Snodgrass R.T., Soo M.D. (1993) Bibliography on spatio-temporal databases. SIGMOD Record 56(6):7–22.

3. Bertino E., Vantini G. (1996) Advanced Database Systems and Geographical Information Systems. In: Proc. of the II Workshop on AITSES, pp. 1–10.
4. Bordogna G., Carrara P., Chiesa S., Spaccapietra S. (2003) A Dual Representation of Uncertain Dynamic Spatial Information. In: Bilgic T., de Baets B., Kaynak O. (Eds.) Fuzzy Sets and Systems – IFSA2003. LNAI, Springer Verlag, Berlin Heidelberg, New York.
5. Bordogna G., Chiesa S. (2002) Fuzzy Querying of Spatial Information. In: Proc. of EUROFUSE, pp. 1–6.
6. Bordogna G., Chiesa S. (2003) A Fuzzy Object-based Data Model for Imperfect Spatial Information Integrating Exact Objects and Fields. Int. J. of Uncertainty, Fuzziness and Knowledge Based Systems 11(1):23–41.
7. Bordogna G., Pasi G. (2002) A semantics for Soft selection conditions in Fuzzy Databases based on a Fuzzy Inclusion. In: Proc. of EUROFUSE, pp. 18–27.
8. Bosc P., Buckles B.B., Petry F.E., Pivert O. (1999) Fuzzy Databases. In: Bezdek J.C., Dubois D., Prade H. (Eds.) The Handbooks of Fuzzy Sets Series, Kluwer Academic Publishers.
9. Bosc P., Prade H. (1994) An Introduction to the Fuzzy Set and Possibility Theory-based treatment of Flexible Queries and Uncertain and Imprecise Databases. In: Motro A., Smets P. (Eds.) Uncertainty Management in Information Systems: from Needs to Solutions, Kluwer Academic Publishers.
10. Burrough P.A., Frank A.U. (1995) Concepts and Paradigms in Spatial Information: are current geographical information systems truly generic? Int. J. Geograph. Inform. Syst. 9(2):101–116.
11. Burrough P.A., Frank A.U. (1996) Geographic Objects with Indeterminate Boundaries. Taylor & Francis.
12. Burrough P.A., McDonnell R. (1998) Principles of geographical information systems. Oxford University Press, Oxford.
13. Clementini E., Di Felice P. (1996) An Algebraic Model for Spatial Objects with Indeterminate Boundaries. In: Burrough P.A., Frank A.U. (Eds.) Geographic Objects with Indeterminate Boundaries, Taylor & Francis.
14. Codd M., Foley H., Petry F., Shaw K. (2000) Uncertainty in Distributed and Interoperable Spatial Information Systems. In: Bordogna G., Pasi G. (Eds.) Recent Issues on Fuzzy Databases, Physica Verlag.
15. Codd M., Petry F., Robinson V. (2000) Uncertainty in Geographic Information Systems and Spatial data. Special issue of Fuzzy Sets and Systems 113(1).
16. Cohn A.G., Gotts N.M. (1996) The "Egg-Yolk" Representation of Regions with Indeterminate Boundaries. In: Burrough P.A., Frank A.U. (Eds.) Geographic Objects with Indeterminate Boundaries, Taylor & Francis.
17. Couclelis H. (1996) Towards an Operational Typology of Geographic Entities with ill-defined Boundaries. In: Burrough P.A., Frank A.U. (Eds.) Geographic Objects with Indeterminate Boundaries, Taylor & Francis.
18. Cross V. (1999) Using the Fuzzy Object Model for geographical Information Systems. In: Proc. of the 18th NAFIPS conference, pp. 814–818.
19. De Caluwe R. (1997) Fuzzy and Uncertain Object-Oriented databases, Concepts and Models, World Scientific.
20. De Caluwe R., De Tré G., Van der Cruyssen B., Devos F., Maesfranckx P. (2000) Time Management in Fuzzy and Uncertain Object-Oriented Databases. In: Pons O., Vila M.A., Kacprzyk J. (Eds.) Knowledge Management in Fuzzy Databases, Studies in Fuzziness and Soft Computing, Physica-Verlag, Heidelberg, Germany.

21. Dragicevic S., Marceau D.J. (2000) An application of fuzzy logic reasoning for GIS temporal modeling of dynamic processes. Fuzzy sets and Systems 113(1):69–80.
22. Dubois D., Lang J., Prade H. (1991) Timed possibilistic logic. Fundamenta Informaticae XV(4):211–234.
23. Dubois D., Prade H. (1988) Possibility Theory: an Approach to Computerised Processing of Uncertainty, Plenum Press, New York.
24. Duckham M., Mason K., Stell J.G., Worboys M.F. (2001) A formal approach to imperfection in geographic information. Computers, Environ. and Urban Syst.
25. Egenhofer M.J., Golledge R.J. (1998) Spatial and Temporal reasoning in geographic information systems. Oxford University Press.
26. Egenhofer M.J., Herring J. (1991) Categorising Binary Topological Relationships between regions, lines, and points in geographic databases. Tech. report, univ. Maine, dep. of Surveying Engineering.
27. Erwig M., Schneider M. (1997) Vague regions. In: Frank A.U., Kuhn W. (Eds.) Advances in Spatial Databases (SSD'97), Springer Verlag.
28. Fisher P. (1997) The pixel: a snare and a delusion. International Journal of Remote Sensing 18(3):679–685.
29. Fisher P. (2000) Sorites paradox and vague geographies. Fuzzy Sets and Systems 113(1):7–18.
30. Frank A. (1992) Theories and methods of spatio-temporal reasoning in geographic space. LNCS 639, Springer Verlag.
31. Geneletti D. (2002) Ecological evaluation for environmental impact assessment. Netherlands Geographical Studies.
32. Goodchild M.F., Gopal S. (1989) Accuracy of spatial databases. Taylor and Francis, London.
33. Guting R.M. et al. (1998) A foundation for representing and querying moving objects. FernUniversitaet Hagen, Informatik-report 238.
34. Janssen L.L.F. (2000) Principles of remote sensing. Enschede: ITC.
35. Mackey D.S., Robinson V.B. (2000) A multiple criteria decision support system for testing integrated environmental models. Fuzzy Sets and Systems 113:53–67.
36. Maling D.H. (1989) Measurements from maps: principles and methods of cartometry, Pergamon press, NY.
37. Morris A., Foster J., Petry F.E. (1999) Providing support for Multiple Collection types in a Fuzzy Object Oriented Spatial Data Model. In: Proc. of the 18th Int. Conf. of the NAFIPS, IEEE press, Piscataway, NJ, pp. 824–828.
38. Papadias D., Karacapilidis N., Arkoumanis D. (1999) Processing fuzzy spatial queries: a configuration similarity approach. Int. J. of Geographical Information Science 13(2):93–118.
39. Parent C., Spaccapietra S., Zimanyi E. (1999) Spatio-temporal conceptual models: data structures + space + time. In: Proc. of the 7th ACM Symposium on Advances in GIS, Kansas City, Kansas.
40. Petry F.E. (1996) Fuzzy Databases, Principles and Applications, Kluwer Academic Publishers.
41. Ratsiatou I., Stefanakis E. (2001) Spatio-Temporal Multicriteria Decision making under Uncertainty. In: Proc. of the 1st Int. Symposium on Robust statistics and fuzzy techniques in geodesy and GIS, Zurich.
42. Richards J.A. (1993) Remote sensing digital image analysis. An introduction. Berlin: Springer Verlag.

43. Robinson V. (2000) On Fuzzy Sets and the Management of Uncertainty in an Intelligent Geographic Information System. In: Bordogna G., Pasi G. (Eds.) Recent Issues on Fuzzy Databases, Physica Verlag.
44. Spaccapietra S., Parent C., Zimanyi E. (1999) MurMur: a research agenda on multiple representations. In: Proc. of the 1999 Int. Symposium on Database Applications in Non-Traditional Environments (DANTE'99), Kyoto.
45. Topaloglou T., Mylopulos J. (1998) Representing Partial Spatial Information in Databases, Geoinformatica.
46. Worboys M.F. (1994) A Unified model for spatial and temporal information. The Computer Journal 37(1):27–34.
47. Worboys M.F. (1998) Imprecision in Finite Resolution Spatial data. Geoinformatica 2:257–280.
48. Yager R.R. (2000) Retrieval from Multimedia databases using Fuzzy temporal Concepts. In: Pons O., Vila M.A., Kacprzyk J. (Eds.) Knowledge Management in Fuzzy Databases, Studies in Fuzziness and Soft Computing, Physica-Verlag, Heidelberg, Germany.
49. Yazici A., Zhu Q., Sun N. (2001) Semantic data modeling of spatiotemporal database applications. Int. J. of Intelligent Systems 16(7):881–904.
50. Yazici A., Akkaya K. (2000) Conceptual Modeling of Geographic Information System Applications. In: Bordogna G., Pasi G. (Eds.) Recent Issues on Fuzzy Databases, Physica Verlag.
51. Zadeh L.A. (1975) The concept of a linguistic variable and its application to approximate reasoning, parts I, II. Information Science 8:199–249,301–357.
52. Zadeh L.A. (1978) Fuzzy Sets as a Basis for a Theory of Possibility. Fuzzy Sets and Systems 1:3–28.
53. Zhan F.B. (1997) Topological relations between Fuzzy regions. In: Proc. of the ACM SAC, pp. 192–196.

# 6

# The Applicability of Generalized Constraints in Spatio-Temporal Database Modelling and Querying

Guy de Tré, Rita de Caluwe, Jörg Verstraete, and Axel Hallez

Department of Telecommunications and Information Processing
Ghent University
Sint-Pietersnieuwstraat 41, B-9000 Gent (Belgium)
{guy.detre,rita.decaluwe,jorg.verstraete,axel.hallez}@ugent.be

## 6.1 Introduction

From a formal point of view, a constraint can be seen as a relationship that has to be satisfied. With respect to data and information modelling, constraints provide a declarative and elegant means to represent data in an efficient way. Moreover, constraints support the definition of data semantics. For example, the set of valid temperatures $t$ on earth can be defined as ranging from a high of 57 degrees Celcius to -70 degrees Celcius, which can be described by the constraint $t \in [-70\,°\mathrm{C}, 57\,°\mathrm{C}]$ or, using linear arithmetic, by $(t \geq -70\,°\mathrm{C}) \wedge (t \leq 57\,°\mathrm{C})$.

These considerations especially hold for temporal and spatial information: temporal data usually consists of time intervals, which on their turn are composed of time points [27], whereas spatial data usually consist of points and line segments [24, 25]. Linear arithmetic constraints are particularly appropriate for representing such kinds of data [14]. For example, if spatial geographical information is modelled, linear arithmetic constraints can be used to define the borders of a country, a city, a region, to define a river, a highway, etc. This is illustrated in Figure 6.1 and in Table 6.1, which respectively represent maps of Europe and Belgium (a real map will be defined by many more constraints, but the basic ideas are the same) and some geometrical descriptions, expressed by means of linear arithmetic constraints.

With respect to database modelling, constraints are additionally considered to be an important and adequate means to enforce data integrity [18, 14, 8]. A database 'record' then belongs to the database in as far as it satisfies all of its defining constraints.

Constraints can also be used to impose selection criteria for information retrieval. In this case, each constraint defines a condition for the database 'records' to belong to the result of the retrieval operation [13, 21]. Every

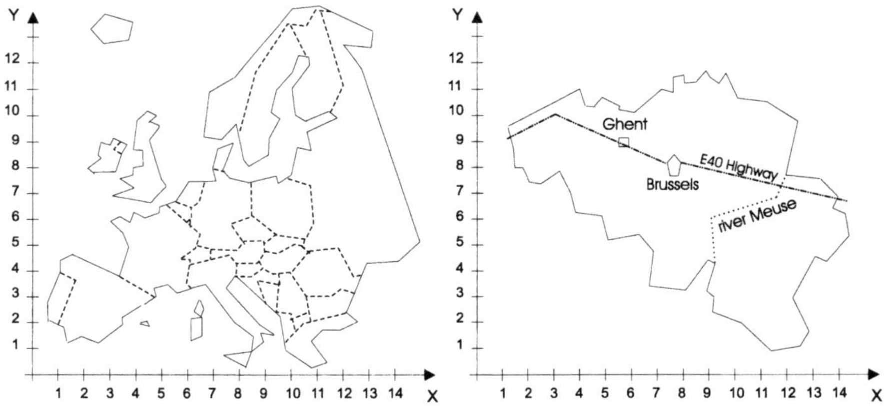


**Figure 6.1.** Modelling geographical information.

**Table 6.1.** Geometrical descriptions using linear constraints.

| Ghent: $(x \geq 5.5) \wedge (x \leq 5.9) \wedge (y \geq 8.8) \wedge (y \leq 9.2)$ |
|---|
| E-40 highway (in Belgium): $((y \geq 9.1) \wedge (1.8y - x = 14.8) \wedge (y \leq 10))$ $\vee ((y \leq 10) \wedge (x + 2y = 23) \wedge (y \geq 8.2))$ $\vee ((y \leq 8.2) \wedge (x + 4.9y = 48.2) \wedge (y \geq 7))$ |
| River Meuse (in Belgium): $((y \geq 4.4) \wedge (x + 0.2y = 10.2) \wedge (y \leq 6))$ $\vee ((y \geq 6) \wedge (2.9y - x = 8.4) \wedge (y \leq 6.9))$ $\vee ((y \geq 6.9) \wedge (x - 0.4y = 8.8) \wedge (y \leq 8))$ |

database 'record' belongs to the result in as far as it satisfies all the imposed criteria. For example, if someone wants to retrieve all the young persons who live in the environment of Ghent, two constraints can be imposed: one to retrieve all young persons and another to retrieve all persons living in the environment of Ghent.

Information, spatio-temporal information in particulary, is for a considerable part only available in an imperfect form [3, 16]. Different types and sources of imperfect information have been distinguished [19, 1, 17]: information can be imprecise, vague, inconsistent, incomplete and/or uncertain. Managing these kinds of information with classical database systems brings a disadvantageous loss of data semantics along. Therefore, advanced database modelling and manipulation techniques are necessary.

In this chapter, a new flexible database modelling and querying technique, that is based on the generalization of the concept 'constraint' and allows to cope with imperfect information, is illustrated. More specifically, it is shown how generalized constraints, as originally introduced by L.A. Zadeh [32, 33], can be integrated in a constraint-based fuzzy object-oriented database model that is extended with spatio-temporal data modelling and data manipulation facilities.

The remainder of the chapter is organized as follows. In Section 6.2, some preliminary definitions with respect to generalized constraints [33] and extended possibilistic truth values (EPTV's) [7] are introduced. In Section 6.3, an algebraic type system for an object oriented database model, that is suited to model both perfect and imperfect information, is described. This type system distinguishes itself from existing type systems by the integration of (generalized) constraints —which are used for the definition of (the semantics of) the data and will also be used for the specification of query criteria— and by its underlying logical framework of EPTV's. The presented type system can be integrated in a database model and thereby support the definition of database schemes [6]. In Section 6.4 the type system is extended with the extra generic data types *Time* and *Space* that allow to model spatio-temporal information in a flexible way, hereby using flexible linear arithmetic constraints. The cases of modelling one-dimensional temporal information, one-dimensional, two-dimensional and $n$-dimensional spatial information and $n$-dimensional spatio-temporal information are discussed. In Section 6.5, the flexible querying of spatio-temporal data is discussed and illustrated by means of some examples. Finally, the achieved results are summarized and some intentions for future developments are given in Section 6.6.

## 6.2 Some preliminaries

In this section some preliminary definitions with respect to generalized constraints and EPTV's are given. In the presented modelling approach, generalized constraints are used for the definition of (spatio-temporal) data, i.e., values of (spatio-temporal) domains, as well as for the definition of (spatio-temporal) restrictions on domains that are associated with attributes in the definition of a type. Whereas, in order to cope with missing information, an underlying many-valued logical framework of EPTV's is used.

### 6.2.1 Generalized constraints

The concept 'generalized constraint' has been introduced by L.A. Zadeh [32, 33] as the basic of a computational approach to meaning and knowledge representation. The introduction of this concept has been motivated by the fact that conventional crisp constraints of the form $X \in C$, where $X$ is a variable and $C$ is a set, are insufficient to represent the meaning of (human) perceptions.

A generalized constraint is, in effect, a family of constraints and can be seen as a generalization of an assignment statement [33].

**Definition 1** *An unconditional generalized constraint on a variable $X$ is defined by*

$$X \textit{ isr } R$$

*where $R$ is the constraining relation and isr is a variable copula in which the discrete-valued variable $r$ defines the way in which $R$ constrains $X$.* □

As specified in [35], the principal generalized constraints are the following:

- *Equality constraint*, $r = e$, i.e., $X$ ise $R$. $X$ equals $R$. For example, the generalized constraint 'Average temperature in Ghent in October=12 °C' denotes that the average temperature in Ghent in October equals 12 °C.
- *Possibilistic constraint*, $r = blank$, i.e., $X$ is $R$. $R$ is the possibility distribution of $X$ [31, 11]. For example, the possibilistic constraint 'Average temperature in Ghent in October is *not_high*', in which *not_high* is a disjunctive fuzzy set with membership function $\mu_{not\_high}$, denotes that $\pi_X(x) = \mu_{not\_high}(x)$, where $x$ is a numerical value of *temperature* and $\pi_X(x)$ is the possibility that the average temperature in Ghent in October is $x$.
- *Veristic constraint*, $r = v$, i.e., $X$ is$v$ $R$. $R$ is the verity distribution of $X$ [34]. For example, the veristic constraint 'Languages spoken in Ghent is$v$ $\{(Dutch, 1), (French, 0.7), (English, 0.5), (German, 0.1)\}$', in which $\{(Dutch, 1), (French, 0.7), (English, 0.5), (German, 0.1)\}$ is a conjunctive fuzzy set which denotes that the verity of the proposition 'Language spoken in Ghent is Dutch' (resp. French, English and German) is 1 (resp. 0.7, 0.5 and 0.1). This expresses that the principal language spoken in Ghent is Dutch, although some people also speak French, English and German.
- *Probabilistic constraint*, $r = p$, i.e., $X$ is$p$ $R$. $R$ is the probability distribution of $X$. For example, the probabilistic constraint 'Number of rainy days in Ghent in October is$p$ $N(14, 4)$' means that this number is a normally distributed random variable with mean 14 and variance 4.
- *Probability-value constraint*, $r = pv$, i.e., $X$ is$pv$ $R$. $X$ is the probability of a fuzzy event [29] and $R$ is its value. For example, the proposition 'it is likely that the temperature in Ghent is not high in October' can be modelled by the probability-value constraint 'Prob(temperature in Ghent in October is *not_high*) is$pv$ likely' in which 'likely' is a fuzzy probability.
- *Random set constraint*, $r = rs$, i.e., $X$ is$rs$ $R$. $R$ is the fuzzy-set-valued probability distribution of $X$. For example, if the temperature in Ghent in October is uncertain and the potential values are modelled by the fuzzy sets '*not_high*', '*around 8* °C' and '*more than 20* °C' with respective probabilities 0.6, 0.3 and 0.1, this can be expressed by the random-set constraint 'temperature in Ghent in October is$rs$ (0.6 $\setminus$ *not_high* + 0.6 $\setminus$ *around 8* °C + 0.1 $\setminus$ *more than 20* °C).
- *Fuzzy graph constraint*, $r = fg$, i.e., $X$ is$fg$ $R$. $X$ is a function and $R$ is its fuzzy graph [33]. For example, if $X$ is a function expressing the

relationship between time of the year and temperatures in Ghent and $X$ is approximated by the fuzzy graph $f^* = winter \times low + spring \times moderate + summer \times rather\_high + autumn \times rather\_low$, this can be expressed by the fuzzy graph constraint '$X$ is$fg$ $f^*$'.

- *Usuality constraint*, $r = u$, i.e., $X$ is*u* $R$. This means that, usually $X$ is $R$. A usuality constraint is a special case of a probability-value constraint. For example, the usuality constraint 'temperature in Ghent in autumn is*u* *cold*' should be interpreted as an abbreviation of *Prob*(temperature in Ghent in autumn is *cold*) is*pv* usually'

### 6.2.2 Extended possibilistic truth values

The concept "extended possibilistic truth value" (EPTV) [7] is defined as an extension of the concept "possibilistic truth value" that was originally introduced in [20] and has been further developed in [4, 5].

EPTV's provide an epistemological representation of the truth of a proposition, which allows to reflect our knowledge about the actual truth and have been specifically designed to deal with those cases where the truth value of a proposition is either unknown or undefined.

The truth value of a proposition is unknown if, e.g., some data in the proposition exist but are not available. For example, the truth value of the proposition '*the average temperature on the top of mountains higher than 3000m in France is -5*°C' is unknown if no information about this temperature is given.

The truth value of a proposition is undefined if, e.g., the proposition can not be evaluated due to the non-applicability of (some of) its elements. For example, the truth value of the proposition '*the average temperature on the top of mountains higher than 3000m in Belgium is -5*°C' is considered to be undefined since the proposition is not applicable because there are no mountains higher than 3000m in Belgium.

**Definition 2 (EPTV)** *With the understanding that $P$ represents the universe of all propositions and $\tilde{\wp}(I^*)$ denotes the set of all ordinary fuzzy sets that can be defined over the universal set $I^* = \{T, F, \perp\}$ of truth values (where $T$ represents 'True', $F$ represents 'False' and $\perp$ represents an undefined truth value), the EPTV $\tilde{t}^*(p)$ of a proposition $p \in P$ is formally defined by means of a mapping*

$$\tilde{t}^* : P \rightarrow \tilde{\wp}(I^*) : p \mapsto \tilde{t}^*(p)$$

*that associates with each $p \in P$ a fuzzy set*

$$\tilde{t}^*(p) = \{(T, \mu_{\tilde{t}^*(p)}(T)), (F, \mu_{\tilde{t}^*(p)}(F)), (\perp, \mu_{\tilde{t}^*(p)}(\perp))\}$$

*The semantics of this associated fuzzy set is defined in terms of a possibility distribution. With the understanding that $t^* : P \rightarrow I^*$ is the mapping function that associates the value $T$ with $p$ if $p$ is true, associates the value $F$ with $p$*

*if p is false and associates the value ⊥ with p if (some of) the elements of p are not applicable, undefined or not supplied, this means that*

$$\forall\, x \in I^* : \pi_{t^*(p)}(x) = \mu_{\tilde{t}^*(p)}(x)$$

*where* $\pi_{t^*(p)}(x)$ *denotes the possibility that the value of* $t^*(p)$ *conforms to* $x$ *and* $\mu_{\tilde{t}^*(p)}(x)$ *is the membership grade of* $x$ *within the fuzzy set* $\tilde{t}^*(p)$. □

Special cases of EPTV's are:

| $\tilde{t}^*(p)$ | interpretation |
|---|---|
| $\{(T,1)\}$ | $p$ is true |
| $\{(F,1)\}$ | $p$ is false |
| $\{(T,1),(F,1)\}$ | $p$ is unknown |
| $\{(\bot,1)\}$ | $p$ is undefined |
| $\{(T,1),(F,1),(\bot,1)\}$ | $p$ is unknown or undefined |

As an example, consider the modelling of the truth value representing 'unknown' by the possibility distribution $\{(T,1),(F,1)\}$, which denotes that it is completely possible that the proposition is true ($T$), but it is also completely possible that the proposition is false ($F$).

New propositions can be constructed from existing propositions, using so-called logical operators, whose definitions are based on the operators of a strong three-valued Kleene logic [22]. An unary operator '$\tilde{\neg}$' is provided for the negation of a proposition and binary operators '$\tilde{\wedge}$', '$\tilde{\vee}$', '$\tilde{\Rightarrow}$' and '$\tilde{\Leftrightarrow}$' are respectively provided for the conjunction, disjunction, implication and equivalence of propositions. The arithmetic rules to calculate the EPTV of a composite proposition and the algebraic properties of extended possibilistic truth values are presented in [7]. The rules for negation, conjunction and disjunction can be summarized as

- *Rule for negation:* $\forall\, p \in P : \tilde{t}^*(NOT\ p) = \tilde{\neg}(\tilde{t}^*(p))$
  where $\tilde{\neg} : \tilde{\wp}(I^*) \to \tilde{\wp}(I^*) : \tilde{V} \mapsto \tilde{\neg}(\tilde{V})$ is defined by
  - $\mu_{\tilde{\neg}(\tilde{V})}(T) = \mu_{\tilde{V}}(F)$
  - $\mu_{\tilde{\neg}(\tilde{V})}(F) = \mu_{\tilde{V}}(T)$
  - $\mu_{\tilde{\neg}(\tilde{V})}(\bot) = \mu_{\tilde{V}}(\bot)$
- *Rule for conjunction:* $\forall\, p, q \in P : \tilde{t}^*(p\ AND\ q) = \tilde{t}^*(p)\tilde{\wedge}\tilde{t}^*(q)$
  where $\tilde{\wedge} : \tilde{\wp}(I^*) \times \tilde{\wp}(I^*) \to \tilde{\wp}(I^*) : (\tilde{U},\tilde{V}) \mapsto \tilde{U}\ \tilde{\wedge}\ \tilde{V}$ is defined by
  - $\mu_{\tilde{U}\tilde{\wedge}\tilde{V}}(T) = \min(\mu_{\tilde{U}}(T), \mu_{\tilde{V}}(T))$
  - $\mu_{\tilde{U}\tilde{\wedge}\tilde{V}}(F) = \max\begin{pmatrix}\min(\mu_{\tilde{U}}(T), \mu_{\tilde{V}}(F)),\\ \min(\mu_{\tilde{U}}(F), \mu_{\tilde{V}}(T)),\\ \min(\mu_{\tilde{U}}(F), \mu_{\tilde{V}}(F)),\\ \min(\mu_{\tilde{U}}(F), \mu_{\tilde{V}}(\bot)),\\ \min(\mu_{\tilde{U}}(\bot), \mu_{\tilde{V}}(F))\end{pmatrix}$
  - $\mu_{\tilde{U}\tilde{\wedge}\tilde{V}}(\bot) = \max\begin{pmatrix}\min(\mu_{\tilde{U}}(T), \mu_{\tilde{V}}(\bot)),\\ \min(\mu_{\tilde{U}}(\bot), \mu_{\tilde{V}}(T)),\\ \min(\mu_{\tilde{U}}(\bot), \mu_{\tilde{V}}(\bot))\end{pmatrix}$

- *Rule for disjunction:* $\forall\, p, q \in P : \tilde{t}^*(p\ OR\ q) = \tilde{t}^*(p)\tilde{\vee}\tilde{t}^*(q)$ where $\tilde{\vee} : \tilde{\wp}(I^*) \times \tilde{\wp}(I^*) \rightarrow \tilde{\wp}(I^*) : (\tilde{U}, \tilde{V}) \mapsto \tilde{U}\ \tilde{\vee}\ \tilde{V}$ is defined by
  - $\mu_{\tilde{U}\tilde{\vee}\tilde{V}}(T) = \max\begin{pmatrix}\min(\mu_{\tilde{U}}(T), \mu_{\tilde{V}}(T)),\\ \min(\mu_{\tilde{U}}(T), \mu_{\tilde{V}}(F)),\\ \min(\mu_{\tilde{U}}(T), \mu_{\tilde{V}}(\perp)),\\ \min(\mu_{\tilde{U}}(F), \mu_{\tilde{V}}(T)),\\ \min(\mu_{\tilde{U}}(\perp), \mu_{\tilde{V}}(T))\end{pmatrix}$
  - $\mu_{\tilde{U}\tilde{\vee}\tilde{V}}(F) = \min(\mu_{\tilde{U}}(F), \mu_{\tilde{V}}(F))$
  - $\mu_{\tilde{U}\tilde{\vee}\tilde{V}}(\perp) = \max\begin{pmatrix}\min(\mu_{\tilde{U}}(F), \mu_{\tilde{V}}(\perp)),\\ \min(\mu_{\tilde{U}}(\perp), \mu_{\tilde{V}}(F)),\\ \min(\mu_{\tilde{U}}(\perp), \mu_{\tilde{V}}(\perp))\end{pmatrix}$

These rules have been obtained by applying Zadeh's extension principle [30] to the operators of the strong three-valued Kleene logic [22]. Kleene logics are truth-functional, which means that according to these systems, the behavior of a logical operator is mirrored in a logical function combining Kleene truth values. Therefore, the extended truth value of every composed proposition can be calculated as a function of the extended truth values of its original propositions.

As illustrated in [9], EPTV's can be used to express query satisfaction in flexible database querying. Hereby, every $o$ object in the result set of a (flexible) query $Q$ has been assigned a calculated EPTV $\tilde{t}^*(o \text{ satisfies } Q)$, where the membership grade of $T$, $F$ and $\perp$ respectively denote the possibility that $o$ satisfies $Q$, the possibility that that $o$ does not satisfy $Q$ and the possibility that $Q$ is not (fully) applicable to $o$.

## 6.3 Types and type system

### 6.3.1 Types

The common characteristics of a data collection can be described by means of a 'type'. For this reason most database models, support some 'type' notion.

In order to give a complete definition of the concept 'type', it is necessary to provide the rules which define its syntax, as well as the rules which define its semantics.

**Definition 3 (Type)** *Each type supported by the database model is defined by its syntax and its semantics.*

- *The syntax of a type. The syntax rules for a type can be formally described by means of some mathematical expressions.*
- *The semantics of a type. The semantic definition of a type $t$ can be fully determined by:*
  - *A set of domains $D_t$.*
  - *A designated domain $dom_t \in D_t$.*
  - *A set of operators $O_t$.*

– *A set of axioms $A_t$.*

□

The designated domain $dom_t$ defines the set of the valid values for the type and is called the domain of the type. In order to deal with cases where a regular domain value does not apply, the assumption has been made that every domain $dom_t$ contains a special, domain specific value $\perp_t$, which is used to represent "undefined" domain values. The set of operators $O_t$ contains the operators, which are defined on the domain $dom_t$. The set of domains $D_t$ consists of the domains that are involved in the definition of the operators of $O_t$, whereas the set of axioms $A_t$ contains the axioms, which are involved in the definition of the semantics of the operators of $O_t$.

### 6.3.2 Type system

A type system for a database model [15] can informally be seen as a collection of rules, which together define all the types supported by that database model.

**Definition 4 (Type system)** *The presented type system $TS$ is defined by the quadruple*

$$TS \equiv [ID, T, \leftrightarrow, \prec]$$

*where*

- *$ID$ is the set of the valid identifiers,*
- *$T$ is the set of valid type definitions,*
- *$\leftrightarrow$ is the partial relation, which is used to define the binary association relationships between object types, and*
- *$\prec$ is the partial ordering relation, which is used to define the inheritance-based type-subtype relationships between object types.*

□

The set $T$ consists of all the type definitions supported by the database model. For each type $t \in T$, a definition as prescribed by Definition 3 is provided. In order to model inapplicability, a type specific element $\perp_t$ has been added to the domain of each type $t \in T$. This element represents an 'undefined' domain value [23].

In order to be consistent with the specifications of the ODMG object model [2], the set $T$ has been partitioned as $T = \{Void\} \cup T_{literal} \cup T_{reference} \cup T_{object}$, where the partitions respectively consists of the definitions of a so-called void type (which is the most primitive type of the system), literal types, reference types (which enable to refer to the instances of object types and are used to formalize binary association relationships between object types in a database scheme) and object types.

### Void type

The *Void* type has been provided do deal with cases where a further specification of a type is not to the purpose or can not be given at that time.

### Literal types

The literal types ($\in T_{literal}$) are further categorized in basic types, collection types, enumeration types and structured types.

The basic types are the basic building blocks of the type system. Different basic types are provided to deal with integer numbers ($Integer$), real numbers ($Real$), binary values ($Boolean$), bytes ($Octet$) and character strings ($String$).

Collection types allow to model multi-valued properties of objects. They are all generic types, designated by a type generator and one or two type parameters. The database model supports collection types for sets ($Set(t)$), bags ($Bag(t)$), lists ($List(t)$), arrays ($Array(t)$) and dictionaries ($Dict(t, t')$), where $t', t \in T_{literal}$ denote the type parameters. Type $t$ is called the significant type of the collection type. In case of nested collection types, the significant type of the innermost collection type is called the most significant type of the collection type.

An enumeration type ($Enum\ id(id_1, id_2, \dots, id_n)$) is used to model a finite set $\{id_1, id_2, \dots, id_n\} \subset ID$ of identifiers in a fixed sequence. $id \in ID$ identifies the enumeration type.

Structured types have been provided to support the modelling of complex objects. A structured type ($Struct\ id(id_1\ isr_1\ t_1; id_2\ isr_2\ t_2; \dots; id_n\ isr_n\ t_n)$) is named by an identifier $id \in ID$ and characterized by a finite set of components that are modelled by means of generalized constraints, as explained further in this section.

### Reference types

Reference types ($\in T_{reference}$) have been introduced to formalize binary association relationships between object types. All reference types are generic types, designated by a type generator and an object type parameter. A binary association relationship between two object types either has a 'one-to-one', a 'one-to-many' or a 'many-to-many' cardinality, which denotes the maximum number of participating domain values of both types. To support the notion of cardinality, a distinction has been made between single-valued reference types ($Ref(t)$), which domains consists of the 'undefined' domain value $\perp_{Ref(t)}$ and of references to regular —composite— objects of type $t$, and multi-valued reference types, which domains consists of an 'undefined' type specific domain value and of collections of references to regular —composite— objects of type $t$. Multi-valued reference types are further subdivided into 'set-of-references'-types ($Set_{Ref}$), 'bag-of-references'-types ($Bag_{Ref}$) and 'list-of-references'-types ($List_{Ref}$). Type $t$ is called the significant type of the multi-valued reference type.

### Object types

Object types are the most elaborated types of the type system. Each object type is characterized by a number of properties (which describe its structure) and a number of explicitly defined operators —also called methods— (which describe its behavior). A property is either an attribute or a binary relationship.

In order to define the binary relationships between object types a partial association relation '$\leftrightarrow$' is defined over the set $T_{object}$. ($id_1 \leftrightarrow id_2$ denotes that 'object type $id_1$ is binary related to object type $id_2$'.) An object type can inherit properties and methods from its parent types [26]. In order to define the inheritance-based type-subtype relationships between object types, a partial ordering relation '$\prec$' is defined over the set $T_{object}$. ($\widehat{id} \prec id$ denotes that 'object type $id$ inherits all characteristics of object type $\widehat{id}$'.)

An object type $(Class\ id : \widehat{id}_1, \widehat{id}_2, \dots, \widehat{id}_m(id_1\ isr_1\ s_1; id_2\ isr_2\ s_2; \dots ; id_n\ isr_n\ s_n))$ is named by an identifier $id \in ID$, the identifiers $\widehat{id}_i \in ID$, $1 \leq i \leq m$ denote the supertypes of the object type (if existent). Like the components of a structured type, the characteristics[1] of an object type are modelled by generalized constraints.

## 6.3.3 Generalized constraints in structured types and object types

Both the components of a structured type and the characteristics of an object type are modelled as generalized constraints. A characteristic or component

$$id\ isr\ s$$

is a (generic) generalized constraint on a variable $id$ with associated specification $s$. In a component of a structured type, $s$ specifies a literal type or reference type, whereas in a characteristic of an object type $s$ specifies a literal type, a reference type or method signature. For literal types and reference types, the generalized constraint restricts the valid values of the variable $id$. For method signatures, the generalized constraint puts restrictions on the valid return values of the method identified by $id$.

In this chapter the considered generalized constraints are limited to equality constraints ($r = e$), possibilistic constraints ($r = blank$) and veristic constraints ($r = v$). Their semantics are specified as follows:

- If $isr = ise$, the valid values of $id$ are restricted to the values of the domain $dom_s$ of the associated type $s$ (or to the values of the domain of the type of the return values of the method that has been specified in the method signature).

[1] The term characteristic is used to denote both the properties (structure) —i.e., the attributes and relationships— and the operators (behavior) of the object type.

- If $isr = is$, $id$ is interpreted as a disjunctive (possibilistic) variable, whose valid values are restricted as follows:
  - If $s$ is not a collection type or a multi-valued reference type, the valid values are restricted to fuzzy sets that are defined over domain $dom_s$ (or to fuzzy sets that are defined over the domain of the type of the return values of the method).
  - If $s$ is a collection type or a multi-valued reference type, the valid values are restricted to collections of fuzzy sets that are defined over the domain $dom_{s'}$ of the (most) significant type $s'$ of $s$.

  The membership grades in these fuzzy sets are interpreted as degrees of possibility.
- If $isr = isv$, $id$ is interpreted as a conjunctive (veristic) variable, whose valid values are restricted as follows:
  - If $s$ is not a collection type or a multi-valued reference type, the valid values are restricted to fuzzy sets that are defined over domain $dom_s$ (or to fuzzy sets that are defined over the domain of the type of the return values of the method).
  - If $s$ is a collection type or a multi-valued reference type, the valid values are restricted to collections of fuzzy sets that are defined over the domain $dom_{s'}$ of the (most) significant type $s'$ of $s$.

  The membership grades in these fuzzy sets are interpreted as degrees of verity.

The domain of a structured type

$$Struct\ id(id_1\ isr_1\ t_1; id_2\ isr_2\ t_2; \ldots; id_n\ isr_n\ t_n)$$

consists of the 'undefined' domain value $\perp_{id}$ and of composite values that contain a value $id\ isr'\ v$, with $isr' \in \{ise, is\}$, for each of the components $id\ isr\ t$, $t \in T_{literal} \cup T_{reference}$ of the structured type.

Similarly, the domain of an object type

$$Class\ id : \widehat{id}_1, \widehat{id}_2, \ldots, \widehat{id}_m(id_1\ isr_1\ s_1; id_2\ isr_2\ s_2; \ldots; id_n\ isr_n\ s_n)$$

contains the 'undefined' domain value $\perp_{id}$ and the 'undefined' domain values $\perp_{\widehat{id}_i}$, $1 \leq i \leq m$ of its subtypes (if existent). Each other domain value is composite and contains a value $id\ isr'\ v$, with $isr' \in \{ise, is\}$, for each of the (inherited) properties $id\ isr\ s$, $s \in T_{literal} \cup T_{reference}$ of the object type.

Each value $id\ isr'\ v$ in a composition, corresponding with a component or property $id\ isr\ s$, is thus on its turn described by a generalized constraint. Hereby, the semantics are specified as follows:

- If $isr' = ise$, the value for component $id$ equals $v$.
- If $isr' = is$, the value for component $id$ is uncertain and described by possibility distribution $v$.

Combining the generalized constraint of the type specification —denoted by the copula *isr*— with the generalized constraint of the domain value —denoted by the copula *isr'*— results in the following possible cases for the value $v$:

- $isr = ise$
  - If $isr' = ise$, then $v \in dom_s$. The value of *id* is perfectly described.
  - If $isr' = is$, then $v \in \tilde{\wp}(dom_s)$, where $\tilde{\wp}(dom_s)$ denotes the fuzzy powerset[2] of $dom_s$. The value of *id* is uncertain. All candidate values are perfectly described.
- $isr = is$
  - $isr' = ise$
    - If $s$ is not a collection type or a multi-valued reference type, then $v \in \tilde{\wp}(dom_s)$. The value of *id* is vague or imprecise.
    - If $s$ is a collection type or a multi-valued reference type, then $v$ is a collection of vague or imprecise values, all specified by fuzzy sets over the domain $dom_{s'}$ of the (most) significant type $s'$ of $s$.
  - $isr' = is$
    - If $s$ is not a collection type or a multi-valued reference type, then $v \in \tilde{\wp}(\tilde{\wp}(dom_s))$. Where $\tilde{\wp}(\tilde{\wp}(dom_s))$ denotes the set of all level-2 fuzzy sets that can be defined over $dom_s$ [12]. The value of *id* is uncertain, what is described by the membership grades in the 'outer-level' fuzzy set. Candidate values can be fuzzy or imprecise, what is described by the 'inner-level' fuzzy sets [10].
    - If $s$ is a collection type or a multi-valued reference type, then $v$ is uncertain and is a fuzzy set of collections of vague or imprecise values, which on their turn are all specified by fuzzy sets over the domain $dom_{s'}$ of the (most) significant type $s'$ of $s$.
- $isr = isv$
  - $isr' = ise$
    - If $s$ is not a collection type or a multi-valued reference type, then $v \in \tilde{\wp}(dom_s)$. The value of *id* is veristic.
    - If $s$ is a collection type or a multi-valued reference type, then $v$ is a collection of veristic values, all specified by fuzzy sets over the domain $dom_{s'}$ of the (most) significant type $s'$ of $s$.
  - $isr' = is$
    - If $s$ is not a collection type or a multi-valued reference type, then $v \in \tilde{\wp}(\tilde{\wp}(dom_s))$. The value of *id* is uncertain, what is described by the membership grades in the 'outer-level' fuzzy set. Candidate values can be veristic, what is described by the 'inner-level' fuzzy sets.
    - If $s$ is a collection type or a multi-valued reference type, then $v$ is uncertain and is a fuzzy set of collections of veristic values, which

[2] The fuzzy powerset of a universe $U$ is the set of all fuzzy sets that can be defined over $U$.

on their turn are all specified by fuzzy sets over the domain $dom_{s'}$ of the (most) significant type $s'$ of $s$.

### 6.3.4 Instances of types

The instances of a reference type, a literal type and an object type are respectively called reference instances, literals and objects, whereas the *Void* type cannot have instances at all.

**Definition 5 (Reference instance)** *Every reference instance $r$ is defined as a pair $[t, v]$, with $t \in T_{reference}$ and $v \in dom_t$.* □

**Definition 6 (Literal)** *Every literal $l$ is defined as a pair $[t, v]$, with $t \in T_{literal}$ and $v \in dom_t$.* □

Depending on its lifetime, an object can either be transient or persistent.

**Definition 7 (Transient object)** *A transient object $o$ is defined as a triple*

$$[t, v, \tilde{t}^*(\text{"}o \text{ is an instance of } t\text{"})]$$

*in which*

- *$t \in T_{object}$ is the type of the object,*
- *$v \in dom_t$ is the state of the object and*
- *$\tilde{t}^*$("o is an instance of t") is the EPTV that expresses the truth value of the proposition "o is an instance of object type t".*

□

**Definition 8 (Persistent object)** *A persistent object $o$ is defined as a quintuple*

$$[oid, N, t, v, \tilde{t}^*(\text{"}o \text{ is an instance of } t\text{"})]$$

*in which*

- *$t \in T_{object}$ is the type of the object,*
- *$v \in dom_t$ is the state of the object,*
- *oid is a unique object identifier,*
- *$N$ is a (finite) set of object names and*
- *$\tilde{t}^*$("o is an instance of t") is the EPTV that expresses the truth value of the proposition "o is an instance of object type t".*

□

The unicity of the object identifier has to be guaranteed over the whole database. The object identifier *oid* is used to refer to the (state of the) object. The set of object names $N$ can be empty.

The set of all the instances of an object type $t \in T_{object}$ is written as $V_t^{instance}$. If $t$ is a subtype of another object type $\widehat{t}$, then $V_t^{instance} \subseteq V_{\widehat{t}}^{instance}$. The extent of an object type $t$ is written as $V_t^{extent}$ and defined as the set of all the persistent objects of $t$ within a particular database. If $t$ is a subtype of another object type $\widehat{t}$, then $V_t^{extent} \subseteq V_{\widehat{t}}^{extent}$.

## 6.4 (Generalized) constraints as a basis for the modelling of spatio-temporal information

The presented type system allows to model both perfect and imperfect information in a more flexible and natural way than traditional type systems do. It can therefore be used as basic component in flexible database models. In this section the type system is extended with a new literal type *Time* and a new generic literal type *Space*. This is done in order to support the modelling of both temporal and spatial information. The domains of these new types contain all subsets of points of a given geometrical space, which on its turn is defined by a finite number of axes, that all have only one point (the origin) in common. For the *Time* type this geometrical space is one-dimensional and represents a time dimension. For the *Space* type the geometrical space is in general $n$-dimensional, where each axis represents a spatial dimension. Linear arithmetic constraints are used to describe the domain values of both types. The newly obtained type system provides a basis for spatio-temporal database modelling.

In next Subsection 6.4.1 the focus is on the modelling of one-dimensional temporal information. The modelling of spatial information is discussed in Subsection 6.4.2. The cases of one-dimensional, two-dimensional and $n$-dimensional spaces are handled. In Subsection 6.4.3 the use of generalized constraints in structured types and object types is reconsidered and discussed in view of the extra types. The modelling of spatio-temporal information is handled in Subsection 6.4.4.

### 6.4.1 Modelling temporal information

In order to model temporal information, a new literal type $Time \in T_{literal}$ is introduced.

The domain of *Time* is defined by

$$dom_{Time} = \wp(\{(x)|x \in \mathbb{R}\}) \cup \{\bot_{Time}\}$$

where $\wp(U)$ denotes the powerset[3] of $U$. Hereby, the set $\mathbb{R}$ of real numbers is used as a representation of the continuum of physical time points [3]. The extra value $\bot_{Time}$ represents an 'undefined' domain value and is used to denote inapplicability of temporal data.

Because by definition all the 'regular' elements of the domain are sets, the operators of the type are the binary operators $=$, $\neq$, $\subset$, $\supset$, $\subseteq$, $\supseteq$, $\cup$, $\cap$ and $\setminus$ and a null-ary operator $\bot$, which always results in an 'undefined' domain value. When restricted to the set $dom_{Time} \setminus \{\bot_{Time}\}$, all binary operators have the same semantics as their counterparts in classical set theory. For the bottom value $\bot_{Time}$, the semantics are:

[3] The powerset of a universe $U$ is the set of all subsets that can be defined over $U$.

$$\forall\, x \in dom_{Time} : op\,(x, \bot_{Time}) = op\,(\bot_{Time}, x) = \bot_{Time}$$

where $op \in \{=, \neq, \subset, \supset, \subseteq, \supseteq, \cup, \cap, \backslash\}$.

The 'regular' elements of $dom_{Time}$ are described by linear arithmetic constraints. The set operators can be straightforwardly translated using the logical operators $\wedge$, $\vee$ and $\neg$.

**Example 1** *For example, the temporal information consisting of time point 60 and time interval from time point 100 until time point 120, i.e.*

$$\{(60)\} \cup \{(t) | t \in [100, 120]\}$$

*can be described with linear constraints as*

$$(t = 60) \vee ((t \geq 100) \wedge (t \leq 120))$$

◇

### 6.4.2 Modelling spatial information

A new data type *SpaceDim* is defined for the modelling of spatial information. This data type has not been included directly in the type system, but is necessary for the definition of the type *Space*.

The domain of *SpaceDim* is defined by

$$dom_{SpaceDim} = \mathbb{R}$$

The considered operators are the binary operators $=$, $\neq$, $<$, $>$, $\leq$, $\geq$, $+$, $-$, $*$ and $/$, which all have the same semantics as their counterparts in $\mathbb{R}^2$.

The type *SpaceDim* has been employed to model a dimension of a multidimensional spatial space, hereby using the set $\mathbb{R}$ of real numbers as a representation of a spatial axis.

The type *Space* is structured and consists of a finite number of components, each representing a spatial dimension. A distinction is made between one-dimensional, two-dimensional and $n$-dimensional spatial data.

#### One-dimensional spatial data

The type *Space* is suited for the modelling of multi-dimensional spatial data. One-dimensional spatial data can be handled by considering one (spatial) component. The specification of *Space* in this case becomes:

$$Space\ id(id_1 : SpaceDim)$$

where $id$ is the identifier of the type and $id_1$ is the identifier of the spatial dimension (with associated type *SpaceDim*). The domain of type $Space\ id(id_1 : SpaceDim)$, shortly written as $dom_{id}$, is defined by

$$dom_{id} = \wp(\{(x)|x \in dom_{SpaceDim}\}) \cup \{\perp_{Space}\}$$

where $\wp(U)$ denotes the powerset of $U$ and $\perp_{Space}$ represents an 'undefined' domain value. With this definition, each 'regular' value of $dom_{id}$ is a set that is defined over the continuum of points on the axis with identifier $id_1$. The operators of the type are the binary operators $=$, $\neq$, $\subset$, $\supset$, $\subseteq$, $\supseteq$, $\cup$, $\cap$ and $\setminus$ and a null-ary operator $\perp$, which always results in an 'undefined' domain value. When restricted to the set $dom_{id} \setminus \{\perp_{Space}\}$, all binary operators have the same semantics as their counterparts in classical set theory. For the bottom value $\perp_{Space}$, the semantics are:

$$\forall\, x \in dom_{Space} : op\,(x, \perp_{Space}) = op\,(\perp_{Space}, x) = \perp_{Space}$$

where $op \in \{=, \neq, \subset, \supset, \subseteq, \supseteq, \cup, \cap, \setminus\}$. The 'regular' elements of $dom_{id}$ are described by linear arithmetic constraints, completely analogous to the temporal case that has been discussed in the previous subsection.

### Two-dimensional spatial data

Two-dimensional spatial data can be modelled with two (spatial) components in the type *Space*, i.e. by considering the specification:

$$Space\ id(id_1 : SpaceDim;\ id_2 : SpaceDim)$$

where $id$ is the identifier of the type and $id_1$ and $id_2$ are the identifiers of the spatial dimensions. In the two-dimensional case, the domain $dom_{id}$ is defined by

$$dom_{id} = \wp(\{(x_1, x_2)|x_1, x_2 \in dom_{SpaceDim}\}) \cup \{\perp_{Space}\}$$

where $\wp(U)$ denotes the powerset of $U$ and $\perp_{Space}$ represents an 'undefined' domain value. Each 'regular' value of $dom_{id}$ is a set that is defined over the continuum of points in the plane defined by the two spatial axes with identifiers $id_1$ and $id_2$. The operators are the same as in the one-dimensional case. The 'regular' elements of $dom_{id}$ are described by linear arithmetic constraints.

**Example 2** *For example, in the two-dimensional space*

$$Space\ id(x : SpaceDim;\ y : SpaceDim)$$

*of the right map of Figure 6.1, the subset of points denoting the city of Ghent, i.e.*

$$\{(x, y)|(x \in [5.5, 5.9]) \wedge (y \in [8.8, 9.2])\}$$

*can be described by linear constraints as*

$$(x \geq 5.5) \wedge (x \leq 5.9) \wedge (y \geq 8.8) \wedge (y \leq 9.2)$$

◇

### $n$-dimensional spatial data

In order to model $n$-dimensional spatial data, the type *Space* has been constructed with $n$ spatial components. I.e.

$$Space\ id(id_1 : SpaceDim;\ id_2 : SpaceDim; \ldots ; id_n : SpaceDim) \in T_{literal}$$

Hereby, the domain $dom_{id}$ is defined by

$$dom_{id} = \tilde{\wp}(\{(x_1, x_2, \ldots, x_n)|x_1, x_2, \ldots, x_n \in dom_{SpaceDim}\}) \cup \{\perp_{Space}\}$$

Each 'regular' value of $dom_{id}$ is now a set that is defined over the continuum of points in the space defined by the $n$ axes with identifiers $id_1, id_2, \ldots, id_n$. The operators are the binary operators $=$, $\neq$, $\subset$, $\supset$, $\subseteq$, $\supseteq$, $\cup$, $\cap$ and $\setminus$ and a null-ary operator $\perp$, which all have the same semantics as described in the one-dimensional and two-dimensional cases. All 'regular' elements of $dom_{id}$ are described by linear arithmetic constraints.

## 6.4.3 The use of generalized constraints revisited

The new types *Time* and $Space\ id(id_1 : SpaceDim;\ id_2 : SpaceDim; \ldots ; id_n : SpaceDim)$ are both literal types and can be used as associated type of a component of a structured type or as associated type of an attribute of an object type. As has been described in Subsection 6.3.3, both the components of a structured type and the attributes of an object type are modelled as generalized constraints. In this subsection, the impact of generalized constraints on the domain values of the new types is discussed.

Let

$$id\ isr\ s, \text{ where } isr \in \{ise, is, isv\}$$

be a (generic) generalized constraint on the identifier $id$ of a component or attribute with associated type $s \in \{Time\} \cup \{Space\ id(id_1 : SpaceDim;\ id_2 : SpaceDim; \ldots ; id_n : SpaceDim)|id, id_1, id_2, \ldots, id_n \in ID\}$ and consider a regular composite domain value in which

$$id\ isr'\ v, \text{ where } isr' \in \{ise, is\}$$

is the value corresponding with this component or attribute.

### Equality constraints

If $isr = ise$, the valid values of the component or attribute contain data that is perfectly described.

- If $isr' = ise$, then $v \in dom_s$. The value of $id$ is described by traditional linear arithmetic constraints.
- If $isr' = is$, then $v \in \tilde{\wp}(dom_s)$, where $\tilde{\wp}(dom_s)$ denotes the fuzzy powerset of $dom_s$. The value of $id$ is uncertain. All candidate values are perfectly described by means of traditional linear arithmetic constraints.

#### Possibilistic constraints

If $isr = is$, the valid values of the component or attribute contain data that might be imprecise or vague. Identifier $id$ identifies a disjunctive (possibilistic) variable.

- If $isr' = ise$, then the valid values for $v$ are restricted to fuzzy sets that are defined over the points of the temporal or spatial space that is modelled by type $s$. The membership grades in these fuzzy sets are interpreted as degrees of possibility.
  - If $s$ is the temporal type $Time$, this means that the valid values of the component or attribute are the elements of

    $$\tilde{\wp}(\{(x)|x \in \mathbb{R}\} \cup \{\perp_{Time}\})$$

    where $\tilde{\wp}(U)$ denotes the fuzzy powerset of $U$.
  - If $s$ is a spatial type $Space\ id(id_1 : SpaceDim;\ id_2 : SpaceDim; \ldots;$ $id_n : SpaceDim)$, this means that the valid values of the component or attribute are the elements of

    $$\tilde{\wp}(\{(x_1, x_2, \ldots, x_n)|x_1, x_2, \ldots, x_n \in dom_{SpaceDim}\} \cup \{\perp_{Space}\})$$

- If $isr' = is$, then the value of $id$ is uncertain and all candidate values might be imprecise or vague.
  - If $s$ is the temporal type $Time$, this means that

    $$v \in \tilde{\wp}(\tilde{\wp}(\{(x)|x \in \mathbb{R}\} \cup \{\perp_{Time}\}))$$

  - If $s$ is a spatial type $Space\ id(id_1 : SpaceDim;\ id_2 : SpaceDim; \ldots;$ $id_n : SpaceDim)$, this means that

    $$v \in \tilde{\wp}(\tilde{\wp}(\{(x_1, x_2, \ldots, x_n)|x_1, x_2, \ldots, x_n \in dom_{SpaceDim}\} \cup \{\perp_{Space}\}))$$

  The membership grades in the 'outer-level' fuzzy set express the uncertainty on the candidate values, whereas the inner-level fuzzy sets express the imprecision or vagueness.

#### Veristic constraints

If $isr = isv$, the valid values of the component or attribute contain data that might be veristic. Identifier $id$ identifies a conjunctive (veristic) variable.

- If $isr' = ise$, then the valid values for $v$ are restricted to fuzzy sets that are defined over the points of the temporal or spatial space that is modelled by type $s$. The membership grades in these fuzzy sets are interpreted as degrees of verity.

- If $s$ is the temporal type *Time*, this means that the valid values of the component or attribute are the elements of

$$\tilde{\wp}(\{(x)|x \in \mathbb{R}\} \cup \{\bot_{Time}\})$$

- If $s$ is a spatial type *Space id*($id_1$ : *SpaceDim*; $id_2$ : *SpaceDim*; ...; $id_n$ : *SpaceDim*), this means that the valid values of the component or attribute are the elements of

$$\tilde{\wp}(\{(x_1, x_2, \ldots, x_n)|x_1, x_2, \ldots, x_n \in dom_{SpaceDim}\} \cup \{\bot_{Space}\})$$

- If $isr' = is$, then the value of $id$ is uncertain and all candidate values might be veristic.
  - If $s$ is the temporal type *Time*, this means that

$$v \in \tilde{\wp}(\tilde{\wp}(\{(x)|x \in \mathbb{R}\} \cup \{\bot_{Time}\}))$$

  - If $s$ is a spatial type *Space id*($id_1$ : *SpaceDim*; $id_2$ : *SpaceDim*; ...; $id_n$ : *SpaceDim*), this means that

$$v \in \tilde{\wp}(\tilde{\wp}(\{(x_1, x_2, \ldots, x_n)|x_1, x_2, \ldots, x_n \in dom_{SpaceDim}\} \cup \{\bot_{Space}\}))$$

  The membership grades in the 'outer-level' fuzzy set express the uncertainty on the candidate values, whereas the inner-level fuzzy sets express the verity.

In order to describe the fuzzy sets of $\tilde{\wp}(\{(x)|x \in \mathbb{R}\} \cup \{\bot_{Time}\})$ and of $\tilde{\wp}(\{(x_1, x_2, \ldots, x_n)|x_1, x_2, \ldots, x_n \in dom_{SpaceDim}\} \cup \{\bot_{Space}\}))$ the concept linear arithmetic constraint has been generalized.

In order to come to the definition of a generalized linear arithmetic constraint, the definitions for the comparison operators $=$, $\leq$ and $\geq$ are generalized. Traditionally, these comparison operators allow to describe crisp subsets of an n-dimensional space (cf. examples 1 and 2). For the generalization, a normalized fuzzy set $\tilde{V}$ has been associated with each operator. This fuzzy set is defined over the universe of valid distances, i.e., the set $\mathbb{R}^*$ of positive real numbers, and the boundary condition $\mu_{\tilde{V}}(0) = 1$ must hold for it.

**Definition 9 (Generalized comparison operators)** *If*

$$d((x_1, x_2, \ldots, x_n), (x'_1, x'_2, \ldots, x'_n))$$

*denotes the Euclidean distance between the regular elements* $(x_1, x_2, \ldots, x_n)$ *and* $(x'_1, x'_2, \ldots, x'_n)$ *of the considered space, i.e.*

$$d((x_1, x_2, \ldots, x_n), (x'_1, x'_2, \ldots, x'_n)) = \sqrt{(x_1 - x'_1)^2 + (x_2 - x'_2)^2 + \cdots + (x_n - x'_n)^2}$$

*then the membership functions of the fuzzy sets described by the generalized operators* $=_{\tilde{V}}$, $\leq_{\tilde{V}}$ *and* $\geq_{\tilde{V}}$ *are defined as follows:*

- *Temporal type Time.* $\forall\,(x) \in dom_{Time} \setminus \{\bot_{Time}\}, \forall\, t \in \mathbb{R}:$
  - $\mu_{x =_{\tilde{V}} t}((x)) = \mu_{\tilde{V}}(d')$, *where*
  $$d' = \min\{d((x),(x'))|(x') \in dom_{Time} \wedge x' = t\}$$
  - $\mu_{x \leq_{\tilde{V}} t}((x)) = \mu_{\tilde{V}}(d')$, *where*
  $$d' = \min\{d((x),(x'))|(x') \in dom_{Time} \wedge x' \leq t\}$$
  - $\mu_{x \geq_{\tilde{V}} t}((x)) = \mu_{\tilde{V}}(d')$, *where*
  $$d' = \min\{d((x),(x'))|(x') \in dom_{Time} \wedge x' \geq t\}$$
- *Spatial type*
  $$Space\ id(id_1 : SpaceDim;\ id_2 : SpaceDim; \ldots ; id_n : SpaceDim)$$
  $\forall\,(x_1, x_2, \ldots, x_n) \in dom_{id} \setminus \{\bot_{id}\}, \forall\,(a_1, a_2, \ldots, a_n, b) \in \mathbb{R}^{n+1}:$
  - $\mu_{a_1x_1 + a_2x_2 + \cdots + a_nx_n =_{\tilde{V}} b}((x_1, x_2, \ldots, x_n)) = \mu_{\tilde{V}}(d')$, *where*
  $$d' = \min\{d((x_1, x_2, \ldots, x_n), (x_1', x_2', \ldots, x_n'))| \\ (x_1', x_2', \ldots, x_n') \in dom_{id} \wedge a_1x_1' + a_2x_2' + \cdots + a_nx_n' = b\}$$
  - $\mu_{a_1x_1 + a_2x_2 + \cdots + a_nx_n \leq_{\tilde{V}} b}((x_1, x_2, \ldots, x_n)) = \mu_{\tilde{V}}(d')$, *where*
  $$d' = \min\{d((x_1, x_2, \ldots, x_n), (x_1', x_2', \ldots, x_n'))| \\ (x_1', x_2', \ldots, x_n') \in dom_{id} \wedge a_1x_1' + a_2x_2' + \cdots + a_nx_n' \leq b\}$$
  - $\mu_{a_1x_1 + a_2x_2 + \cdots + a_nx_n \geq_{\tilde{V}} b}((x_1, x_2, \ldots, x_n)) = \mu_{\tilde{V}}(d')$, *where*
  $$d' = \min\{d((x_1, x_2, \ldots, x_n), (x_1', x_2', \ldots, x_n'))| \\ (x_1', x_2', \ldots, x_n') \in dom_{id} \wedge a_1x_1' + a_2x_2' + \cdots + a_nx_n' \geq b\}$$

□

Figure 6.2 illustrates the membership functions that result from the application of the generalized comparison operators $\leq_{\tilde{V}}$, $\geq_{\tilde{V}}$ and $=_{\tilde{V}}$ to a given fuzzy set $\tilde{V}$ in a one-dimensional temporal space.

The definition of a linear arithmetic constraint is then obtained by replacing all regular comparison operators by appropriate generalized comparison operators and by replacing the regular logical operators $\wedge$, $\vee$ and $\neg$ by their fuzzy counterparts $\tilde{\wedge}$, $\tilde{\vee}$ and $\tilde{\neg}$, which semantics have been defined as follows:

- the impact of the $\tilde{\wedge}$ operator is reflected by applying Zadeh's (standard) intersection operator [28] onto the fuzzy sets described by the arguments of the operator, i.e. with arguments $\tilde{A}$ and $\tilde{B}$, the membership grade of an element $(x_1, x_2, \ldots, x_n)$ in the resulting fuzzy set $\tilde{A}\tilde{\wedge}\tilde{B}$ equals
  $$\min(\mu_{\tilde{A}}((x_1, x_2, \ldots, x_n)), \mu_{\tilde{B}}((x_1, x_2, \ldots, x_n)))$$

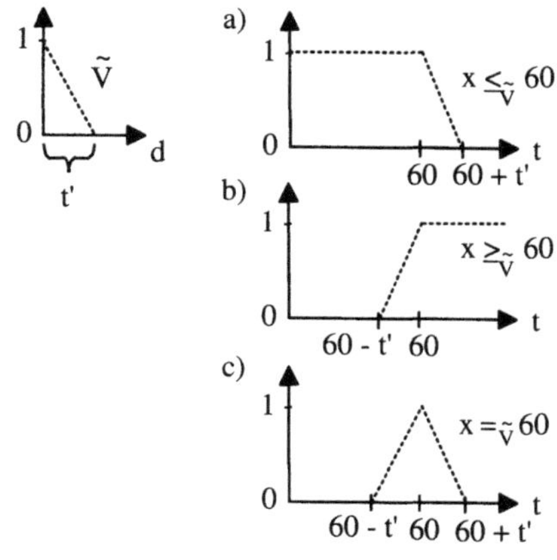


**Figure 6.2.** Application of the generalized comparison operators.

- the impact of the $\tilde{\vee}$ operator is reflected by applying Zadeh's (standard) union operator [28], i.e. with arguments $\tilde{A}$ and $\tilde{B}$, the membership grade of an element $(x_1, x_2, \ldots, x_n)$ in the resulting fuzzy set $\tilde{A}\tilde{\vee}\tilde{B}$ equals

$$\max(\mu_{\tilde{A}}((x_1, x_2, \ldots, x_n)), \mu_{\tilde{B}}((x_1, x_2, \ldots, x_n)))$$

- the impact of the $\tilde{\neg}$ operator is reflected by applying Zadeh's (standard) complement operator [28], i.e. with argument $\tilde{A}$, the membership grade of an element $(x_1, x_2, \ldots, x_n)$ in the resulting fuzzy $\tilde{\neg}(\tilde{A})$ set equals

$$1 - \mu_{\tilde{A}}((x_1, x_2, \ldots, x_n))$$

**Definition 10 (Generalized linear arithmetic constraints)** *The generalized counterpart $\tilde{c}$ of a given linear arithmetic constraint c is obtained by replacing all comparison operators in c by an appropriate generalized comparison operator $=_{\tilde{V}}$, $\leq_{\tilde{V}}$ or $\geq_{\tilde{V}}$ and by replacing all logical operator $\wedge$, $\vee$ and $\neg$ in c by their fuzzy counterparts $\tilde{\wedge}$, $\tilde{\vee}$ and $\tilde{\neg}$.* □

Operators have been provided for the handling of fuzzy sets. Among the considered operators are: $\tilde{\cap}$ (t-norm), $\tilde{\cup}$ (t-conorm), *co* (complement), *normalize*, *support*, *core*, $\alpha - cut$ and $\bar{\alpha} - cut$. Each operator preserves its usual semantics. Additionally, a null-ary operator $\perp$, which always results in an 'undefined' domain value, is added.

**Example 3** *With appropriate fuzzy sets $\tilde{U}$, $\tilde{V}$ and $\tilde{W}$, the fuzzy temporal information* 'around *time point 60 or from time point* about *100 until time point* about *120' can be described as:*

$$(t =_{\tilde{U}} 60) \;\tilde{\vee}\; ((t \geq_{\tilde{V}} 100) \;\tilde{\wedge}\; (t \leq_{\tilde{W}} 120))$$

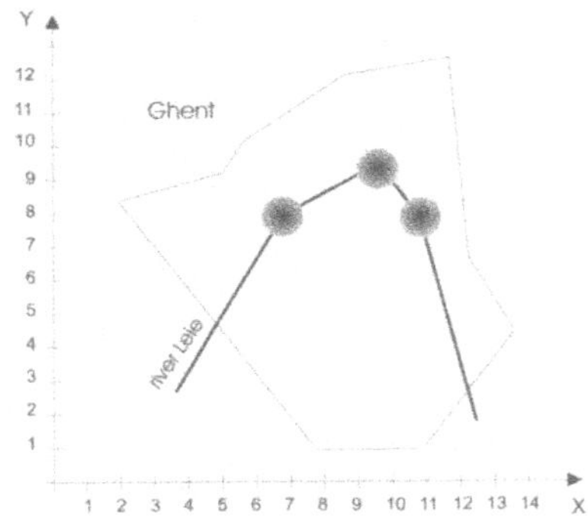


**Figure 6.3.** Illustration of the modelling of imperfect two-dimensional spatial data.

◇

**Example 4** *In the two-dimensional space*

$$Space\ id(x : SpaceDim;\ y : SpaceDim)$$

*of the map of Ghent in Figure 6.3, the fuzzy set of points representing 'In Ghent, in the environment of a winding of the river Leie' can be described by the generalized linear constraint*

$$(\tilde{c}_1 \tilde{\wedge} \tilde{c}_2 \tilde{\wedge} \tilde{c}_3 \tilde{\wedge} \tilde{c}_4 \tilde{\wedge} \tilde{c}_5 \tilde{\wedge} \tilde{c}_6 \tilde{\wedge} \tilde{c}_7 \tilde{\wedge} \tilde{c}_8 \tilde{\wedge} \tilde{c}_9) \tilde{\wedge} (\tilde{c'}_1 \tilde{\vee} \tilde{c'}_2 \tilde{\vee} \ldots \tilde{\vee} \tilde{c'}_n)$$

*where the constraints*

- $\tilde{c}_1 = (y \geq_{\tilde{U}} 1)$,
- $\tilde{c}_2 = (y + 1.24x \geq_{\tilde{U}} 10.68)$,
- $\tilde{c}_3 = (y - 0.21x \leq_{\tilde{U}} 7.78)$,
- $\tilde{c}_4 = (y - 3.25x \geq_{\tilde{U}} -8.65)$,
- $\tilde{c}_5 = (y - 0.63x \leq_{\tilde{U}} 6.57)$,
- $\tilde{c}_6 = (y - 0.17x \leq_{\tilde{U}} 10.67)$,
- $\tilde{c}_7 = (y + 21.67x \leq_{\tilde{U}} 272.7)$,
- $\tilde{c}_8 = (y + 1.15x \leq_{\tilde{U}} 20.39)$ *and*
- $\tilde{c}_9 = (y - 1.42x \geq_{\tilde{U}} -14.65)$

*with appropriate fuzzy set* $\tilde{U}$ *are used to model Ghent and the constraints* $c'_i$, $i = 1, 2, \ldots, n$ *are used to model the environments of the windings of the river Leie. For example, with an appropriate fuzzy set* $\tilde{V}$, *the three windings depicted in the figure can be described by the generalized linear constraints*

- $\tilde{c'}_k = (x =_{\tilde{V}} 7.2 \tilde{\wedge} y =_{\tilde{V}} 7.8)$,
- $\tilde{c'}_l = (x =_{\tilde{V}} 9.8 \tilde{\wedge} y =_{\tilde{V}} 9)$ *and*
- $\tilde{c'}_m = (x =_{\tilde{V}} 11 \tilde{\wedge} y =_{\tilde{V}} 7.8)$

◇

### 6.4.4 The modelling of spatio-temporal information

A fortiori, the presented extension of the type system described in Section 6.3 allows to handle spatio-temporal information. By adding extra, separate literal types for the modelling of temporal and spatial information to the type system, the modelling capacities of both the structured types and the object types have been enriched.

Both the components of a structured type and the characteristics of an object type can have a temporal type or a spatial type as associated type. Furthermore, components with temporal types, spatial types and 'regular' can be grouped and combined within one single structured type or object type.

**Example 5** *With the object types City and Street being specified by*

$$\begin{aligned} &Class\ City(PostalCode\ ise\ String;\\ &\qquad Name\ ise\ String;\\ &\qquad Area\ isv\ Space\ Map(x : SpaceDim; y : SpaceDim);\\ &\qquad Streets\ ise\ Set_{Ref}(Street)) \end{aligned}$$

*and*

$$\begin{aligned} &Class\ Street(Name\ ise\ String;\\ &\qquad Location\ is\ Space\ Map(x : SpaceDim; y : SpaceDim);\\ &\qquad Buildings\ ise\ Set(Struct\ Building(Number\ ise\ Integer;\\ &\qquad\qquad Location\ is\ Space\ Map(x : SpaceDim; y : SpaceDim);\\ &\qquad\qquad Functionality\ isv\ String));\\ &\qquad Traffic\ ise\ Set(Struct\ Sample(Time\ ise\ Time;\\ &\qquad\qquad NumberOfVehicles\ is\ Integer))) \end{aligned}$$

*one could model information about cities and the traffic load and buildings in the streets of these cities.*

*An example of a persistent instance of class City is (cf. Definition 8)*

$$\begin{aligned} &[oid_{C1}, \emptyset, City,\\ &(PostalCode\ ise\ \text{'B-9000'};\\ &\quad Name\ ise\ \text{'Ghent'};\\ &\quad Area\ ise\ (\tilde{c}_1 \tilde{\wedge} \tilde{c}_2 \tilde{\wedge} \tilde{c}_3 \tilde{\wedge} \tilde{c}_4 \tilde{\wedge} \tilde{c}_5 \tilde{\wedge} \tilde{c}_6 \tilde{\wedge} \tilde{c}_7 \tilde{\wedge} \tilde{c}_8 \tilde{\wedge} \tilde{c}_9);\\ &\quad Streets\ ise\ Set(oid_{S1}, oid_{S2}, \ldots, oid_{Sm})), \{(T, 1)\}] \end{aligned}$$

*where* $(\tilde{c}_1 \tilde{\wedge} \tilde{c}_2 \tilde{\wedge} \tilde{c}_3 \tilde{\wedge} \tilde{c}_4 \tilde{\wedge} \tilde{c}_5 \tilde{\wedge} \tilde{c}_6 \tilde{\wedge} \tilde{c}_7 \tilde{\wedge} \tilde{c}_8 \tilde{\wedge} \tilde{c}_9)$ *are the constraints describing Ghent, as given in Example 4, and* $\{oid_{S1}, oid_{S2}, \ldots, oid_{Sm}\}$ *is a finite set of object identifiers of street objects, describing the streets of Ghent.*

*As indicated by the EPTV* $\{(T,1)\}$, *this persistent object completely belongs to the extent of type City.*

*An example of a persistent Street object is*

$$
\begin{aligned}
&[oid_{S7}, \emptyset, Street,\\
&(Name\ ise\ \text{`Sint-Pietersnieuwstraat'};\\
&Location\ ise\ (y - 3.2x =_{\tilde{V}} -23.8)\ \tilde{\wedge}\ (x \geq_{\tilde{V}} 8.5)\ \tilde{\wedge}\ (x \leq_{\tilde{V}} 9);\\
&Buildings\ ise\ Set(Building(Number\ ise\ 1;\\
&\quad Location\ ise\ (x =_{\tilde{U}} 8.5\ \tilde{\wedge}\ y =_{\tilde{U}} 3.4);\\
&\quad Functionality\ is\ \{(\{(Home,1)\},1),(\{(Student\_rooms,1)\},0.8)\}),\\
&\quad\quad Building(Number\ ise\ 41;\\
&\quad Location\ ise\ (x =_{\tilde{U}} 8.84\ \tilde{\wedge}\ y =_{\tilde{U}} 4.5);\\
&\quad Functionality\ ise\ \{(labo,1),(school,1),(office,0.8)\}),\ldots);\\
&Traffic\ ise\ Set(Sample(Time\ ise\ (t = 15/09/2006);\\
&\quad NumberOfVehicles\ ise\ \text{`about 1200'}),\\
&\quad\quad Sample(Time\ ise\ (t = 15/10/2006);\\
&\quad NumberOfVehicles\ ise\ \text{`more than 3500'}),\\
&\quad\quad Sample(Time\ ise\ (t = 15/11/2006);\\
&\quad NumberOfVehicles\ ise\ \text{`3204'}),\ldots)),\{(T,1)\}]
\end{aligned}
$$

*Hereby,* $\tilde{V}$ *(resp.* $\tilde{U}$*) is the fuzzy set associated with the comparison operators for the modelling of the location of the street (resp. location of the buildings).*

*The functionality of the building with number 1 is uncertain: it is completely possible that the building is a home and less possible that the building is used to house students. The functionality of the building with number 41 is certain, but is multiple-valued (veristic): the building is multi-functional and used for research activities (laboratory), teaching activities (school) and to a lesser extent also for administrative activities (office).*

*The counting of vehicle traffic is provided at accurate, certain time intervals (days), but the counts themselves might be imprecise and are therefore modelled by means of possibilistic fuzzy sets (resp. denoted by the labels 'about 3000', 'more than 3500' and '3204').*

◇

## 6.5 Flexible querying of spatio-temporal information

In order to support flexible querying an underlying many-valued logical framework of EPTV's[7] will be used. Traditional Boolean logic no longer suffices since with the modelling of impreciseness, vagueness and uncertainty query satisfaction is no longer bivalent, but becomes a matter of degree. This is because the objects in the result set of a query will typically belong to this set to a calculated degree. In the presented approach, this degree will be expressed by the EPTV that is part of the definition of transient and persistent objects (cf. Definitions 7 and 8).

At the basis of the manipulation of spatio-temporal information are the set operators $\cap$, $\cup$ and $\setminus$ (which can be modelled by means of linear constraints, using the logical operators $=$, $\neq$, $<$, $>$, $\leq$, $\geq$, $\wedge$, $\vee$ and $\neg$), the fuzzy set operators $\tilde{\cup}$, $\tilde{\cap}$ and *co* (which can be modelled by means of generalized linear constraints, using the generalized logical operators $=_{\tilde{V}}$, $\leq_{\tilde{V}}$ and $\geq_{\tilde{V}}$ and Zadeh's operators $\tilde{\wedge}$, $\tilde{\vee}$ and $\tilde{\neg}$ [28]), the fuzzy set operators *normalize*, *support*, *core*, $\alpha - cut$ and $\bar{\alpha} - cut$, and the null-ary operator $\perp$.

For flexible querying purposes, logical operators has been provided. All these operators result in an EPTV and together with the negation ($\tilde{\neg}$), conjunction ($\tilde{\wedge}$) and disjunction ($\tilde{\vee}$) operators for EPTV's allow to calculate the degree with which a given database object satisfies the criteria imposed by a the query.

### 6.5.1 Logical operators for perfectly described information

In the case of perfectly described information, the considered sets are all conjunctive, that is all points together represent the spatio-temporal entity.

Therefore, the traditional logical operators for sets $\subset$, $\supset$, $\subseteq$ and $\supseteq$ can be applied and have semantics as has been described in Subsections 6.4.1 and 6.4.2.

Furthermore, the basic topological relations *equal*, *disjoint*, *overlap*, *inside*, *contains* and *meet* are provided with similar semantics, i.e. if at least one of the operands of an operator contains an 'undefined' element, the operator results in the EPTV $\{(\perp, 1)\}$.

If one of the above mentioned operators results in the Boolean truth value *True*, the associated EPTV is $\{(T, 1)\}$, otherwise if the operator results in the Boolean truth value *False*, the corresponding EPTV is $\{(F, 1)\}$.

### 6.5.2 Logical operators for imperfect information

For imperfect spatio-temporal information a distinction has to be made between conjunctive fuzzy sets, which represent veristic data and disjunctive fuzzy sets, which represent possibilistic data.

#### Veristic interpretation

In the case of conjunctive data, the considered operators are the generalizations $\widetilde{equal}$, $\widetilde{disjoint}$, $\widetilde{overlap}$, $\widetilde{inside}$, $\widetilde{contains}$ and $\widetilde{meet}$ of the basic topological relations, which have been defined by:

**Definition 11 (Generalized topological relation)** *With the understanding that $\tilde{A}$ and $\tilde{B}$ are two veristic fuzzy subsets over an n-dimensional space*

*$S$ and $\perp_S$ stands for $\perp_{Time}$ if $S$ is the one-dimensional temporal space and $\perp_S$ stands for $\perp_{Space}$ if $S$ is a spatial space,*

- *(the EPTV of) $\tilde{equal}(\tilde{A},\tilde{B})$ is defined by:*
  - $\mu_{\tilde{equal}(\tilde{A},\tilde{B})}(T) = 1 - \max(|\mu_{\tilde{A}}((x_1,x_2,\ldots,x_n)) - \mu_{\tilde{B}}((x_1,x_2,\ldots,x_n))|)$ *where* $(x_1,x_2,\ldots,x_n) \in S$
  - $\mu_{\tilde{equal}(\tilde{A},\tilde{B})}(F) = \max(|\mu_{\tilde{A}}((x_1,x_2,\ldots,x_n)) - \mu_{\tilde{B}}((x_1,x_2,\ldots,x_n))|)$ *where* $(x_1,x_2,\ldots,x_n) \in S$
  - $\mu_{\tilde{equal}(\tilde{A},\tilde{B})}(\perp) = \max(\mu_{\tilde{A}}(\perp_S), \mu_{\tilde{B}}(\perp_S))$
- *(the EPTV of) $\tilde{disjoint}(\tilde{A},\tilde{B})$ is defined by:*
  - $\mu_{\tilde{disjoint}(\tilde{A},\tilde{B})}(T) = 1 - \max(\mu_{\tilde{A}\tilde{\cap}\tilde{B}}((x_1,x_2,\ldots,x_n)))$ *where* $(x_1,x_2,\ldots,x_n) \in S$
  - $\mu_{\tilde{disjoint}(\tilde{A},\tilde{B})}(F) = \max(\mu_{\tilde{A}\tilde{\cap}\tilde{B}}((x_1,x_2,\ldots,x_n)))$ *where* $(x_1,x_2,\ldots,x_n) \in S$
  - $\mu_{\tilde{disjoint}(\tilde{A},\tilde{B})}(\perp) = \max(\mu_{\tilde{A}}(\perp_S), \mu_{\tilde{B}}(\perp_S))$
- *(the EPTV of) $\tilde{overlap}(\tilde{A},\tilde{B})$ is defined by:*
  - $\mu_{\tilde{overlap}(\tilde{A},\tilde{B})}(T) = \max(\mu_{\tilde{A}\tilde{\cap}\tilde{B}}((x_1,x_2,\ldots,x_n)))$ *where* $(x_1,x_2,\ldots,x_n) \in S$
  - $\mu_{\tilde{overlap}(\tilde{A},\tilde{B})}(F) = 1 - \max(\mu_{\tilde{A}\tilde{\cap}\tilde{B}}((x_1,x_2,\ldots,x_n)))$ *where* $(x_1,x_2,\ldots,x_n) \in S$
  - $\mu_{\tilde{overlap}(\tilde{A},\tilde{B})}(\perp) = \max(\mu_{\tilde{A}}(\perp_S), \mu_{\tilde{B}}(\perp_S))$
- *(the EPTV of) $\tilde{inside}(\tilde{A},\tilde{B})$ is defined by:*
  - $\mu_{\tilde{inside}(\tilde{A},\tilde{B})}(T) = 1 - \max(\mu_{\tilde{A}}((x_1,x_2,\ldots,x_n)) - \mu_{\tilde{A}\tilde{\cap}\tilde{B}}((x_1,x_2,\ldots,x_n)))$ *where* $(x_1,x_2,\ldots,x_n) \in S$
  - $\mu_{\tilde{inside}(\tilde{A},\tilde{B})}(F) = \max(\mu_{\tilde{A}}((x_1,x_2,\ldots,x_n)) - \mu_{\tilde{A}\tilde{\cap}\tilde{B}}((x_1,x_2,\ldots,x_n)))$ *where* $(x_1,x_2,\ldots,x_n) \in S$
  - $\mu_{\tilde{inside}(\tilde{A},\tilde{B})}(\perp) = \max(\mu_{\tilde{A}}(\perp_S), \mu_{\tilde{B}}(\perp_S))$
- *(the EPTV of) $\tilde{contains}(\tilde{A},\tilde{B})$ is defined by:*

$$\tilde{contains}(\tilde{A},\tilde{B}) = \tilde{inside}(\tilde{B},\tilde{A})$$

- *(the EPTV of) $\tilde{meet}(\tilde{A},\tilde{B})$ is defined by:*
  - $\mu_{\tilde{meet}(\tilde{A},\tilde{B})}(T) = \sup_{\alpha\in[0,1]\ \wedge\ meet(\alpha-cut(\tilde{A}),\alpha-cut(\tilde{B}))=T}(1-\alpha)$
  - $\mu_{\tilde{meet}(\tilde{A},\tilde{B})}(F) = 1 - \sup_{\alpha\in[0,1]\ \wedge\ meet(\alpha-cut(\tilde{A}),\alpha-cut(\tilde{B}))=T}(1-\alpha)$
  - $\mu_{\tilde{meet}(\tilde{A},\tilde{B})}(\perp) = \max(\mu_{\tilde{A}}(\perp_S), \mu_{\tilde{B}}(\perp_S))$

□

### Possibilistic interpretation

To query disjunctive data, a so-called 'IS' operator has been provided. This operator allows to build 'IS' propositions that allow to compare stored (spatio-temporal) labels with (spatio-temporal) labels provided by the user.

**Definition 12 (IS operator)** *With the understanding that $\pi_{A_i}$ is the possibility distribution function characterizing the value of a possibilistic attribute $A_i$, $\mu_L$ is the membership function of the disjunctive fuzzy set that models the label L that is given by the user and $\perp_{A_i}$ is the 'undefined' domain value of the domain of the type that is associated with $A_i$, the EPTV of the proposition "$A_i$ IS L" is defined by*

- $\mu_{\tilde{t}^*(\text{“}A_i \text{ IS } L\text{”})}(T) = \sup_{p \in dom_{A_i}} \min(\pi_{A_i}(p), \mu_L(p))$
- $\mu_{\tilde{t}^*(\text{“}A_i \text{ IS } L\text{”})}(F) = \sup_{p \in dom_{A_i} \setminus \{\perp_{A_i}\}} \min(\pi_{A_i}(p), 1 - \mu_L(p))$
- $\mu_{\tilde{t}^*(\text{“}A_i \text{ IS } L\text{”})}(\perp) = \min(\pi_{A_i}(\perp_{A_i}), 1 - \mu_L(\perp_{A_i}))$

□

### 6.5.3 Some examples

In order to illustrate (some of) the potentials of the proposed querying framework, the following spatio-temporal queries are discussed.

### Temporal querying

Query 1. *Find the periods of the year when traffic in Sint-Pietersnieuwstraat is very busy.*

Hereby, 'very busy' can be modelled by means of a possibility distribution, over the valid values for the component $NumberOfVehicles$ of the $Traffic$ attribute of class *Street* that is defined by the fuzzy set with membership function

$$\mu_{very\_busy}(x) = \begin{cases} 0 & \text{if } x < 1000, \\ \frac{1}{2000}(x - 1000) & \text{if } 1000 \le x \le 3000, \\ 1 & \text{if } x > 3000 \end{cases}$$

This query is an example of the flexible querying of crisp data. For each traffic sample in the street object representing 'Sint-Pietersnieuwstraat', the IS-operator (cf. Definition 12) will be applied to calculate the EPTV $\{(T, \mu_T), (F, \mu_F), (\perp, \mu_\perp)\}$ of the proposition '*NumberOfVehicles* IS *very_busy*'. Next, the value(s) of the associated time indication *Time* will be assigned a membership grade

$$\frac{1 + (\mu_T - \mu_F)(1 - \frac{\mu_\perp}{2})}{2}$$

Finally, the resulting possibilistic fuzzy set over $dom_{Time}$ is calculated as the fuzzy union ($\tilde{\cup}$) of all these time indications.

Query 2. *What was the traffic load in Sint-Pietersnieuwstraat around the beginning of the academic year in 2003?*

Suppose that 'Around the beginning of the academic year in 2003' is described by the generalized linear constraint

$$t =_{\tilde{V}} \text{'October 1st, 2003'}$$

where $\tilde{V}$ is the fuzzy set with membership function

$$\mu_{\tilde{V}}(x) = \begin{cases} -\dfrac{x}{20 \text{ days}} + 1 & \text{if } x \leq 20 \text{ days} \\ 0 & \text{if } x > 20 \text{ days} \end{cases}$$

For each traffic sample in the street object representing 'Sint-Pietersnieuwstraat', it has to be checked wether the sample is taken within the considered period. This can be done using the $\widetilde{inside}$ operator. For each traffic sample, for which the application of the $\widetilde{inside}$ operator results in an EPTV that differs from $\{(F,1)\}$, the time indication $Time$, $NumberOfVehicles$ and corresponding EPTV will be returned.

## Spatial querying

Query 3. *Give all streets in Ghent, which are in the environment of a winding of the river Leie.*

Let 'the environments of the windings of the river Leie in Ghent' be described by the generalized linear constraint

$$(\tilde{c}_1 \tilde{\wedge} \tilde{c}_2 \tilde{\wedge} \tilde{c}_3 \tilde{\wedge} \tilde{c}_4 \tilde{\wedge} \tilde{c}_5 \tilde{\wedge} \tilde{c}_6 \tilde{\wedge} \tilde{c}_7 \tilde{\wedge} \tilde{c}_8 \tilde{\wedge} \tilde{c}_9) \tilde{\wedge} (\tilde{c'}_1 \tilde{\vee} \tilde{c'}_2 \tilde{\vee} \ldots \tilde{\vee} \tilde{c'}_n)$$

as given in Example 4.

For each street object related to the city object that represents Ghent, it has to be checked wether the location of the street overlaps with the environments of the windings of the river Leie in Ghent. Hereby, the $\widetilde{overlap}$ operator can be used. Each street object, for which the application of the $\widetilde{overlap}$ operator results in an EPTV that differs from $\{(F,1)\}$, is an element of the answer set of the query.

Query 4. *Give all school environments in Ghent where traffic is very busy.*

Suppose again that 'very busy' traffic is modelled by means of a possibility distribution with membership function $\mu_{very\_busy}$ as given in Query 1.

Furthermore, suppose that 'environment' of a school building is modelled by means of a fuzzy set $\tilde{V}$ of valid distances that is defined by the membership function

$$\mu_{\tilde{V}}(x) = \begin{cases} -\dfrac{x}{200m} + 1 & \text{if } x \leq 200m \\ 0 & \text{if } x > 200m \end{cases}$$

For each street object related to the city object that represents Ghent, it has to be checked wether there are buildings in the street with a school *functionality.* For each such building, the environment of the building should be described. This can be done by using the fuzzy set $\tilde{V}$ as associated fuzzy set for the generalized comparison operators in the generalized linear constraints that describe the location of the building. For example, since the location of building 41 in Sint-Pietersnieuwstraat in Example 5 is described by $(x =_{\tilde{U}} 8.84 \tilde{\wedge} y =_{\tilde{U}} 4.5)$, its environment will be modelled by $(x =_{\tilde{V}} 8.84 \tilde{\wedge} y =_{\tilde{V}} 4.5)$. Next, all streets within the environment can be find by taking the intersection $(\tilde{\wedge})$ of the location of the school environment and the locations of all streets in Ghent. Finally, the traffic sample set for the street can be checked for samples with very busy traffic using the IS-operator, i.e., by evaluating the predicate '*NumberOfVehicles* IS *very_busy*'.

### Spatio-temporal querying

Query 5. *Localize person X, given that it is now 13h40min, X is in Ghent, at noon X is going to have lunch in an Italian restaurant, after that she is going to buy shoes and X is moving on foot.*

Let 'noon' be defined by the fuzzy set with membership function

$$\mu_{noon}(t) = \begin{cases} 0 & \text{if } t < 11h45min, \\ \dfrac{1}{15min}(x - 11h45min) & \text{if } 11h45min \leq t \leq 12h00min, \\ 1 & \text{if } 12h00min > t > 13h00min, \\ -\dfrac{x}{60min} + 14 & \text{if } 13h00min \leq t \leq 14h00min, \\ 0 & \text{if } t > 14h00min \end{cases}$$

Then $\mu_{noon}(13h40min) = 0.33$. By applying the IS-operator, it is obtained that the EPTV

$$\tilde{t}^*(\text{'13h40min IS } noon\text{'}) = \{(T, 0.33), (F, 0.66)\}$$

which can be seen as the truth value of the proposition '$X$ is in an Italian restaurant in Ghent', whereas the negation

$$\tilde{\neg}(\tilde{t}^*(\text{'13h40min IS } noon\text{'})) = \{(T, 0.66), (F, 0.33)\}$$

can be seen as the truth value of the proposition '$X$ is buying shoes'. Since both options are still possible, the database must be queried for Italian restaurants in Ghent that are within walking distance of a shoe shop and for shoe shops in Ghent that are within walking distance of an Italian restaurant. Hereby, the *Location* and *Functionality* components of the *Buildings* attribute of the *Street* class can be used in order to find the Italian restaurants and shoe shops in Ghent. Furthermore, with the understanding that 'walking distance' is defined by the fuzzy set with membership function

$$\mu_{walking\_distance}(d) = \begin{cases} 1 & \text{if } d < 3000m, \\ -\dfrac{x}{3000m} + 2 & \text{if } 3000m \leq d \leq 6000m, \\ 0 & \text{if } d > 6000m \end{cases}$$

the distance between each Italian restaurant and each shoe shops can be calculated and checked for consistency with the concept 'within walking distance'. Hereby, an EPTV $\tilde{t}^*$('calculated_distance IS *walking_distance*') is calculated.

All locations of Italian restaurants, shoe shops and streets connecting a restaurant with a shop are potential locations of $X$. Their associated degree of satisfaction can be calculated from the conjunction of the EPTV's, $\tilde{t}^*$('13h40min IS *noon*') and $\tilde{t}^*$('calculated_distance IS *walking_distance*').

## 6.6 Conclusions

In this chapter a database modelling technique, based on constraints, has been presented. On the one hand, it has been shown that generalized constraints, as presented by L.A. Zadeh, are an adequate means to model database semantics and, more specifically, geographical database semantics. On the other hand, is has been discussed and illustrated how generalized linear constraints can be used to describe both temporal an spatial information. An underlying many-valued logic of EPTV's supports flexible querying and serves as a basis for spatio-temporal reasoning. The potentials of the proposed techniques have been illustrated and discussed by means of some spatio-temporal queries.

In the presented approach only the equality, possibilistic and veristic generalized constraints have been incorporated in the type system. In future work, also the use of the other generalized constraints, presented by L.A. Zadeh, will be studied. Furthermore, it seems to be interesting to extend the framework with the more general concept of protoforms. Another research topic will be the incorporation of more advanced topological relations which, for example, allow to deal with relative positions of objects, indicated by wind directions.

## References

1. Bosc P., Prade H. (1993) An Introduction to Fuzzy Set and Possibility Theory Based Approaches to the Treatment of Uncertainty and Imprecision in Database

Management Systems. In: Proc. of the 2nd Workshop on Uncertainty Management in Information Systems: From Needs to Solutions, Catalina, California, USA.

2. Cattell R.G.G., Barry D. (2000) The Object Data Standard: ODMG 3.0. Morgan Kaufmann Publishers, San Francisco, USA.
3. de Caluwe R., de Tré G., Van der Cruyssen B., Devos F., Maesfranckx P. (2000) Time management in Fuzzy and Uncertain Object-Oriented Databases. In: Pons O., Vila M.A., Kacprzyk J. (Eds.) Knowledge Management in Fuzzy Databases, Studies in Fuzziness and Soft Computing, Physica-Verlag, Heidelberg, Germany, pp. 67–88.
4. de Cooman G. (1995) Towards a possibilistic logic. In: Ruan D. (Ed.) Fuzzy Set Theory and Advanced Mathematical Applications, Kluwer Academic Publishers, Boston, USA, pp. 89–133.
5. de Cooman G. (1999) From possibilistic information to Kleene's strong multivalued logics. In: Dubois D., Klement E.P., Prade H. (Eds.) Fuzzy Sets, Logics and Reasoning about Knowledge, Kluwer Academic Publishers, Boston, USA, pp. 315–323.
6. de Tré G., de Caluwe R., Van der Cruyssen B. (2000) A Generalised Object-Oriented Database Model. In: Bordogna G., Pasi G. (Eds.) Recent Issues on Fuzzy Databases, Studies in Fuzziness and Soft Computing, Physica-Verlag, Heidelberg, Germany, pp. 155–182.
7. de Tré G. (2002) Extended Possibilistic Truth Values. International Journal of Intelligent Systems 17:427–446.
8. de Tré G. (2003) Enhancing the ODMG model: A framework based on constraints. In: Mastorakis N.E., Antoniou G.E., Manikopoulos C., Bojkovic Z., Gonos I.F. (Eds.) Recent Advances in Communications and Computer Science, Electrical and Computer Engineering Series, WSEAS press, pp. 398–406.
9. de Tré G., de Caluwe R. (2003) Modelling Uncertainty in Multimedia Database Systems: An Extended Possibilistic Approach. International Journal of Uncertainty, Fuzziness and Knowledge-Based Systems 11(1):5–22.
10. de Tré G., de Caluwe R. (2003) Level-2 fuzzy sets and their usefulness in object-oriented database modelling. Fuzzy Sets and Systems 140(1):29–49.
11. Dubois D., Prade H. (1988) Possibility Theory. Plenum Press, New York, USA.
12. Gottwald S. (1979) Set theory for fuzzy sets of higher level. Fuzzy Sets and Systems 2(2):125-151.
13. Kanellakis P.C., Kuper G.M., Revesz P.Z. (1995) Constraint Query Languages. Journal of Computer and System Sciences 51:26–52.
14. Kuper A., Libkin L., Paredaens J. (Eds.) (1993) Constraint Databases. Springer-Verlag, Berlin, Germany.
15. Lausen G., Vossen G. (1998) Models and Languages of Object-Oriented Databases. Addison-Wesley, Harlow, England.
16. Morris A. (2001) Why Spatial Databases Need Fuzziness. In: Proc. of Joint 9th IFSA and 20th NAFIPS International Conference, Vancouver, Canada, pp. 2446–2451.
17. Motro A. (1995) Management of Uncertainty in Database Systems. In: Kim W. (Ed.) Modern Database Systems, The object model, interoperability and beyond, Addison-Wesley Publishing Company, Reading, Massachusetts, USA.
18. Paredaens J., De Bra P., Gyssens M., Van Gucht D. (1989) The Structure of the Relational Database Model. Springer-Verlag, Berlin, Germany.

19. Parsons S. (1996) Current Approaches to Handling Imperfect Information in Data and Knowledge Bases. IEEE Transactions on Knowledge and Data Engineering 8(3):353–372.
20. Prade H. (1982) Possibility sets, fuzzy sets and their relation to Lukasiewicz logic. In: Proc. of the 12th International Symposium on Multiple-Valued Logic, pp. 223–227.
21. Prade H. (2001) From flexible queries to case-based evaluation. In: Proc. of ECSQARU-2001 workshop on Management of uncertainty and imprecision in multimedia information systems, Toulouse, France.
22. Rescher N. (1969) Many-Valued Logic. Mc.Graw-Hill, New York, USA.
23. Riedel H., Scholl M.H. (1997) A Formalization of ODMG Queries. In: Spaccapietra S., Maryanski F. (Eds.) Proc. of the 7th Working Conference on Database Semantics (DS-7), Leysin, Swiss, pp. 63–90.
24. Rigaux P., Scholl M., Voisard A. (2002) Spatial Databases with Application to GIS. Morgan Kaufmann Publishers, San Francisco, USA.
25. Shekhar S., Chawla S. (2003) Spatial Databases: A Tour. Pearson Education Inc., New Jersey, USA.
26. Taivalsari A. (1996) On the Notion of Inheritance. ACM Computing Surveys 28(3):438–479.
27. Tansel A., Clifford J., Gadia S., Jajodia S., Segev A., Snodgrass R. (1993) Temporal Databases: Theory, Design, and Implementation. The Benjamin/Cummings Publishing Company Inc., Redwood City, USA.
28. Zadeh L.A. (1965) Fuzzy sets. Information and Control 8(3):338–353.
29. Zadeh L.A. (1968) Probability measures of fuzzy events. Journal of Mathematical Analysis and Applications 23:421–427.
30. Zadeh L.A. (1975) The concept of linguistic variable and its application to approximate reasoning Parts I, II and III. Information Sciences 8:199–251,301–357 9:43–80.
31. Zadeh L.A. (1978) Fuzzy sets as a basis for a theory of possibility. Fuzzy Sets and Systems 1(1):3–28.
32. Zadeh L.A. (1986) Outline of a computational approach to meaning and knowledge representation based on a concept of a generalized assignment statement. In Thoma M., Wyner A. (Eds.) Proc. of the International Seminar on Artificial Intelligence and Man-Machine Systems, Springer, Heidelberg, Germany, pp. 198–211.
33. Zadeh L.A. (1997) Toward a theory of fuzzy information granulation and its centrality in human reasoning and fuzzy logic. Fuzzy Sets and Systems 90(2):111–127.
34. Zadeh L.A. (1999) From computing with numbers to computing with words—from manipulation of measurements to manipulation of perceptions. IEEE Transactions on Circuit Systems 45:105–119.
35. Zadeh L.A. (2002) Toward a preception-based theory of probabilistic reasoning with imprecise probabilities. Journal of Statistical Planning and Inference 105:233–264.

# 7

# An Error Tolerant Parser for Natural Language Queries in GISs

Fangju Wang

Department of Computing and Information Science
University of Guelph
Guelph, Ontario, Canada N1G 2W1
fjwang@cis.uoguelph.ca

## 7.1 The Necessity of a Natural Language Interface

Geographic information systems (GISs) have established as an important branch of information system technology. GISs have been applied in almost all the areas where maps are used. The major application areas include environmental studies, agriculture, transportation and traffic control, oil and mining industries, military, tourism, archaeology, and many others. Users in a large number of areas apply GISs as tools to support their needs for geographic information.

The nature of GIS applications decides the nature of GIS users, who are very diversified in terms of their professional background, technical education, and computing skills. For such a user community, GIS query languages have become an issue affecting system accessibility. It is widely recognized that artificial query languages (e.g. SQL) are difficult to learn and use, especially by occasional and novice users. The language issue has been realized by GIS researchers and practitioners. For example, as early as 1989, in its research plan, the National Center for Geographic Information and Analysis (NCGIA) suggested that "future GIS must include user interfaces that are easy to learn, appear natural to the user and do not depend on the internal structure of the system" [15].

The currently used artificial query languages have an additional limitation when used in querying geographic data. In such languages, query conditions are specified in formula systems based on Boolean logic. A Boolean condition places a clear-cut boundary between the data satisfying and unsatisfying it. The formula systems are suitable for expressing precise query conditions, but unsuitable for vague and imprecise conditions. Humans use numerous vague concepts in dealing with geographic information, for example, far, high, mild, northeast, and so on. Obviously, such concepts are difficult to be expressed in the Boolean formula systems. Simple translation from a vague condition into a precise one may often lead to unreasonable or even ridiculous results.

A natural language may provide a better user interface to a GIS. The interface can be used with minimum knowledge about the underlying data structures and program coding, and requires little training and computing skills. It is easier and more natural to use a natural language to express the imprecise concepts that are commonly used by end users. Since 1980's, great efforts have been made in creating natural language interfaces for GISs. Research has been conducted in theoretical studies and experimental systems. Related work included studies of GIS query language issues [9], design and evaluation of user interfaces [17], knowledge-based system approaches to support natural language interfaces [18], incorporations of natural language words into artificial query languages [13, 19], a fuzzy object model to handle spatial data storage and querying [14], a fuzzy approach to optimize file I/O for answering imprecise queries [6], processing fuzzy directions [7], and handling ambiguity in parsing GIS queries [20].

While research was going on to create natural language user interfaces, great progress had been achieved in building graphical user interfaces (GUIs). With the rapid development of graphics hardware and software in 1990's, GUIs became powerful and popular. Compared with text query languages, they do not have complex syntax and other rules, and thus are much easier to learn and use. For many information system applications, work in GUIs has solved many of the problems that natural language interfaces were expected to solve, and as a result, in recent years the pace in developing database natural language interfaces became slow in general [16].

However, in dealing with geographic data, GUIs cannot entirely solve the problems that natural language interfaces may solve. A GUI has difficulties in expressing vague search conditions. For example, when looking for suitable areas to plant a certain crop, the end user may require that the soil rooting depth to be "deep", "very deep", "not very deep", "moderately deep", "excessively deep", "somewhat deep", "not very shallow", or others. Using different colors to distinguish areas of the possible soil rooting depth "values" is impractical, and some of the "values" are overlapped. Creating a button for each of the "values" may make the interface difficult to use. Furthermore, when performing multi-criteria search, it may be a tedious task to graphically pick all the geographic areas and examine combinations of their properties [21].

The above discussion suggests that we will still need natural language interfaces to improve GIS accessibility. A natural language may provide a more flexible query interface. The flexibility enables a user interface to better satisfy the needs of different user groups, and to better express users' original data requests. However, it is the flexibility that presents challenges in developing natural language interfaces. The challenges are caused by errors, ambiguity, and impreciseness in processing natural language queries.

## 7.2 Dealing with Errors, Ambiguity, and Impreciseness

Traditionally, geographic data are stored in a database as precise digital values. An artificial query language precisely specifies to the database system what data to be retrieved. A most commonly used approach for developing a natural language interface for an existing GIS is to build a language converter on the top of the existing query interface. The converter processes natural language queries, and generates queries in the artificial query language. Grammatical errors, ambiguity in parsing, and imprecise words may create great difficulties in generating artificial language queries.

Natural language queries may contain errors including mis-spelled words and grammatical errors. In our research, we address grammatical errors only. By a grammatical error, we mean a violation of a grammar rule. Grammatical errors are very common in natural language sentences, even if the language is the author's native language. It is more difficult to always write perfect natural language sentences than to always write perfect artificial language commands. If a language converter is not able to deal with grammatical errors properly and fails whenever an error occurs, its usefulness would be largely discounted. To make things worse, for most of the natural languages, there are no international nor national standards.

Grammatically correct sentences may be ambiguous. There are many causes of ambiguity. Most commonly, ambiguity occurs when a sentence can be parsed in different ways. For example, a grammatical constituent can be associated with different other constituents. When the constituent is translated into a selection condition, it may be used to select different classes of spatial objects when different associations are chosen in parsing the sentence. For example,

It is an advantage for a natural language interface to accommodate imprecise words for querying geographic data. However, imprecise words cannot be directly used to retrieve precise geographic data stored in computer systems. The gap between imprecise words and precise search conditions in artificial languages presents challenges in developing natural language interfaces for GISs.

In our research, we developed techniques for solving the problems of grammatical errors, ambiguity in parsing, and imprecise search conditions. Since the research reported in this chapter is a continuation of our previous work, we briefly describe those techniques in the following. The techniques were based on fuzzy sets, fuzzy grammars and possibility theory, which will be presented in more details in the next section.

In our previous research, for grammatical errors we addressed incomplete sentences [21]. In an incomplete sentence, one or more grammatical constituents are missing. For parsing incomplete sentences, we modified grammar rules. In each rule, we assigned labels to grammatical constituents. When parsing a sentence, a grammar rule can be fired even though some of the expected constituents are missing from the sentence. When a rule is fired, a plausibility value is calculated based on the label of the constituent that is considered to be missing. When a parse tree is generated from a sentence, a total plausibility value is calculated. When all the possible parse trees are generated, the one with the highest plausibility value is chosen as the most possible parsing structure. The missing constituents can be identified from the structure and the sentence can be repaired into a complete one. The technique is robust and effective in repairing incomplete sentences.

The technique we developed for dealing with ambiguity in parsing is based on domain knowledge [20]. A GIS has a definite application domain. If the knowledge needed for eliminating ambiguity in parsing can be properly represented, the "knowledge base" should have a manageable size. We represented the knowledge as a fuzzy compatibility matrix. When a sentence is ambiguous, more than one parse tree may be created. The knowledge stored in the matrix is used to calculate each tree's plausibility value. Then the most plausible parse structure can be selected. This technique and the technique for "repairing" incomplete sentences both involve calculating plausibility values. However, the former is based on domain knowledge while the latter based on knowledge about a language's grammar.

In the technique for translating vague search conditions into Boolean formulae, we developed the approach of inverse membership functions of fuzzy formulae [19]. A fuzzy formula may have truth values in [0,1]. It contains one or more variables, and is associated with an inverse membership function. When given a membership grade, it returns the variable values that have grades equal to or higher than the given grade. In our technique, a system can use an inverse membership function, at a required confidence level, to select the data that satisfy a vague condition.

We combine the three techniques into a three-step procedure for converting natural language sentences into artificial language queries. The three techniques work together smoothly, partly because they are all based on fuzzy set theory, as well as the fuzzy grammar theory and possibility theory derived from it. Since the fuzzy grammar theory and possibility theory are also the underlying approaches of the new technique we report in this chapter, we briefly describe them in the following section.

## 7.3 The Fuzzy Set Approaches

The grammar of a language is a key in processing sentences in the language. In most language understanding techniques, the first step is to parse sentences. In parsing a sentence, the grammar is used to decide if a string of words is a valid sentence of the language, and, if so, to identify grammatical constituents in the sentences, including phrases and clauses. In a language that does not allow errors and ambiguity (e.g. an artificial query language), a valid sentence produces a unique parse structure, and a string of words is rejected if it is not a valid sentence.

When a language allows errors and ambiguity, we need a more flexible grammar. The grammar defined in our technique is based on fuzzy grammar. A fuzzy grammar is usually represented as a quadruple $G = (V_N, V_T, P, s)$ where $V_T$ is a set of *terminal* symbols that represent language primitives or classes of primitives in input strings, $V_N$ is a set of *nonterminal* symbols that usually represent groups of primitives, for example phrases or clauses ($V_T \cap V_N = \phi$), $P$ is a set of *grammar rules*, and $s \in V_N$ is the *starting* symbol. The rules in $P$ are of the form $\alpha \xrightarrow{\rho} \beta$ where $\alpha$ and $\beta$ are strings in $(V_T \cup V_N)^*$, which is the set of finite strings obtained by concatenation of elements of $V_T$ and $V_N$. Rule $\alpha \xrightarrow{\rho} \beta$ means that $\alpha$ is defined as $\beta$, and $\rho$ is a context-dependent function that estimates the plausibility of the definition. Let $\alpha_1$, $\alpha_2$, ..., $\alpha_m$ be strings in $(V_T \cup V_N)^*$ $(m \geq 2)$. If there are $\alpha_1 \rightarrow \alpha_2$, ..., $\alpha_{m-1} \rightarrow \alpha_m$, then $\alpha_m$ is said to be *derivable* from $\alpha_1$ in $G$. The sequence of derivations $\alpha_1 \rightarrow \alpha_2$, ..., $\alpha_{m-1} \rightarrow \alpha_m$ is referred to as a *derivation chain* from $\alpha_1$ to $\alpha_m$. Such a derivation chain is usually written in a short form as $s \Rightarrow S$, where $S$ is a string in $(V_T \cup V_N)^*$.

When a fuzzy grammar is used in a bottom-up parsing algorithm, the central data structure is a parse stack. Usually, the stack contains the grammatical symbols that have been recognized from the input string, and need to be further processed. When the $\beta$ part of a grammar rule match the symbol(s) at the top of the stack, the rule is fired and the $\alpha$ part of the rule replaces the symbol(s) in the stack that the $\beta$ matches. We say that the $\beta$ symbols in the stack are *reduced* into $\alpha$. Note that the directions of reduction and derivation are opposite. Let $S$ be the input string. When the starting symbol $s$ is the only symbol remaining in the stack, the parsing of $S$ is successful. We represent this fact as $s \Rightarrow S$ and $S \in L(G)$, which means respectively that $S$ can be derived from $s$, and $S$ is a valid sentence in the language defined by grammar $G$, denoted as $L(G)$.

*Possibility theory* was initially developed for estimating the possibility for an event to occur in a given condition [22, 4]. In the theory, an event can be represented as an expression containing *variables*, a condition can be represented by using a *restricting fuzzy set*, and the possibility for the event to occur under the condition can be estimated by a *possibility distribution function.*

Let $U$ be a *universe* of discourse, whose generic elements are denoted as $u$. A *fuzzy set* $F$ defined over $U$ is characterized by a *membership function* $\mu_F$ that relates with each $u \in U$ a membership grade in [0,1]. Let $x$ be a variable taking values in $U$, and $F$ be the name of a fuzzy set. Variable $x$ may be restricted by the fuzzy set $F$. We denote such a restriction as $\Re(x, F)$ and call $F$ the restricting fuzzy set of $x$. $\Re(x, F)$ associates a *possibility distribution* with $x$. The possibility distribution function $\pi_F^x(u)$ estimate the possibility for $x$ to take value $u$ under the restriction of $F$. Numerically, the distribution function of $x$ under the restriction of $F$ is defined to be equal to the membership function of $F$, that is

$$\pi_F^x(u) = \mu_F(u) \quad \forall u \in U. \tag{7.1}$$

In our research, we use possibility theory to find the most possible events in certain conditions. For example, in eliminating ambiguity in parsing we use the possibility theory to find the most possible, i.e. the most plausible parse structure when paring a given natural language query against a given database.

## 7.4 The Error-Tolerant Parsing

Developing an error-tolerant parser is a difficult task. Such a parser must be tolerant to various errors that may occur in input word strings. It must be able to identify an error when it occurs, handle it and generate the right parse structure from the erroneous sentence. In other words, when parsing an erroneous sentence, the parser should not simply reject the sentence, instead, it should find and handle the error(s), and create a parse structure as the basis for subsequent processing.

Presently, the well-developed techniques for syntactical, semantic, and pragmatical analyses are mainly based on the assumption that input sentences contain no errors [1, 10, 12]. They are developed for parsing grammatically correct sentences, or for processing parse structures that are assumed to be correct. The development of the techniques for handling grammatical errors have lagged far behind other language processing techniques. According to a survey [2], "most of the existing NLIDBS can handle only simple cases of extragrammatical utterances" (NLIDBS – natural language interfaces to databases). A more recent survey [16] indicated that the situation did not improve much. According to the survey, in the existing systems queries with errors could hardly be parsed, and "syntactic errors gave the user more difficulty".

The problem of input strings outside the systems' lexical or syntactic coverage is usually termed "undergeneration" in natural language processing. According to a survey [5] conducted in 1997, the most commonly used techniques for addressing undergeneration is to define "error" rules. The early work included strategies of manual specification of error rules and rule relaxation. Such strategies involved adding some extra rules into language grammars. They were labor-intensive and suffered from the difficulty of predicting all the possible types of errors. The more recent attempts have been made to develop grammar rules (including the rules for errors) by using statistic induction to "learn" the grammar for a corpora of data. A major problem of such techniques is that it is often difficult to converge to a useful grammar, especially "from modesty sized corpora" [3]. Another major problem with the above mentioned techniques is poor parse performance. The cost for parsing a sentence is proportional to the number of grammar rules, especially when errors and ambiguity in input are allowed. The techniques usually generated a large number of grammar rules, which may result in high parsing costs. To the author's knowledge, in the most recent years there have been no significant advances in the techniques for addressing undergeneration.

In this research, we develop a technique that is aimed at dealing with more types of grammatically errors, which may often occur in GIS natural language queries. The errors are:

- *Omission*: A grammatical constituent is missing. "Show me the properties are greater than one acre" is an example of this type of errors, in which the "that" before "are" is missing. This type of errors has been addressed in our previous work. In this research, omission errors are handled in a more general framework.
- *Repetition*: A word is entered twice. For example, "the" is entered twice in a sentence. Such errors are common when users enter sentences using key boards.
- *Insertion*: An extra word is entered, for example, the "did" in "Tell me when did the owner buy the property".
- *Erroneous word order*: Words are entered in a wrong order. In this research, we only deal with the cases in which two grammatical constituents are reversed. For example, "Tell me who the property's agent is" is entered as "Tell me who is the property's agent".

In our technique, we use the fuzzy grammar approach with a "normal" grammar that is defined for parsing correct English sentences. No "error rules" are added. In each parse step, each rule is applied once to match the grammatical constituents at the top of the parse stack. Approximate match is allowed and each parse alternative is assigned a plausibility value. Plausibility values are calculated based on the possibilities for different errors to occur. Finally, the parse alternative with the highest plausibility value is chosen as the parse structure of the input sentence.

Approximate match may lead to a very large search space, which in turn may result in high costs in computing time and memory space. We use a GIS query model to eliminate parse structures that are impossible to be converted into meaningful queries. We call such parse structures "spurious" structures. A modified dynamic programming parsing algorithm is used for high performance. Also, a set of data structures are developed to save memory space.

The approximate match method, the parsing algorithm, and the data structures will be presented in the sections of "The Grammar and Approximate Match" and "The Parsing Strategy, Algorithm, and Data Structures". Before presenting the technical details, we first describe the GIS query model that we use for semantic representation and spurious structure elimination. The description may provide a more complete picture of the natural language query processing, for better understanding the technique.

## 7.5 The GIS Query Model

Most language processing techniques use a two-step method. The first step is syntactical analysis, in which the grammatical structure of an input string is analyzed and (usually) represented as a parse tree. The second step is semantic analysis, in which the semantics (meaning) of the input is extracted and represented. The representation method for semantics is application dependent. In building a natural language interface, it is an important task to develop a semantic representation model suitable for subsequent operations. In our research, we use a GIS query model for this purpose. After a parse tree is created from an input query, it is converted into the query model.

In terms of object terminology, a GIS stores classes of objects. The objects can be non-spatial objects (e.g. property owners), or spatial objects (e.g. geographic areas). For non-spatial objects, a GIS stores non-spatial attributes (e.g. names of property owners). For spatial objects, a GIS stores non-spatial attributes (e.g. zoning regulations for geographic areas) as well as spatial representations (e.g. X and Y coordinates in a 2-D vector GIS).

A GIS query is used to request information about spatial or non-spatial objects that satisfy certain search conditions. The objects may belong to one or more classes. A GIS query is evaluated by performing a sequence of operations. The operations can be non-spatial or spatial. A non-spatial operation is performed on non-spatial attributes only, A spatial operation is performed on spatial representations. An operation can be unary, which is performed on single objects, or can be binary, which is performed on object pairs. For example, finding property owners whose names are Smith is a unary non-spatial operation, while finding the land parcels adjacent to a certain conservation area is a binary spatial operation. Some spatial operations do not create new objects, but some others create new objects. Operations in the former group are mainly used to examine spatial relationships between spatial objects, like overlap, within, and cross. For example, an operation for "overlap" can be used to test if two geographic areas overlap. Operations in the latter group include the operations of overlay, view-sheding, and buffering. An overlay operation can be used to combine two precisely registered polygon maps of the same geographic area, by placing one on top of the other. The resultant map has the attributes of the two input maps. An operation of view-sheding can be used to create a visibility map for given viewing positions. A buffering operation is for mapping buffer zones surrounding spatial objects of points, lines, and polygons.

Based on the above analysis, we created a graph model for GIS queries. A query can be modeled as a directed graph $(N, E)$, where $n \in N$ represents object classes involved in the query, and $e \in E$ the binary operations. Formally, a node $n \in N$ is structured as

$$n = \{C, (p)^*\} \tag{7.2}$$

where $C$ is a stored object class, and $(p)^*$ denotes zero or more unary operations on objects in the class. An edge $e \in E$ is

$$e = \{< op >, (n_i, n_j)\} \tag{7.3}$$

where $< op >$ is the binary operation it represents, and $n_i$ and $n_j$ are the two nodes (representing two object classes) the operation is performed on. The direction of the operation is from $n_i$ to $n_j$.

We use a recursive algorithm to convert a parse tree into a query graph. The algorithm starts at the root, and progresses in pre-order. Informally speaking, for a subtree representing a noun phrase, if the noun represents a stored class of objects, a graph node is created. For each complement of the noun (that may be a preposition phrase, a relative clause, or an adjective phrase), if the complement has a noun phrase and the noun phrase can be converted into a graph node, an edge is created between the two nodes. The preposition, verb, or adjective in the complement is used to determine the operation for the edge. For example, we can determine spatial operations from words "within", "on", "to", "contain", "pass", "intersect", "overlap", "connect", "adjacent", "close to", and so on, which are commonly used in natural language queries in GISs. If a noun complement contains a noun phrase that cannot be converted into a graph node, the system checks if the noun phrase corresponds/relates to an attribute in the object class (created from the noun modified). If so, the complement (a subtree) is converted into a predicate associated with the node (that is for the parent of the subtree). If a noun phrase can be converted into a node and it has an adjective phrase or relative clause that does not have a noun phrase, the adjective phrase or relative clause is converted into the predicate of a unary operation of the node. The target object classes are identified as those that do not appear in noun complements.

Once a query graph has been created, the translation into an artificial language query is straightforward and unique. For example, if the GIS supports an object-relational SQL [11], the nodes can be translated into relations in the "from" clause, the unary operations and graph edges can be converted into operations in the "where" clause (with necessary defuzzification), and the target object classes are those in the "select" clause. The translated query can then be executed by the database software of the GIS.

## 7.6 The Grammar and Approximate Match

Now we discuss the technical details of the approximate match, the parsing algorithm, and the efficient data structures. As discussed before, a *fuzzy grammar* can be represented as a quadruple $G = (V_N, V_T, P, s)$. A fuzzy grammar has a plausibility function $\rho$ associated with each rule in $P$:

$$\alpha \xrightarrow{\rho} \beta. \tag{7.4}$$

The plausibility function $\rho$ indicates the plausibility for substituting $\beta$ with $\alpha$ in a reduction step. Its value range is [0,1]. The grammar rules that we use in this research are listed in the appendix. They are modified from the version published in [20]. This set of rules has a good coverage and is easy to implement. It is suitable for building experimental systems of natural language interfaces for databases.

Differing from the techniques of adding error rules, our technique addresses grammatical errors using approximate match. Let

$$A \xrightarrow{\rho} XYZ$$

be a rule in $P$, where $A \in V_N$, and $X, Y, Z \in V_N \cup V_T$. The approximate match is informally illustrated in (7.6) – (7.9), while (7.5) represents an exact match:

$$s \Rightarrow \alpha_1 A \alpha_2 \rightarrow \alpha_1 XYZ \alpha_2 \tag{7.5}$$

$$s \Rightarrow \alpha_1 A \alpha_2 \rightarrow \alpha_1 XZ \alpha_2 \tag{7.6}$$

$$s \Rightarrow \alpha_1 A \alpha_2 \rightarrow \alpha_1 XYYZ \alpha_2 \tag{7.7}$$

$$s \Rightarrow \alpha_1 A \alpha_2 \rightarrow \alpha_1 XWYZ \alpha_2 \tag{7.8}$$

$$s \Rightarrow \alpha_1 A \alpha_2 \rightarrow \alpha_1 XZY \alpha_2 \tag{7.9}$$

where $\Rightarrow$ denotes a derivation chain, $\rightarrow$ the derivation step concerned, $\alpha_1, \alpha_2 \in (V_T \cup V_N)^*$, and $W$ are in $V_T \cup V_N$, i.e. a single terminal or nonterminal symbol. The approximate match illustrated by (7.6) through (7.9) can be used to deal with the four types of errors listed in the section of "The Error-Tolerant Parsing". After a reduction step, the string on the left hand side of $\rightarrow$ becomes the top of parse stack, and the parse tree of the string is expanded by incorporating the following subtree into it:

$$\overbrace{XYZ}^{A} \tag{7.10}$$

In node $A$, the information about the match and the detected possible error is recorded. The information will be taken into account in subsequent parsing steps. For example, if the rule is fired through (7.6) and thus the $Y$ is artificially introduced, this fact must be considered when assigning a plausibility value when $Y$ (possibly along with other symbols) is reduced to another symbol.

In an unambiguous artificial query language, for a valid sentence $S$, there is a unique derivation chain $s \Rightarrow S$. However, in a natural language, for sentence $S$ that may be erroneous and/or ambiguous strings, the derivation from $s$ to $S$ cannot be represented as a unique chain. Instead, it has to be represented as a structure that we call *derivation tree*, because at a point of time, there may be more than one rule that approximately matches the symbol(s) at the top of the parse stack. The root of the tree is $S$, all the leaf nodes are the starting symbol $s$, and the interior nodes are strings in $(V_N \cup V_T)^*$, representing intermediate parse results. Let $\alpha_i \in (V_N \cup V_T)^*$ be a string of grammatical constituents recognized from $S$. If the approximate match allows $n$ rules to fire when parsing $\alpha_i$, we will have $\alpha_{i+1} \to \alpha_i$, $\alpha_{i+2} \to \alpha_i$, ... $\alpha_{i+n} \to \alpha_i$. Thus in the derivation tree, the node for $\alpha_i$ has $n$ children: $\alpha_{i+1}$, $\alpha_{i+2}$, ... $\alpha_{i+n}$. The path from a leaf node $s$ to the root $S$ is a derivation chain, which represents one of the parse alternatives of sentence $S$.

For user input $S$, we calculate a plausibility value for each parse alternative. The plausibility values play a key role in the selection of the parse structure that most likely represent what the user wants to express. We use the possibility theory to guide the selection. To apply the theory, we define variable $t$ and fuzzy set $F_S$. The domain of $t$ is the universe of all parse trees and the elements in $F_S$ are parse trees of sentence $S$. When the process of parsing $S$ is completed, $t$ contains *the* parse tree of $S$.

Since we have the statement "$t$ contains *the* parse tree of $S$", the fuzzy set of the parse trees of $S$, i.e. $F_S$, is used to restrict $t$. That is, we have restriction $\Re(t, F_S)$. The restriction associates $t$ with possibility distribution function $\pi^t_{F_S}$. Let $\tau$ be a parse tree (instance) created from $S$. When $t$ takes $\tau$ as its value, $\pi^t_{F_S}(\tau)$ returns the possibility value for $\tau$ to be *the* parse tree of $S$.

As discussed before, numerically the distribution function of $t$ under the restriction of $F_S$ is defined to be equal to the membership function of $F_S$, that is

$$\pi^t_{F_S}(\tau) = \mu_{F_S}(\tau) \quad \forall \tau. \tag{7.11}$$

To define $\mu_{F_S}$, we associate each parse alternative (that will be converted into a parse tree) created from $S$ with a membership grade. The semantics of the membership grade is defined as the plausibility for $S$ to be parsed into $\tau$. The plausibility is calculated from the plausibility functions of all the grammar rules used in creating $\tau$.

Now we briefly discuss the plausibility functions associated with grammar rules. When a rule is fired, the function returns a plausibility value between 0 and 1 inclusive. The value is calculated based on how close the right hand side of the rule matches the top of the parse stack, and how common the possible error is (if the match is not exact). If the match is exact, it returns 1. If the match is approximate, possibly an error is detected and it returns a value less than 1. Roughly speaking, if the error is a common one, the plausibility value should be high, but if the error is rare, the value should be low. The plausibility function of a rule has a "switch" structure of five cases, which respectively correspond to the five different kinds of matches illustrated in (7.5) through (7.9). When a match enables the rule to be fired, the code in the corresponding case is executed.

Examples are given in the following to show how plausibility values are decided. If Rule 14.1 is fired through (7.9) and a "BE" followed by a predicate is seen at the top of the parse stack, possibly a "erroneous word order" is detected. Since such an error is common, the plausibility function returns a high value. If Rule 14.1 is fired through (7.6) and no predicate is seen at the top of the stack, possibly an "omission" is detected. Since such an error is rare, the function returns a low value. If Rule 14.3 is fired through (7.8) and a "DO" is seen before a noun phrase or pronoun, possibly an "insertion" is detected. Since such an error is common, the function returns a high value.

Defining the plausibility functions is difficult but feasible. We currently mainly code our "subjective" knowledge into the functions. Since the system is experimental, it allows us to use an incremental and adaptive method to define the functions. Many errors are dealt with when encountered and some plausibility values are adjusted when found improper. Our technique is based on reasoning, whereas the "error rules" techniques are based on enumeration. Our technique is closer to the way in which humans detect and handle grammatical errors. It is easier to incorporate other techniques, e.g. machine learning and statistic analysis, in our technique for better parsing results. We are considering to incorporate a machine learning component into the system for automatic adjusting plausibility functions.

As mentioned before, when parsing sentence $S$, each time a rule is fired, a node is added to the derivation tree. We attach the plausibility value of the rule with the tree node. For a derivation chain (a path from a leaf to the root), we thus have a sequence of plausibility values. The total plausibility value of the chain is calculated as the arithmetic product of the sequence of values. When a sentence has two or more errors, arithmetic multiplication allows all of them to contribute to the total plausibility value. This method gives more reasonable results than the commonly used method of taking minimum, in which only the lowest of the plausibility values contributes and the number of errors does not affect the plausibility for choosing a parse tree.

For the parse tree $\tau$ generated from the chain, we use the total value as its membership grade in $F_S$, the fuzzy set of parse trees of $S$. According to (7.11), the value can be used to evaluate the plausibility for the parse tree to be *the* parse tree of $S$.

## 7.7 The Parsing Strategy, Algorithm, and Data Structures

The fuzzy grammar and the approximate match approach enable us to use "normal" grammar rules to detect and handle many grammatical errors. The power and flexibility naturally relate to high costs in computing time and memory space. Without an efficient parsing strategy and data structures, the technique can only be a "conceptual" one. In this section, we describe the parsing strategy and data structure that make the technique practical. In fact, they should be considered as an integrated part of the technique.

For a natural language query, to find the parse tree that most possibly represent the user's request, we may have two different methods. One is the exhaustive generation method that generate all the parse alternatives, and then chooses the one with the highest plausibility value. Another is to use the dynamic programming method to generate one alternative that has the highest value. The former is computationally expensive. The latter is more efficient but sometimes the parse result does not represent a meaningful query. This is because the plausibility values are calculated mainly based on syntactical knowledge, a syntactically plausible parse alternative is not necessarily highly semantically plausible. If this happens, it is costly to discard a result and generate another one with the dynamic programming method. In contrast, choosing another alternative is relatively easy when using the exhaustive method.

In our research, we use a strategy that combines the two methods. To present our strategy, we first briefly describe the dynamic programming method. Dynamic programming is a family of algorithms that are characterized by the guaranteed best results and high performance [8]. A dynamic programming algorithm solves a problem by solving its sub-problems. It produces the best result of the problem based on the best results of its sub-problems. The high performance is achieved by avoiding repeated processing of sub-problems.

Jurafsky and Martin described an effective dynamic programming algorithm for parsing ambiguous sentences [12]. The input of the algorithm includes the grammar for parsing $G = (V_N, V_T, P, s)$ and a string of words $w_1w_2...w_n$ that is the sentence to be parsed. The non-terminal symbols in the grammar are indexed from 1 to $|V_N|$. The starting symbol $s$ has index 1.

A central data structure of the described algorithm is a 3-D dynamic programming table $\Pi$. The table element $\Pi[i, j, a]$ holds the maximum score (e.g. plausibility or probability value) and other necessary information for a symbol with non-terminal index $a$ to span words $w_iw_{i+1}...w_j$. The final result corresponds to $\Pi[1, n, 1]$: the parse tree with the highest score whose root is the starting symbol $s$ and which spans the entire string of words $w_1w_2...w_n$. The process of calculating and assigning scores for the elements in $\Pi$ starts with assigning scores in $\Pi[i, i, a]$ for $1 \leq i \leq n$ and $1 \leq a \leq |V_N|$, which are for sub-strings of length 1, then scores for sub-strings of length 2 are calculated and assigned, and so on, until $\Pi[1, n, 1]$. That is, all the elements are calculated and assigned. The score for a string is calculated based on the highest scores of its sub-strings. To obtain $\Pi[1, n, 1]$, a rule $s \rightarrow \beta_1...\beta_m$ must be fired, where $\beta_1...\beta_m \in V_T \cup V_N$ $(m \leq n)$. Since each of $\beta_1$,...,$\beta_m$ spans a shorter string of the $n$ words, and the scores in the corresponding $\Pi$ elements have been assigned, the values can be used to obtain the highest score for $\Pi[1, n, 1]$.

In our technique, we combine the methods of exhaustive generation and dynamic programming parsing. When generating the alternatives, we use GIS query models to eliminate the parse alternatives that lead to no meaningful queries. Thus the number of parse alternatives is largely reduced. (The elimination using GIS query models will be discussed in the next section.) We use a dynamic programming table to store information about parsing sub-strings. Since different parse alternatives may more or less share the same parse structures of sub-strings, the information in the dynamic programming table helps avoid repeated calculation, and thus dramatically reduce parsing costs.

Differing from the dynamic programming method described in [12], our method employs a 4-D array $\Pi'$. Element $\Pi'[i, j, a, m]$ stores the information about applying the $m$'th match to fire the rule that defines the $a$'th non-terminal symbol to cover words $w_i w_{i+1} ... w_j$. Another major difference is that we calculate table elements in a "lazy" way. When parsing a sub-string, the system first checks the table to see if the parsing was done before, if so, the information stored in the table is used. Otherwise, the system parses the sub-string, and then assigns the parsing information to the corresponding table element. We do not calculate and assign the values for all the table elements, for the calculation is expensive and most of the values may not be needed at all. This strategy has proved very efficient. It does not require much extra costs to maintain the dynamic programming table, whereas the table helps minimize the costs for generating the parse alternatives. This strategy has the major advantages of both methods: It guarantees to find the most plausible parse structure, avoids repeated parsing of sub-strings, and, if the first choice structure has to be discarded, the second best is ready to use.

Now we discuss the data structure. The parsing process can be conceptually represented as a derivation tree, in which a parse alternative is a path from a leaf node to the root. (Note a derivation tree is different from a parse tree, in which the root is starting symbol $s$ and the leaves are words.) When the approximate match is applied, for a sentence containing errors, each interior node in the tree may have several children. Thus the number of parse alternatives may be very large. In our experiments, it is common that the numbers may be as high as several hundreds. If a separate parse stack is maintained for each parse alternative, the high costs in memory usage (as well as memory allocation and de-allocation) may make the technique unaffordable. This problem may become worse when the technique is used in multi-user environment. An efficient data structure is a key that makes the technique feasible.

We developed a tree structure for our technique, which contains the stacks of all the parse alternatives of a sentence. The structure is aimed at minimizing space usage.

The basic data structures are for word units (WU), stack elements (SE), and tree nodes (TN). A WU object (instance) contains the value of a word when it is recognized by a lexical analyzer. A word is associated with only one WU object. An SE object represents a terminal symbol or a non-terminal symbol. When it represents a terminal symbol, it points to a WU object, and when it represents a non-terminal symbol, it points to all the SE objects representing the defining symbols. A TN object contains an array of SE objects.

At a point of time during a parsing process, the path from a leaf TN object to the tree root TN object can be used as a parse stack. The bottom is at the root and top is at the (current) leaf. When a rule is fired at the top of a stack, a new TN object is created and attached to the current top node (the leaf TN object) as its child. A new SE object is created for the symbol on the left hand side of the rule, and "pushed" onto the new TN node. The new SE object points to the SE objects that define it. Those defining SE objects are in the parent or ancestor TN objects. In addition, the new SE object carries the plausibility value returned by the rule's plausibility function, as well as other necessary information. When a new word is read in but no rule fires, a new SE object is created for the word, and "pushed" onto the current top TN object.

Differing from the "conventional" parse stacks, such a stack does not pop out the symbols that have been reduced. When tracing from the leaf node (stack top) to the root node (stack bottom), if we find that an SE object is pointed by another (the defined) SE object, we consider the former as "reduced". In a stack, if the only un-reduced symbol is the starting symbol $s$, the parsing alternative is completed.

This structure has good space efficiency. When more than one rule fires at the top of a stack (a leaf node of the derivation tree) and thus a derivation chain has to be "forked", the different parsing alternatives share their common SE objects, and no duplications are needed.

Another advantage is with the way for dealing with "reduced" SE objects. To be shared by different parsing alternatives, an SE object reduced by one rule should not be removed from the stack nor be labeled. In a stack, since the defined SE object is always "higher" than the defining ones, the pointers from the former to the latter provide sufficient information for distinguishing "reduced" and "un-reduced" SE objects in *that* stack. In this way, one stack's information does not cause confusion in another stack. By retaining the "reduced" SE objects in a stack, all the information required for selecting the most plausible parsing alternative for a sentence can be obtained in the tree, and the parse tree of that alternative can be easily generated by following the pointers from defined to defining SE objects. In fact, a path in the derivation tree is a linear representation of a parse tree, with the parse tree root at the derivation tree leaf on the path.

In the whole process of parsing a sentence, besides the dynamic programming table, the tree is the only data structure to maintain, and its size can be handled by most of the computers available nowadays.

## 7.8 Eliminating Spurious Parse Structures by the GIS Query Model

When a sentence with errors or/and ambiguity is parsed, a very large number of parse alternatives may be created. Most of them are spurious because they will not generate meaningful GIS queries. Such parse structures should be deleted at the earliest possible time to save computing time and space. We use the GIS query model discussed in the section of "The GIS Query Model" to eliminate spurious parse structures.

Recall the translation from a parse tree to a query graph. For a parse tree to be converted into a meaningful query in an artificial query language, nouns in noun phrases must be converted into stored object classes or something that can be derived from attributes of the classes. If a parse tree contains a noun phrase that cannot be converted in this way, the tree cannot be converted into a valid query.

We use the knowledge about the relationships between noun phrases and stored object classes in eliminating spurious parse structures. Whenever symbols at a parse stack are reduced into a noun phrase, the noun phrase is examined for the possibility to be converted into a node or a node predicate in a query graph. If the phrase cannot be converted into a node or a node predicate, the TN (tree node) object is marked as "dead-end" to indicate the parse alternative is discarded.

Also recall that verbs, prepositions, and adjectives in noun complements are converted into spatial or non-spatial operations, which can be unary operations attached to nodes or binary operations between nodes in query graphs. In a query graph, even though all the nodes represent stored object classes, some unary or binary operations may not "make sense" when performed on given objects. For example, it is meaningless to test if a polygon object is within a point object. We also consider parse structures with such operations spurious structures and eliminate them using the query model. Whenever a noun complement is parsed, the verb, preposition, or adjective is examined. If it represents an impossible operation in that context, the parse structure is marked "dead-end".

In addition, the query model is used in eliminating duplicate parse structures. Approximate match may allow different parse processes to produce identical parse trees. Duplicates should be eliminated to reduce parsing costs. Only one "copy" should go for further processing. Whenever a reduction is performed on a path in the derivation tree (a stack) and a new symbol is added, the path is compared with all the "active" paths. If the path is found to be the same as another one in creating parse trees, the leaf node of the path is marked as "dead-end".

## 7.9 A Sample Query

In this section, an erroneous query is used as an example to illustrate the basic idea we discussed in this chapter. The query is "Find the houses that close to a school", which has a grammatical error. The query is processed in the context of a GIS for real estate purposes.

When the approximate match is used, there may exist a number of parse alternatives. The following list five of them:

1. The string "that close to a school" is parsed as a relative clause using Rules 14.2 and 17, and the `VERB` in Rule 17 is considered missing. The relative clause is the complement of "the houses".
2. The string "that close to a school" is parsed as a relative clause using Rules 14.2 and 17, but the "to" is considered as an insertion. The relative clause is the complement of "the houses".
3. The string "that close to a school" is parsed as an adjective phrase using Rule 2, and the "that" is considered as an insertion. The adjective phrase is the complement of "the houses".
4. The string "that close" is parsed as a relative clause using Rule 17, and the string "to a school" is parsed as a preposition clause using Rule 13. Both are complements of "the houses".
5. The string "that close" is parsed as a relative clause using Rule 17, and the string "to a school" is parsed as a preposition clause using Rule 13. Both are complements of "the houses". When Rule 17 is applied, the objective noun phrase is considered missing.

Of the five alternatives, the last one is assigned a very low plausibility value, because it is very unlikely that an objective noun is missing. In the second alternative, the operation that "houses" close "a school" is meaningless (because "close" means "shut" or "stop" when used as a verb). Thus the alternative is eliminated. In the fourth alternative, the relationship "to" between "houses" and "a school" is meaningless. (We usually look for the roads *to* a school.) Therefore this structure is treated as a spurious structure. The first and the third alternatives both represent meaningful operations that can be used to find houses close to a school. Since the two alternatives can be converted into the same query graph, one of them is eliminated and only the other remains for further processing.

## 7.10 Concluding Remarks

An effective error-tolerant parser must satisfy the following criteria:

1. The "correct" parse structure must be one of the parse alternatives generated.
2. All spurious parse structures should be eliminated as early as possible.
3. The parser must be able to select the "correct" parse structure from the alternatives.
4. The parser must be efficient in computing time and memory usage.

We have implemented the technique as an experimental system. The machine used for the experiments is an HP workstation of 2.3GHz Pentium processor and 256 MB RAM. The operating system is Red Hat Linux 7.1. The program is coded in gcc. By the time of this writing, about 150 queries in English have been tested, a few are grammatically correct, and most have various kinds of grammatical errors. The queries are applied on a GIS for real estate applications, which stores objects of houses, schools, hospitals, shopping centers, parks, roads, zoning areas, property vendors, etc.

The experiments show that the criteria (1), (2) and (4) have been well satisfied. For almost all the test queries, the "correct" parse structures are generated. The query model based approach can eliminate all the spurious parse alternatives, and eliminate spurious only. The parsing is efficient. The longest parsing time recorded is about 3.54 seconds, which can be considered acceptable. No swapping is observed, which indicates memory usage is efficient.

Criterion (3) is the most difficult to achieve. Currently the accuracy is slightly lower than 80%. This rate is encouraging and indicates the success of the overall technique. A key for higher accuracy rate is to improve plausibility functions. As the work in the near future, we are planning to use a machine learning approach that learns from users' feedback and automates the adjustment of plausibility functions, which is tedious work when done manually.

## Addendum

```
1.   <adjective complement>:-
1.1       <preposition phrase> |
1.2       [FOR {<noun phrase> | PRONOUN>}] <infinitive> |
1.3       THAT <declarative phrase>

2.   <adjective phrase>:-
          [<adverb group>] ADJECTIVE [<adjective complement>]

3.   <adjective phrase list>:-
3.1       <adjective phrase> |
3.2       <adjective phrase list> <adjective phrase>

4.   <adverb group>:-
4.1       ADVERB |
4.2       <adverb group> ADVERB

5.   <adverb phrase>:-
          ADVERB [<noun phrase>]

6.   <declarative sentence>:-
6.1       {<noun phrase> | PRONOUN} [AUXILIARY]
               BE <predicate> '.'|
6.2       {<noun phrase> | PRONOUN} [AUXILIARY]
               <verb phrase> '.'
```

```
7.  <imperative sentence>:-
         [PLEASE] [DO [NOT]]
              {BE <predicate> | <verb phrase>} '.'

8.  <noun phrase>:-
         [DETERMINER] [<adjective phrase list>] NOUN
              [<noun complement list>]

9.  <noun complement list>:-
9.1      <noun complement> |
9.2      <noun complement list> <noun complement>

10. <noun complement>:-
10.1     <preposition phrase list> |
10.2     <relative clause list> |
10.3     <adjective phrase list>

11. <predicate>:-
11.1     <adjective phrase> |
11.2     <adverb phrase> |
11.3     <noun phrase> |
11.4     <preposition phrase>

12. <preposition phrase list>:-
12.1     <preposition phrase> |
12.2     <preposition phrase list> <preposition phrase>

13. <preposition phrase>:-
         PREPOSITION <noun phrase>

14. <relative clause>:-
14.1     {THAT | WHICH | WHO | WHERE} <predicate> BE |
14.2     {THAT | WHICH | WHO} <verb phrase> |
14.3     {THAT | WHICH | WHO | WHERE}
              {<noun phrase> | PRONOUN}
              [AUXILIARY] <verb phrase>

15. <relative clause list>:-
15.1     <relative clause> |
15.2     <relative clause list> <relative clause>

16. <sentence>:-
16.1     <declarative sentence> |
16.2     <imperative sentence> |
```

```
16.3      <yes-no interrogative sentence> |
16.4      <wh interrogative sentence>

17. <verb phrase>:-
          VERB [PRONOUN] [<noun phrase>]

18. <wh interrogative sentence>:-
18.1      {<what phrase> | <how phrase> | <where phrase>} BE
            {<noun phrase> | PRONOUN} '?' |
18.2      {<what phrase> | <how phrase> | <where phrase>} BE
            <preposition phrase> '?'

19. <yes-no interrogative sentence>:-
19.1      BE {<noun phrase> | PRONOUN} <predicate> '?'|
19.2      DO {<noun phrase> | PRONOUN} <verb phrase> '?'
```

## References


1. Allen J. (1995) Natural Language Understanding (second edition). The Benjamin/Cummings Publishing Company Inc, Redwood City, California, USA.
2. Androutsopoulos I., Pitchie G.D., Thanisch P. (1995) Natural Language Interfaces to Databases – An Introduction. Natural Language Engineering 1(1):29–81.
3. Ballim A., Pallotta V., Lieske C. (1999) Robust Text analysis: an Overview. http://lithwww.epfl.ch/ pallotta/rota.ps.gz.
4. Dubois D., Prade H. (1988) Possibility Theory. Plenum Press, New York, USA.
5. Carroll J., Briscoe T. (1997) Robust Parsing – A Brief Overview. Report CSRP435, COGS, University of Sussex.
6. Cobb M.A., Petry F.E., Kevin B. (2000) Fuzzy spatial relationship refinements based on minimum bounding rectangle variations. Fuzzy Sets and Systems 113(1) 111–120.
7. Cobb M.A., Petry F.E., Wen L. (2001) Modifications of fuzzy directional queries for spatial databases. In: Proc. of the Joint IFSA World Congress and 20th NAFIPS international Conference, Vancouver, B.C. Canada.
8. Cormen T.H., Leiserson C.E., Rivest R.L. (1990) Introduction of Algorithms. The MIT Press, Cambridge, MT, USA.
9. Frank A.U., Mark D.M. (1991) Language Issues for GIS. In: Maguire D.J. et al. (Eds.) Geographical Information Systems Principles and Applications, Longman Scientific and Technical, Essex, England.
10. Hausser R. (1999) Foundations of Computational Linguistics. Springer-Verlag, Berlin.
11. ISO/IEC (1999) Information Technology – Database languages – SQL Multimedia and Application Packages – Part 3: Spatial. The International Organization of Standardization.
12. Jurafsky D., Martin J.H. (2000) Speech and Language Processing. Prentice Hall, Upper Saddle River, NJ, USA.

13. Kollias V.J., Voliotis A. (1991) Fuzzy reasoning in the development of geographical information systems FRSIS: a prototype soil information system with fuzzy retrieval capabilities. International Journal of Geographical Information Systems 5:209–223.
14. Morris A., Petry F.E., Cobb M. (1998) Incorporating Spatial Data into the Fuzzy Object Oriented Data Model. In: Proc. of the seventh International Conference on Information Processing and Management of Uncertainty, Paris, pp. 604–611.
15. NCGIA (National Center for Geographic Information and Analysis) (1989) The Research Plan of the National Center for Geographic Information and Analysis. International Journal of Geographical Information Systems 3(2):117–136.
16. Ogden W.C., Bernick P. (1997) Using Natural Language Interfaces. In: Helander M.G. et al. (Eds.) Handbook of Human-Computer Interaction Second, Completely Revised Edition. Elsevier, Amsterdam.
17. Raper J.R. (1991) User Interface. In: Masser I., Blakemore M. (Eds.) Handling Geographical Information: Methodology and Potential Applications. Longman Scientific and Technical, Essex, England.
18. Smith T.U., Yiang J. (1991) Knowledge-Based Approaches in GIS. In: Maguire D.J. et al. (Eds.) Geographical Information Systems Principles and Applications. Longman Scientific and Technical, Essex, England.
19. Wang F. (1994) Towards a Natural Language User Interface: an Approach of Fuzzy Query. International Journal of Geographic Information Systems 8:143–162.
20. Wang F. (2000) A Fuzzy Grammar and Possibility Theory Based Natural Language User Interface for Spatial Queries. Fuzzy Sets and Systems 113(1):147–159.
21. Wang F. (2003) Handling Grammatical Errors, Ambiguity and Impreciseness in GIS Natural Language Queries. Transactions in GIS 7(1):103–122.
22. Zadeh L.A. (1978) PRUF – A Meaning Representation Language for Natural Languages. International Journal of Man–Machine Studies 10:395–460.

# 8

# An MPEG-7 Based Video Database Management System

Adnan Yazici[1], Öznur Yavuz[1], and Roy George[2]

[1] Dept. of Computer Engineering
METU, 06531, Ankara, Turkey
yazici@ceng.metu.edu.tr

[2] Dept. of Computer Science
Clark Atlanta University
Atlanta, GA 30314, U.S.A.

## 8.1 Introduction

The availability of higher storage capacities and faster processing speeds have made it possible to store and operate on multimedia data resources such as images, graphics, 3D models, audio, speech, and video. The quantity of multimedia data available in digital archives, on the World Wide Web (WWW), and in personal and professional databases is growing rapidly. There is a need for systems that can index, store, and retrieve these media items for applications such as digital libraries, geographical information systems, video on demand services, teleconferencing, etc. However, developing these mechanisms is a difficult task due to the high volume and complexity of multimedia material, and a lack of adequate indexing standards. Since traditional database systems are inadequate in handling multimedia data, the development of multimedia databases has drawn both research and industry efforts. Although searching and querying on image databases have been studied intensively it is only in recent years, video database management systems have been researched.

In this research we develop a web-based video database management system that supports the semantic description of video content, and facilitates content-based querying on video data. Video entities are classified as Objects (cars, balls, etc.), Agent Objects (persons, teams, etc.), Activities (eating, shouting, etc.), and Events (eating an apple by John.). These artifacts and the spatial, temporal, and spatio-temporal properties of these entities are expressed using the description tools of MPEG-7 standard. The Data Description Language of MPEG-7 is based on the XML schema language [11, 15, 16] and is used to define multimedia objects in the video database system. Additionally, new description schemas may be defined by extending these description tools to address specific requirements of the system, especially to deal with the temporal, spatial, and spatio-temporal queries (object trajectories) that have to be supported by the system. Defining the spatial relationships between objects and agents objects crisply is an oversimplification and is not sufficient for deployable systems. We define a flexible spatial and spatio-temporal query model that retrieves the results based on fuzzy similarity measures. This generalized query model also retrieves exact results as a subset of the larger set of similarity based results. XML documents containing MPEG-7 descriptions are stored on text large object type columns as attributes of videos. All queries are performed using XML Path Language (XPath), which is a query notation for addressing and filtering the elements and text of XML documents. With our system it is possible to query multiple related videos defined under a single video group name.

There are a number of innovate contributions presented in this chapter. A video database system based on the spatio-temporal video model [21] was developed during this research. This model differs from the other extant ones in several ways. First, all entities and their properties are described using MPEG-7 descriptions. The XML documents containing MPEG-7 descriptions are stored in tree like DOM structure [7]. Queries are performed using XPath expressions which are applied to the nodes of DOM tree [7]. Second, this model supports a richer and more flexible query model in comparison to extant video models. The enhanced query model includes temporal relationships queries (starts, finishes, meets, equals, overlaps, during, and before) between events and directional trajectory queries (e.g., trajectories of an object starting from the left of a region and ending at the bottom of this region) of objects and agents. Querying on video groups and the definition of still and moving regions as inputs to the query model is also facilitated by the system.

The organization of the rest of the chapter is as follows: Section 8.2 explains the concepts and data structures of AVIS model [1], and the spatio-temporal video extensions to this model. Section 8.3 briefly overviews the MPEG-7 description tools used in the system and Section 8.4 gives the implementation details of the system. Query processing mechanisms and some of the XPath expressions used by the system are included in this section. Finally, Section 8.5 investigates possible extensions and concludes the study.

## 8.2 Related Work

There have been several previous research efforts on developing video models that permit advanced querying methodologies. These efforts include annotation-based models, physical level video segmentation approaches and object-based modeling approaches. Annotation-based video models allow free text or attribute/keyword annotations in order to describe semantic and spatial attributes of video data. The physical level video segmentation approach does not address semantic concepts, such as, objects, events, roles, players, etc., rather the video data is described as a stream of small segments with application specific temporal and spatial properties. The object based models focus on the modeling of semantic content of the video data using object-oriented modeling techniques, describing components of the video in terms of the actors, objects, locations, etc. In this section we briefly describe several systems that have features relevant to our effort.

AVIS (Advanced Video Information System) [1] is a formal object-based video data model. AVIS models the semantic entities as objects, activities, events, and temporal properties, enabling users to query semantic video content. AVIS uses a number of array structures to store video entities, and their properties. Temporal properties of entities are given with a set of frame sequences, and these frame sequences are stored in a frame segment tree structure. Temporal queries (object occurrence, activity occurrence, and event occurrence queries) can be performed with AVIS. Köprülü et al. [21] extends the video model of AVIS to support modeling of spatio-temporal properties of object entities. That study replaces the frame sequences of objects on frame segment tree with frame sequences of object-region pairs. Array structures of AVIS is also modified to store spatial properties of objects. That is, spatial and spatio-temporal queries are supported. Spatial relations (left, right, top, bottom, topleft, topright, bottomleft, bottomright, and overlap) between object entities (spatial queries) and trajectories of object entities (spatio-temporal queries) can also be queried.

VARS (Video Archival and Retrieval System) [8] proposes a stratification-based video model where video frames are modeled as multi-layered descriptions. Each context layer, stratum, describes the temporal properties of a simple concept, and the video is the union of all strata present. A wide range of entities like objects, events, locations, and categories, audio dialog, and structured data (dates and times) are modeled as strata. Strata are extracted automatically in VARS. Content based browsing of video data is supported by the system. Like AVIS [1], video entities are defined with a set of frame sequences in VARS. Audio dialogs are modeled and indexed at the paragraph level. Paragraphs are defined with start and end frames of them. Complex queries are performed by using AND and OR logical operations on frame sequences. The frame sequences on query results are extended to ensure that they have good visual and audio continuity. Fast-forwarding directly using strata information is also supported.

BilVideo [14] is a web-based video database management system that includes a fact-base and a set of rules implemented in Prolog. It uses an object-relational database system. Query processor of BillVideo interacts with both of the knowledge base and object-relational database. The rule-based approach of the study eliminates the need for the computation of relations at the time of query processing. Object occurance, external-predicate, trajectory-projection, similarity-based object trajectory are the types of spatio-temporal queries supported by the system. Semantic queries based on keywords, events, activities and categories are implemented. Queries on low-level concepts like color, shape, and texture are also supported. The system has an SQL-like textual query language and a visual query language.

The VideoText model [18, 19] and Algebraic Video Model [30] are video models based on the textual annotations assigned to the frame sequences of videos. Since free text or keyword annotations are used to describe semantic attributes of frame sequences, video annotations can be added, removed, and modified independently from the raw video data. VideoText and and Algebraic Video Model systems support querying temporal relationships between frame sequences. Theodosis et al. [29] describes indexing schemas on spatio-temporal properties of objects. They categorize spatio-temporal relationships of objects as inclusive and exclusive. In inclusive relationships, objects overlap temporally and spatially. In exclusive relationships, they are temporally and spatially disjoint. This study uses a 3D R-tree whose third dimension is time. OVID [25] is a content-based object-oriented video data model. The interval-inclusion based inheritance feature of the model enables common descriptional data to be shared among objects. A query language, VideoSQL, that extends SQL for video databases is defined for the mode. An interesting feature of the system is that it permits users to specify the number of frames they would like to see in a presentation. The Logical HyperVideo Data Model [20] is an effort that supports spatial, temporal, and semantic queries on the objects that are described with their semantic associations to other objects. These associations are defined as video hyperlinks.

Several MPEG-7-based projects are currently under development. AMOS [31], a video object segmentation and search system models object entities on videos by using a semi-automatic approach to define video objects. An object is modeled as a set of regions with corresponding visual features like color, texture and spatio-temporal relations. Textual annotations can also be entered for the objects. In AMOS the MPEG-7 standard is used to describe video segments, which are decomposed into moving regions spatio-temporally. Moving regions are described and the regions are decomposed into other moving regions to represent salient regions of objects. Visual, textual annotation and segment relation tools of MPEG-7 standard are also used. The SAMBITS (System for Advanced Multimedia Broadcast and IT services) project [22] has developed access methods for searching AV (Audio-Visual) data using MPEG-7 descriptions in the context of a broadcast application. The system is implemented with HySpirit software development kit. Video shots and spatio-temporal regions on those shots are modeled with SAMBITS. Audio-Visual Segment DS is used to represent spatio-temporal segments of video, StillRegion DS is used to represent spatial regions of images, SegmentDecomposition tools are used to define decomposition of segments into sub segments, and TextAnnotiation DS is used to define annotations of video segments like Whom, What, and When. The prototype system of the study is a web-based application. Retrieval is implemented with standard relational algebra operations like UNION, JOIN, and PROJECT. Query results are ranked from the most relevant to least relevant.

## 8.3 Modeling Video Data

The approach taken to data modeling is similar to that in AVIS (Advanced Video Information System) [1] which models entities on video data like objects, activity types, events, and temporal properties of these entities. Previously, Köprülü et al. [21] extended AVIS to model spatio-temporal properties of object entities. In this section we briefly overview these studies.

### 8.3.1 The Advanced Video Information System (AVIS)

In the Advanced Video Information System (AVIS) model [1], the main focus is on objects, events, and activities, called *entities* , appearing in the video. The video *objects* may be characters in a movie such as Rambo, James Bond, etc. or they may be objects such as a car, a tree, a ball, etc. An *activity* is accepted as the subject of a given frame sequence. It is the type of the action performed in a video (e.g., walking, eating, and kicking a ball). Multiple activity types may occur simultaneously in a frame sequence. An *event* is an instance of an activity type. It consists of a unique activity type, a set of roles in the activity, and objects as the actors of roles in the activity. For example, in the event "Philip is eating a cake", the activity type is eating, the role is eater, and the actor is Philip.

Each entity has a set of frame sequences attached to it. These are the frames in which the entity appears in the video. The collected information about entities is indexed with some special index structures. During query processing, these index structures are used in order to retrieve information about entities, and more importantly, to make correlations between entities.

A frame sequence is a set of contiguous frames containing a semantically important data, like an object or an event. A frame sequence can be considered as a shot defined by segmenting a video stream using a semantic entity. An association map is constructed from frame sequences as a visualization tool to view the segmentation of the video stream.

An association map represents the main idea in this data model. An association map corresponds to horizontal line segments on the Cartesian plane, where x-axis is the time and y-axis is the set of entities. A line segment in the association map represents the occurrence of an entity in the video during the corresponding frame interval. The model provides means to represent entities of interest in a video, frame sequences to identify the occurrence time intervals of entities, and association maps to combine entities and frame sequences into one graph. Thus, association maps can be stored using any method to store such collinear line segments, such as the well-known segment tree.

In AVIS, indexing is based on a segment tree, which is named as the *frame-segment tree* (FST). The FST is constructed from the line segments in the association-map. A great deal of the line segments for different entities overlap in the association map. The overlapping line segments are divided into smaller segments by imaginary vertical lines, so that overlapping sections of line segments for different entities are considered as a separate node in the frame segment tree. A list of entities that appear in the time interval represented by a node is stored along with the node in the tree.

There are three array structures in the AVIS model that provide additional indices to the nodes of the FST. These are *ObjectArray, EventArray,*and *ActivityArray*. Each entity in the database is loaded into the corresponding array depending on the type of the entity, using a hashing method. The ObjectArray is an array whose $i$th element contains an object with hash number $i$. Each element in the array is associated with a list of numbers each of which represents a node that contain object $i$ in the frame segment tree. Each element also contains additional information about the object, like the name of the object.

The EventArray stores event descriptions. Each event has an identity calculated from the unique properties of the event. The EventArray also stores pointers to the corresponding nodes in the frame segment tree.

The ActivityArray stores the activity types together with the events associated with it. Activity type and event entities have $1 : N$ relation in between. Therefore, an activity type may point to more than one event, whereas, an event can point to only one activity type. ActivityArray does not store pointers to the FST like other two arrays. When an access to FST is needed from this array, this is done indirectly over the EventArray. For example, when we want to find the frame intervals of a given activity type, we first find the events pointed to by the activity record, and then we find the frame intervals of each of these events. For all arrays, pointers to the frame segment tree nodes are kept ordered according to the frame interval that they represent.

### 8.3.2 Modeling Spatio-temporal relations in video

This study is also related to the video data model developed by Köprülü et al. [21]. This model extends the video model of AVIS to represent spatial properties of objects. Spatial property of an object is the information about the spatial location of that object on the video frames. Two-dimensional coordinates are used to store spatial properties of objects, and they are represented by the user-specified rectangles. The spatial property of an object $A$ is a tuple $(R, I)$, where, $R$ is a rectangular area which covers all areas in which object $A$ appears during the time interval $I = [t_i, t_f]$. So, $R$ is obviously not the minimum-bounding rectangle (MBR)[3] of $A$. At any time $t$ in $[t_i, t_f]$, object $A$ may be located at anywhere in $R$.

The spatial properties of objects are represented by dynamic rectangles, and so the spatial relationships between objects are computed dynamically. Previously, Li et al. [23] extended Allen's temporal interval algebra [2] to a two-dimensional space in order to define spatial relationships between rectangular areas. They defined two groups of spatial relations: topological and

[3] A minimum bounding rectangle is an imaginary rectangle that is accepted to be the minimum rectangle that covers all parts of an object.

directional. Topological relations refer to the relationships between the rectangles in terms of their topology while directional relations prescribe the orientation of these rectangles with respect to each other. Köprülü et al. [21], use an extended set of directional relations that included intermediate directions (analogous to *northwest*, *southeast*, etc.) and a restricted set of topological relations (i.e., *overlaps*, *equal*, *inside*, *contain*, *touch*, and *disjoint*).

The spatial relationships between objects are defined using fuzzy logic; that is, a certainty level (in fuzzy terms membership value) is used to define the relationships between objects. The membership values ($\mu_i$) of the relationships are calculated using the centers of rectangles. The angle between the x-axis and the line between centers of the rectangles is used to calculate membership values. The data structures used in AVIS are extended to support spatial and temporal properties of objects. The changes are briefly described as follows:

- A unique frame interval in the interval list of an object is used to represent the continuous list of frames in which the related object appears in a single region. Each element in the object array tracks the information for an object together with all the frame interval/region pairs, i.e., an object would have a list of interval/region pairs $(i, r)$ that show the interval $[i_{start}, i_{end}[$ in which the specified object appears in region $r$.
- Changes to frame segment tree are based on the fact that moving entities visit more than one region. In the segment tree of AVIS, this entity is represented by one FST-node, since the frames are continuous. However, in the extended model, this single interval must be divided into sub-intervals, when the region of the object changes. A separate tree node is used to represent each of these sub-intervals. So a tree node will be defined to contain a list of object/region pairs. Semantically, this means that, each $(o_i, r_i)$ pair in the list means that, "object $o_i$ appears in region $r_i$ within the time interval represented by this node".
- The association map is modified so that the y-axis tracks $(object, region)$ pairs and events as video entities.

Several new query types are supported in addition to the basic query types of AVIS:

- Given objects and relation, find frames: "Find the set of time intervals, in which plane is seen at left of towel, with a threshold value of 0.7".
- Given object and interval, find regions: "Give a list of all regions, where the ball appears in, between $5^{th}$ and $10^{th}$ minutes".
- Given object and region, find frames: "List all frame intervals during which a plane appears in the rectangle $R$".
- Find trajectories of objects (spatio-temporal queries): "Show me all frames in which plane moves from rectangle $R_1$ to rectangle $R_2$". The study, the formal definition of a trajectory is as follows:

A trajectory is a set $T = \{(I_n, R_n) \mid n > 1\}$, such that $I_n$ is the time interval $[t_{ns}, t_{nf}]$, $R_n$ is the rectangular area defined for the object in time interval $I_n$, and for any $i$, where $1 < i \leq n$, $t_{(i+1)s} = t_{if} + 1$.
$R_{i+1}$ is a neighbor of R$_i$, neighborhood is defined as follows:

Suppose we have two rectangles $R_1$ and $R_2$, with coordinates $(r_1x_1, r_1x_2, r_1y_1, r_1y_2)$ and $(r_2x_1, r_2x_2, r_2y_1, r_2y_2)$ respectively. $R_1$ is a neighbor of $R_2$ if and only if one of the following conditions is satisfied:

$$r_1x_1 = r_2x_2 \wedge [r_1y_1, r_1y_2] \text{ and } [r_2y_1, r_2y_2] \text{ overlaps}$$
$$r_1x_2 = r_2x_1 \wedge [r_1y_1, r_1y_2] \text{ and } [r_2y_1, r_2y_2] \text{ overlaps}$$
$$r_1y_1 = r_2y_2 \wedge [r_1x_1, r_1x_2] \text{ and } [r_2x_1, r_2x_2] \text{ overlaps}$$
$$r_1y_2 = r_2y_1 \wedge [r_1x_1, r_1x_2] \text{ and } [r_2x_1, r_2x_2] \text{ overlaps}$$

The definition of overlap is taken from Allen's temporal algebra. Since this algebra can be applied to any one-dimensional space, it is used for x-axis and y-axis.

## 8.4 Describing Video Semantic Content with MPEG-7 Description Tools

The concepts developed previously are incorporated into a video database management system modeling semantic content of video data using MPEG-7 description tools. Description tools of MPEG-7 standard are used to describe video entities and their properties instead of using array data structures and frame segment tree as it is performed in AVIS [1] and the extended AVIS [21]. Object, activity type, and event entities, spatio-temporal properties of object entities, and temporal properties of event entities are all modeled in the developed system. The MPEG-7 standard concerns not only semantic content of Audio Visual (AV) data but also other aspects of it, like creation, production and usage information for the data. For instance, MPEG-7 standard provides description tools to define catalog (title, creator, etc.), semantic (who, what, when, and where information about objects and events), and structural (color, shape, texture, etc.) features of audiovisual (AV) data [17]. Only a subset of MPEG-7 description tools is sufficient to describe the semantic content of video data. The MPEG-7 XML Schema used in this study (based on [5] and [9]) is minimized to ignore unused multimedia, audio, visual descriptors, and description schemas.

MPEG-7 description tools are a set of descriptions (namely descriptors —Ds— and description schemas —DSs—) in XML Schema language [11]. MPEG-7 standard includes a number of multimedia descriptors and description schemas pertaining generically to the description of multimedia data, and a number of visual descriptors and description schemas dealing with the motion, and localization visual features of multimedia data that can be used to define video entities and their spatio-temporal properties [24]. In this study, new descriptors and description schemas are also defined in order to suit to the requirements of the video model of the system by extending or simplifying MPEG-7 Ds and DSs. Table 8.1 shows the correspondence between AVIS entities and MPEG-7 descriptors/description schemas used in this system, and lists the MPEG-7 description tools to define spatial and temporal properties of these entities. In the next section we explain in detail how these MPEG-7 description tools are used in the developed system to describe video entities and their spatial and temporal properties.

| **AVIS Entities and Spatio-temporal Properties of Entities** | **MPEG-7 Description Tools Used In the Developed System** |
|---|---|
| Objects | SemanticBase DS<br>Object DS<br>avisObject DS<br>AgentObject DS<br>avisAgentObject DS |
| Activity types | SemanticBase DS<br>avisActivity DS |
| Events | SemanticBase DS<br>EventDSavisEvent DS |
| Temporal properties of entities (frame sequences) | MediaTime DS<br>avisMediaTime DS<br>Segment DS<br>VideoSegment DS |
| Event ⇔ Frame sequence relations<br>Event ⇔ Object relations (Roles and players) | MPEG-7 Semantic Relation Tools:<br>RelationBase DS, Relation DS<br>SemanticRelationBase DS<br>SemanticBaseRelation DS<br>SegmentSemanticBase Relation DS<br>ObjectEventRelation DS |
| Spatio-temporal properties of objects | Segment DS<br>StillRegion DS<br>avisStillRegion DS<br>MovingRegion DS |

**Table 8.1.** Correspondence between AVIS entities and MPEG-7 description tools

### 8.4.1 Describing Objects, and Agent Objects

Object entities on a video may be described by using *Object DS* multimedia description schema of the MPEG-7 standard. In MPEG-7 all semantic content in the multimedia data is described based on the abstract *SemanticBase DS*.

SemanticBase DS is used to describe common properties of semantic entities of any type in narrative world. *Object DS* is a specialized semantic description tool derived from the *SemanticBase DS* to describe a perceivable or abstract object. An object that exists with temporal and spatial properties in a narrative world is a perceivable object. On the other hand, an abstract object is the result of applying abstraction to a perceivable object (e.g. any car) [5].

For the purpose of the video data model it is sufficient to be able to name an object, and to define its spatial and temporal properties. The label element of *Object DS* inherited from the abstract *Semantic Base DS* is suitable to name an object. Spatio-temporal properties of objects are not described with *Object DS*, and the description tools used to define these features of objects are explained in the next sections. The *Object*, and *ObjectRef* elements of Object DS are used to describe or to refer objects resulting from the decomposition of objects into a number of other objects. These elements are not necessary for this video model and a simpler description schema —*avisObject DS*— is defined and used in this study.

In MPEG-7, the *AgentObject DS* extended from the *Object DS* is used to describe a person, an organization (e.g. company), a group of people (e.g. soccer team), or personalized objects (e.g. a talking cup in an animated movie). We use *AgentObject DS* to also define people, companies, and teams on videos. Table 8.2 gives the definitions of elements used in *AgentObject DS*.

| **Name** | **Definition** |
|---|---|
| AgentObjectType | Describes an object that is an agent – a person, an organization, or a group of people in a narrative world. |
| Agent | Describes the agent represented by the object (optional) |
| AgentRef | References the existing description of the agent represented by the object (optional) |

**Table 8.2.** Definitions of elements used in AgentObject DS

The optional *Agent* and *AgentRef* elements of *AgentObject DS* are used to define or to refer an agent representing the object. (In MPEG-7, there is an abstract description schema namely *Agent DS* to define entities responsible for some action like people, or organizations. Other description schemas, like *Person DS*, *PersonGroup DS*, and *Organization DS*, are derived from *Agent DS*, and used to describe features of agents in detail. For example, *Person DS* describes features of a person like first name, last name, and organization.) *AgentObject DS* is defined to simplify the unnecessary schema.

### 8.4.2 Describing Activities and Events

An activity type may be accepted as the subject of a given frame sequence. Elements and attributes of *SemanticBase DS* are enough to describe the ActivityType entity of AVIS. The only requirement to describe activity types is to name activity types, and to define the relationship between activity types and events. However, *SemanticBase DS* is an abstract DS, and abstract description schemas cannot be instantiated on its own. That's why, in this study, a new description schema namely avis *Activity DS* is defined based on the *SemanticBase DS* to describe activities on videos.

Event entities on a video can be described by using Event DS multimedia description schema of MPEG-7 standard. The Event DS is extended from the *SemanticBase DS*, and used to describe a perceivable or abstract event. A perceivable event is a dynamic relation involving one or more objects occurring in a region in time and space of a narrative world [5], (e.g., John is driving the car). An abstract event is the result of applying abstraction to a perceivable event [5], (e.g., anyone is driving the car).

The *Event* and *EventRef* elements of *Event DS* are used to describe or to refer events resulting from the decomposition of the parent event into a number of other events. *SemanticPlace* and *SemanticTime* elements are used to define the place and time of an event semantically. For this study, these four elements are not necessary, and they are not used. That's why, a new and simple description schema namely *avisEvent DS* is defined, and only perceivable events are handled with this DS. The type attribute of *avisEvent DS* determines whether it is perceivable or not, and the default value of this attribute is "perceivable".

*SemanticTime* element of *Event DS* can be seen as a good candidate to define temporal properties of events. However; *SemanticTime* element describes occurrence time of an event semantically, and its type is based on the *SemanticBase DS*. *SemanticTime* does not let you to define the frame sequences of an event with frame rate time units. For the event "Ronaldo has scored a goal on the first half time of the match", "first half time of the match" is the *SemanticTime*. The temporal properties of events are described by using *VideoSegment DS*, and *Relation DS* is used to relate *avisEvent DSs* to those *VideoSegment DSs*.

*AvisActivity* elements are defined under the root element to describe activity type entities on videos. In XML Schema Language, each element has a simple or complex type. *AvisActivity* element has a complex type extended from the *avisActivity DS*. In this extended complex type, *AvisEvent* elements in *avisEvent DS* complex type are defined under the *avisActivity* element in order to relate events to their activity types. *VideoSegment* elements in *VideoSegment DS* complex type that describe temporal properties of events are also defined under the *avisActivity* element. This hierarchical arrangement of elements is useful to query XML descriptions in an efficient and effective way. Definition of *avisActivity* element is given in Figure 8.1.

```
<!--
################################################## -
->
<!-- Definition of avisActivity Element
-->
<!--
################################################## -
->
<element name = "avisActivity">
  <complexType>
    <complexContent>
        <extension
base="mpeg7:avisActivityType"
                <sequence maxOccurs="unbounded">
            <element name="videoSegment"
            <type="mpeg7:VideoSegmentType"
            minOccurs="0"/>
            <element name="avisEvent"
            <type="mpeg7:VideoEventType"
            minOccurs="0"/>
                </sequence>
            </extension>
    </complexContent>
  </complexType>
</element>
```

**Figure 8.1.** Definition of AvisActivity element.

### 8.4.3 Describing Players of Events

In MPEG-7, there exist semantic relation tools to describe relations among semantic entities, between semantic relations and semantic states, between segments and semantic entities, or between analytic models and semantic entities. Figure 8.2 shows the organization of these semantic relation tools. *SemanticRelationBase DS* and *SemanticBaseRelation DS* are the abstract types from which all other tools describing relations between semantic entities are derived. These tools are derived from abstract tools *Relation DS* and *RelationBase DS.*

The semantic relation tool "*ObjectEventRelation DS*" is suitable for defining relations between events and objects (role and player pairs of events). It is extended from the *SemanticBaseRelation DS*, and used to describe a relation between one object and one event corresponding to one of the following relations: $agentOf$, $patientOf$, $experiencerOf$, $stimulusOf$, $causerOf$, $locationOf$, $sourceOf$, $destinationOf$, $beneficiaryOf$, $themeOf$, $resultOf$, $causeOf$, $instrumentOf$, $pathOf$, and $accompanierOf$; and the inverse relations when applicable.

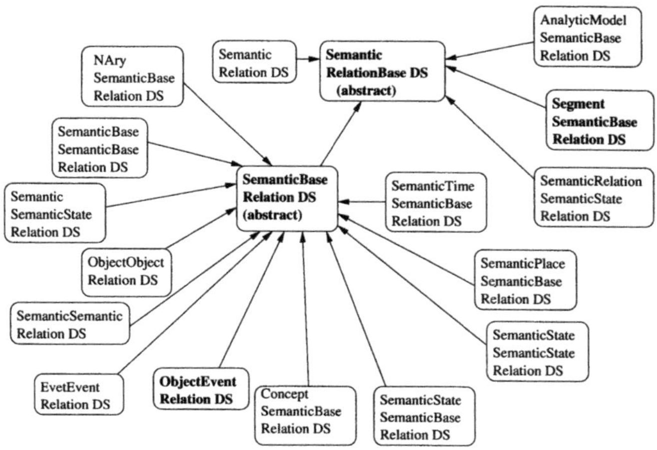


**Figure 8.2.** Organization of semantic relation tools.

The *relation* element of *avisEvent* element inherited from the *SemanticBase DS* can be used to describe relations between this event and other elements like objects, agent objects, and video segments. When the relation between an event and objects/ agent objects (players of events) is defined with the relation element of *avisEvent* element, the source attribute (the first argument) of the *relation* element becomes the event, the *target* attribute (the second argument) of the *relation* element becomes the object (player), and the *name* attribute of the *relation* element corresponds to the role. Since the source is an event, and target is an object, only the inverse relations, *hasCauserOf, hasAgentOf*, and *hasDestinationOf* are meaningful.

### 8.4.4 Describing Temporal Properties of Objects, Agent Objects and Events

Temporal properties of events, objects, and agent objects can be defined with frame sequences. A frame sequence is the set of contiguous frames containing a semantically important data, like an object or an event [21]. It is possible to describe a frame sequence with *MediaTime DS* of MPEG-7 standard. This description schema is used to specify a time point or a time interval in multimedia data. *MediaTime DS* has two elements; a start time point element to describe start time of a frame sequence and the duration to describe duration of a frame sequence. When only a time point has to be specified, the duration can be omitted. A time point or interval can be described in two different ways with *MediaTime DS*:

- Gregorian date and time may be used. In that case, *MediaTimePoint*, or *MediaRelTimePoint* element is used to describe start time point, and *MediaDuration* element is used to describe duration.
- Time units specified by attributes may be counted. For video segments, these time units may be chosen as the frame rate of them. In that case, *MediaRelIncrTimePoint* element is used to describe the start frame, and *MediaIncrDuration* element is used to describe the duration [5].

In this study, we must use the frame rate of the videos as time units. That's why, frame sequences of entities must be described with *MediaTime DS* by selecting *MediaRelIncrTimePoint* element for describing the start time point, and the *MediaIncrDuration* element for describing the duration of the time interval. While Köprülü et al. [21] describe a frame sequence by start and end time points, in this study, duration of a time interval is used instead of the end time point of the interval.

In the developed video database management system, video groups that include a number of related videos, and enable users to define and query them as a unit are handled. Entities like activities, objects, and agent objects may be common for some videos of video groups. In this case, the frame sequence of any entity makes sense with the video id of the video that includes this frame sequence. For this reason, a new description schema namely *avisMediaTime DS* is extended from *MediaTime DS* to include a new attribute, videoid.

In the developed system, to define temporal properties of events, *VideoSegment DS* of MPEG-7 standard is used. *VideoSegment DS* is extended from the abstract *Segment DS*. The *Segment DS* describes a fragment or segment of multimedia content like an image, a partial region in an image, or a moving region in a video, and defines the common properties of segments that apply to the specialized *Segment DSs* like *VideoSegment DS*, *StillRegion DS*, and *MovingRegion DS* [5]. The *VideoSegment DS* is used to describe a temporal interval or a segment of video data. An arbitrary sequence of frames, a single frame, or even the full video sequence can be given as an example for video segments.

The type of *MediaTime* element of the *VideoSegment DS* is changed from *MediaTime* type to *avisMediaTime* type to be able to support video groups. As stated previously, *VideoSegment* elements are defined as child elements of *avisActivity* elements like *avisEvent* elements. The relationship between events and video segments is defined with *SegmentSemanticBaseRelation DS*. This description schema is extended from *SemanticRelationBase DS*, and used to describe a relation between one segment and one semantic entity. The relation element of the *avisEvent* element inherited from *SemanticBase DS* is used to describe the relationship between video segments and events. In that case, the source attribute (the first argument) of the relation element becomes the event, the target attribute (the second argument) of the relation element becomes the video segment, and the name attribute of the relation element becomes *hasMediaPerceptionof*. An example XML document in Figure 8.3

shows the description of an *avisActivity* element, its *avisEvent* elements, and their temporal properties.

The temporal properties of objects/ agent objects are described together with their spatial properties by the help of *StillRegion DS*, and *MovingRegion DS* that are explained in the following section. These description schemas have elements in *avisMediaTime* type to describe frame sequences of objects/ agent objects.

### 8.4.5 Describing Spatio-temporal Properties of Objects, and Agent Objects

The spatio-temporal properties of objects/agent objects are described by the help of *StillRegion*, and *MovingRegion* description schemas of MPEG-7 standard. The *StillRegion DS* is extended from the *Segment DS*, and is used to describe a two dimensional spatial region of an image or a video frame, which can correspond to an arbitrary set of pixels, a single pixel, or even the full image or a video frame [5].

The spatial property of an object is defined as "a tuple $(R, I)$, where, $R$ is a rectangular area which covers all area in which object $A$ appears during the time interval I" in the video model of the study in [21]. This definition implies that the object may be located at anywhere in $R$ during the time interval $I$. However, *StillRegion DS* of MPEG-7 standard defines the spatial property of an object/agent object at a time point not at a time interval. It gives the exact location of the object/agent object at that time point. In this study, a new description schema namely *avisStillRegion DS* is defined to represent the spatial property of an object/agent object in a time interval not at a time point. The difference between *StillRegion DS* and *avisStillRegion DS* is that *avisStillRegion DS* contains a *MediaTime* element in *avisMediaTime* type to represent the time interval of the still region. On the other hand, *StillRegion DS* contains a time point element in *MediaTimePoint*, *MediaRelTimePoint*, or *MediaRelIncrTimePoint* type to represent the time point of the still region. In addition, *StillRegion DS* contains *VisualDescriptors*, *VisualDescriptionSchema*, and *GridLayoutDescriptors* elements that are used to describe visual properties of still regions or their subregions like color, shape, etc. Since visual properties of objects are not handled in this study, *avisStillRegion DS* does not include these elements.

*MovingRegion DS* extended from *Segment DS* is another way to describe the spatio-temporal property of an object/agent object. *MovingRegion DS* of MPEG-7 standard is used to describe a 2D spatio-temporal region of video data, which can correspond to an arbitrary set of pixels from an arbitrary sequence of video frames, one pixel of one video frame, a full video frame, or even the full video sequence. The difference between *avisStillRegion DS* and *MovingRegion DS* is that *MovingRegion DS* describes movements of reference regions during a time interval, and it can be used to describe the set of regions of an object/agent object at different time points during a time interval as

```
<avisActivity id="A1">
  <Label>
      <Name> fuze atma </Name>
  </Label>
  <videoSegment id= "VS1">
      <MediaTime videoid= "1">

    <MediaRelIncrTimePoint> 125 </MediaRelIncrTimePoint>
          <MediaIncrDuration> 5 </MediaIncrDuration>

      </MediaTime>
  </videoSegment>
  <avisEvent>
     <Relation
xsi:type= "SegmentSemanticBaseRelationType"
   name= "HasMediaPerceptionOf" target=<mpeg7:#VS1"/>
     <Relation xsi:type= "ObjectEventRelationType"
name= "HasAgentOf" target= "#ucak"/>
   </avisEvent>
   <videoSegment id="VS2">
     <MediaTime videoid="2">

   <MediaRelIncrTimePoint>   50 </MediaRelncrTimePoint>
            <MediaIncrDuration>  3  </MediaIncrDuration>
     </MediaTime>
   </videoSegment>
   <avisEvent>
      <Relation
 xsi:type= "SegmentSemanticBaseRelationType"
   name= "HasMediaPerceptionOf" target= <mpeg7:#VS2" />
      <Relation xsi:type= "ObjectEventRelationType"
 name= "HasAgentOf" target= "#ucak"/>
   </avisEvent>
   </avisActivity>
```

**Figure 8.3.** Example activity and event description.

a single moving region, however; *avisStillRegion DS* describes a region in which object/agent object appears motionless during a time interval. The most important element of *MovingRegion DS* is the *SpatioTemporalLocator* element in *SpatioTemporalLocator* type. *SpatioTemporalLocator* is a visual descriptor of MPEG-7 standard that defines spatio-temporal regions in a video sequence. It describes moving regions in multiple frames by one or several sets of a reference region and its motion.

In the developed system, spatio-temporal properties of objects and agent objects are kept with object/agent object descriptions. *AavisObject* and *avisAgentObject* elements are defined in a complex type extended from the *avisObject* and *avisAgentObject DS.* The *StillRegion* and *MovingRegion* elements are defined under the *avisObject* and *avisAgentObject* elements.

For an XML document in XML Schema Language, a root element is necessary. In this study, root element namely mpeg7Video is designed for efficient retrieval of query results. The definition of root element is given in Figure 8.4. Filename attribute of *mpeg7Video* element is used to store filename of a video file described, and a video group attribute is used to define whether the MPEG-7 descriptions belong to a single video or to a number of related videos (video group).

```
<!--##########################################################
>
<!-- Definition of root element
-->
<!--##########################################################
>
<element name= "mpeg7Video" type= "mpeg7:MPEG7VideoType" >

<complexType name=    "\MPEG7VideoType">
<sequence minOccurs="0">
    <sequence minOccurs="0" maxoccurs="unbounded">
        <choice>
            <element name= "avisObject" \>
             <element name="avisAgentObject" >
        </choice>
    </sequence>
    <sequence minOccurs="0" maxoccurs="unbounded">
        <element name= "avisActivity">
    </sequence>
</sequence>
<attribute name=  "filename" type="string"
use="required/>
<attribute name=  "videogroup"  type="boolean"
default=\false"/>
</complexType>
```

**Figure 8.4.** Definition of root element.

## 8.5 A Video Database Management System Based on MPEG-7 Standard

In this study, a web-based video database management system based on the MPEG-7 standard is developed. In the developed system, the semantic content of video data including object, agent object, activity, event entities, spatial, temporal, and spatio-temporal properties of these entities are described using MPEG-7 standard. MPEG-7 multimedia and visual description tools can be used to describe not only the semantic content of video data but also different aspects of video, and other multimedia data.
The features of this system are as follows:

- Semantic content of video data is described using MPEG-7 standard.
- This system supports spatial, temporal, and spatio-temporal queries on the semantic content of video data. In spatial queries, spatial relationships between objects/agent objects, and spatial properties of objects/agent objects (locations of objects/agent objects on certain frame intervals), are queried. In temporal queries, temporal relationships between events are queried. In spatio-temporal queries, trajectories and directional trajectories of objects/ agent objects are queried.
- Fuzziness is included in spatial and spatio-temporal queries. The spatial properties of objects/ agent objects, the spatial relationships between objects/agent objects, and trajectories of objects/agent objects are given with certainty levels.
- The only data structure used in the prototype system is the tree-like XML Document Object Model (DOM).
- All MPEG-7 descriptions of video data produced by the system are validated against the MPEG-7 XML schema.
- The prototype system is a web-based dynamic HTML (DHTML) application. Every user of this video database management system has a userid and a password. Any user registers or signs on the system, and defines, updates or queries videos in the database.
- The authorization information about users, MPEG-7 descriptions of videos and raw video data are kept in a database. MPEG-7 descriptions are stored in the database as XML in a text large object columns. Raw video data is kept in the database as a BLOB (binary large object).
- MPEG-7 descriptions (XML documents) are queried using the XPath language. XPath is a general-purpose query notation for addressing and filtering the elements and text of an XML document.
- All queries can be performed on multiple videos. The system enables users to define related videos as a group so that entities on these videos can be shared, the semantic contents of these videos can be described, accessed, and queried like a single video.

In this section, the overall architecture of the prototype system developed by this study is explained, usage and some of the implementation details of the system are given, and query types and query mechanisms are described.

### 8.5.1 Overview of the System

A DHTML (Dynamic HTML) application is developed as the prototype system of this study. A DHTML application is a set of additions to HTML (Hyper Text Markup Language) that permit page authors and developers to change the style and attributes of elements on an HTML page dynamically. In addition, users can insert, delete, or modify elements and the text after a page is loaded. DHTML applications reside on the browser, interpret and respond to actions. Users interact with the browser by clicking a button to load an image or passing the mouse over a certain part of the page. Since the majority of processing is performed on the client with DHTML, the application can respond to user actions quickly without making time-consuming trips to the server [32].

In the developed system, users download the DHTML application from the server, and then perform video entry, video update and video query operations on the client machine using this application. Applications connect to the database on the server to load XML documents containing MPEG-7 descriptions, or to query over these descriptions. The architecture of the system is given in Figure 8.5, and the details of the architecture are explained below.

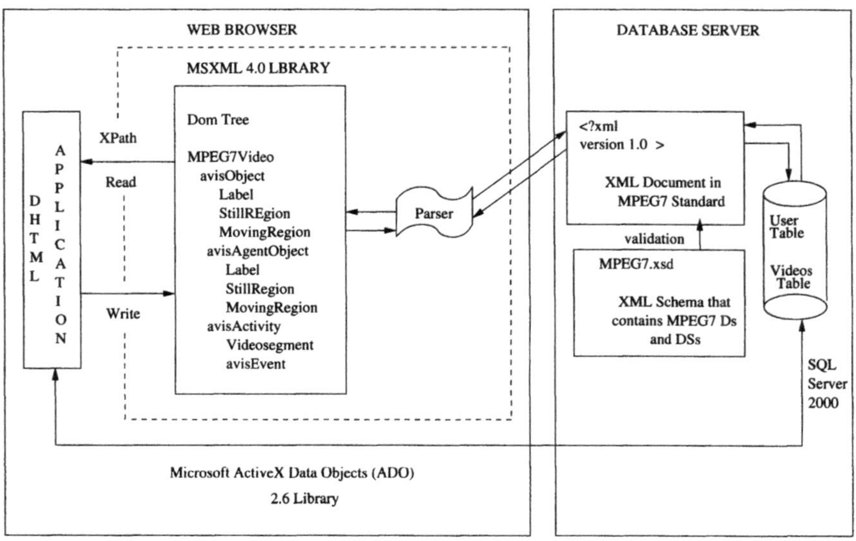


**Figure 8.5.** The Architecture of the system

### System Description

The DHTML application works using a web browser interface. A user registers or signs on the system. At that time, a connection to the database on the server is done to check his/her authorization information. If authorized the user may perform video entry, video update, and video query operations on the system. The following processing is performed in the web browser during video entry and video update operations:

- Video entities (objects, agent objects, activities, and events), and their spatial, temporal, and spatio-temporal properties (video segments, still regions, and moving regions) are entered by the help of the GUI (Graphical User Interface) of the application.
- These entries are stored in an XML DOM (Document Object Model) structure using the MSXML Parser 4.0 library (MSXML 4.0).
- MPEG-7 descriptions stored in the DOM structure are stored in the database on the server when the user wants to save descriptions after video entry or update processing. Validation of the descriptions is performed at that time.

MPEG-7 descriptions may be encoded in two ways: with binary representation and with textual representation. In this study, the textual representation is selected. When MPEG-7 descriptions are encoded textually, the DDL (Description Definition Language) is used. DDL syntax is based on the XML Schema Language of W3C (World Wide Web Consortium). XML schema language defines the structure of XML documents containing MPEG-7 descriptions. XML parsers supporting XML Schema Definition (XSD) are used to read these XML documents and provide access to their content and structure. MSXML 4.0 enables XSD support. It is also a validating XML parser that can compare the XML document containing MPEG-7 descriptions to the corresponding MPEG-7 XML schema, and reports errors [33].

MSXML 4.0 library provides two different application-programming interfaces (API) to parse an XML document: Document Object Model (DOM), and Simple API for XML (SAX). The DOM creates an object tree that is managed by the MSXML parser. Developers can use the logic built into MSXML for managing XML content, rather than having to create their own. On the other hand, the SAX parser reads an XML document and generates events based on specific symbols that it encounters in the document. The SAX parser does not create a tree structure in memory for the document. Developers who need to build applications for specific purposes may find it more efficient to use SAX to read the document and then manage the information in their internal data structures [33]. For the following reasons, the DOM approach is more suitable for our application domain:

- SAX is designed for reading XML documents, not writing. DOM is a better choice for modifying an XML document and saving the changed document to memory.

- To perform complex XML Path Language (XPath) filtering and retain complex data structures that hold context information, DOM provides an enhanced tree structure of that retains this context information automatically. With SAX, the context data structures to store the information must be created programmatically.
- Random access to information is crucial in a database application and DOM provides data structures that facilitate this type of access.
- SAX support is not built into Microsoft Internet Explorer.

In the system, MSXML parser reads an XML document containing MPEG-7 descriptions, and parses its contents into a set of abstract information containers called nodes. When the MSXML parser loads an XML document into DOM, it reads it from start to finish and creates a logical model of nodes from the structures and content within the XML document. The document itself is considered a single node that contains all of the other nodes, including a node representing the root element, which, in turn, contains all of the elements, attributes, and text nodes in the document.
Processing a video query in our system is done as follows:

- The user selects a video/video group.
- The XML document containing MPEG-7 descriptions for the selected video/video group is taken from the database, it is parsed by the help of MSXML 4.0 library, and a DOM tree is constructed.
- The DOM tree is traversed to present all objects, agent objects, activities, and events to the user in the GUI of the query page of the system so that he/she can perform queries.
- XML Path Language, which is a general-purpose query notation for addressing and filtering the elements and text of XML documents, is used to perform the queries. The system implements XPath as described in the XML Path Language (XPath) Version 1.0 Recommendation by the W3C.

### Query Types and XPath Expressions of the System

In our prototype system, all queries are performed using XPath expressions. XPath expressions are applied to the nodes of DOM trees that store XML documents containing MPEG-7 descriptions. XPath is implemented through the DOM extensions *selectNodes* and *selectSingleNode* in the system. *SelectNodes* extension applies the specified XPath expression to any node in the DOM tree and returns the list of matching nodes as a node list. Similarly, the *SelectSingleNode* extension applies the specified XPath expression to any node in the DOM tree and returns the first matching node. Context node of an expression gives the node of the DOM tree to which this XPath expression is applied.

In XPath, it is possible to retrieve a specific node within a set of nodes by indexing. Because collections returned by XPath queries preserve document order. In the system objects, agent objects, and activities are stored in the order that they are defined by the user so that they can be reached using their entry sequence numbers to the system with the indexing property of XPath expressions. For example, if "plane" object is entered to the system as the second object, this object can be reached with indexing on XPath expressions like "avisObject" [2] instead of using its name element or id attribute. In the implemented system, this property of the XPath language is used to make queries more efficient. The following sections include only some of the query types supported in the system, and give XPath expressions to perform those query types.

*"From Object/Agent Object to Frame Intervals" Type Queries*

In this query type, an object/agent object is given, and frame intervals in which this object/agent object appears are retrieved. An example to this type of query is: *"Find frame intervals in which a computer engineer is seen"*

*XPath expression 1:*
"mpeg7:mpeg7Video/avisObject[*object-index*]/StillRegion/MediaTime
mpeg7:mpeg7Video/avisObject[*object-index*]/
MovingRegion/SpatioTemporalLocator/FigureTrajectory/Time"
*Context node of the Expression 1*: root element

*XPath Expression 2:*
"mpeg7:mpeg7Video/avisAgentObject[*agentobject-index*]/
StillRegion/MediaTime
mpeg7:mpeg7Video/avisAgentObject[*agentobject-index*]/
MovingRegion/ SpatioTemporalLocator/FigureTrajectory/Time"

*Context node of the Expression 2*: root element
Objects and agent objects are described with *avisObject*, and *avisAgentObject* elements in the system, and spatio-temporal properties of them are described with *StillRegion* and *MovingRegion* elements. *MediaTime* element of *StillRegion* element gives a frame interval during which the object/agent object appears on a still region. In the system, *FigureTrajectory* element of *MovingRegion* element is used to describe the moving rectangular regions of an object/agent object during a frame interval. Time element of *FigureTrajectory* element keeps this frame interval. That's why, to find frame intervals of objects/agent objects, we need to reach *MediaTime* elements of their *StillRegion* elements and *Time* elements of their *MovingRegion* elements. That's why, XPath expressions 1 and 2 are used to perform this query type. As it is seen from expressions, the object/agent object is reached with indexing using of its entry sequence number to the system.

*"From Activities to Frame Intervals" Type Queries*

In this query type, an activity is given, and frame intervals, in which this activity is performed, are retrieved. An example to this type of query is *"Find frame intervals in which someone is driving a car"*

*XPath expression 3:*
"mpeg7:mpeg7Video/avisActivity[*activity-index*]/
videoSegment/MediaTime"

*Context node of the Expression 3*: root element
Activities are described with *avisActivity* elements in the system. *VideoSegment* elements of *avisActivity* element contain video segments in which an instance of this activity (an event) is occurred, and *MediaTime* elements of *VideoSegment* elements give the frame intervals during which this activity is performed. That's why, XPath expression 3 is used to perform this query type. As it is seen from the expression, the activity is reached with indexing using its entry sequence number to the system.

*"From Events to Frame Intervals" Type Queries*

In this query type, an event is given (events are given with their activity type, and role-player pairs), and frame intervals in which this event is occurred are retrieved. An example to this type of query is *"Find frame intervals in which Nigar is playing a car"*

*XPath expression 4:*
"mpeg7:mpeg7Video/avisActivity[Label[Name="activity-name"]]"
*Context node of the Expression 4*: root element

*XPath expression 5:* "videoSegment"
*Context node of the Expression 5:* avisActivity element

*XPath expression 6:*
"avisEvent[Relation[@xsi:type= *"SegmentSemanticBaseRelationType"* and
@target= *"mpeg7:videosegment-id"*] and
Relation[@xsi:type= *"ObjectEventRelationType"* and
@name = *"role1-name"* and @target= *"player1-name"*] and
Relation[@xsi:type= *"ObjectEventRelationType"* and
@name = *"role2-name"* and @target= *"player2-name"*] and
Relation[@xsi:type= *"ObjectEventRelationType"* and
@name = *"roleN-name"* and @target= *"playerN-name"*] ]"

*Context node of the Expression 6:* avisActivity element
Events are described with *avisEvent* elements in the system. In this query, events are given with their activity types, and role-player pairs. *AvisEvent* elements and their temporal properties — *VideoSegment* elements— are defined as child elements of an *avisActivity* element. That's why, to reach an *avisEvent*

element, its parent *avisActivity* element must be found. Since the activity of the event is given with its name, indexing can't be used on XPath expressions. However, the Label element of *avisActivity* element can be used to reach the activity with its name. XPath Expression 4 is used for this purpose. For each *VideoSegment* element of the *avisActivity* element, a relation element whose type attribute is *SegmentSemanticBaseRelationType* is defined as the child element of its corresponding *avisEvent* element, and target attribute of this relation element keeps the id of the *VideoSegment* element. XPath expression 5 is used to find video segment ids of all VideoSegments for the *avisActivity* element found with XPath expression 4. *AvisEvent* elements also have relation elements whose type attributes are *ObjectEventRelationType*, and these relation elements keep role-player pairs for the event. XPath Expression 6 can be used to reach *avisEvent* elements with their corresponding video segment ids and role-player pairs. When the role-player pairs are matched, *MediaTime* element of the corresponding *VideoSegment* element gives one of the frame intervals of the given event.

*"Find Trajectory" Type Queries*

In this query, an object/agent object, two rectangular regions namely $R1$ and $R2$, and a threshold value are given, and the trajectories of the given object that start at region $R1$, and end at region $R2$ is found. In the query, start and end regions of the resulting trajectories and the given regions $R1$ and $R2$ are not equal exactly. These equalities can be given with a certainty level. The given threshold value in the query determines the requested certainty level by the user. An example to this type of query is *"Find trajectories of the plane starting from the rectangular region* $R1$ *given by pixel values* $[50, 60, 50, 120]$ *and ending at rectangular region* $R2$ *given by pixel values* $[80, 90, 80, 150]$ *with a threshold value of 0.5"*

*XPath expression 7:*
"mpeg7:mpeg7Video/avisObject[*object-index*]/MovingRegion/
SpatioTemporalLocator/FigureTrajectory"
*Context node of the Expression 7*: root element

*XPath expression 8:*
"mpeg7:mpeg7Video/avisAgentObject[agentobject-index]/
MovingRegion/
"SpatioTemporalLocator/FigureTrajectory"
*Context node of the Expression 8:* root element

*XPath expression 9:* "vertices[0]/KeyValue"

*Context node of the Expression 9:* FigureTrajectory element

*XPath expression 10:* "vertices[1]/KeyValue"
*Context node of the Expression 10:* FigureTrajectory element

*XPath expression 11:* "vertices[2]/KeyValue"
*Context node of the Expression 11:* FigureTrajectory element

*XPath expression 12:* "vertices[3]/KeyValue"
*Context node of the Expression 12:* FigureTrajectory element

*XPath expression 13:* "vertices[0]/KeyTimePoint/
MediaRelIncrTimePoint"
*Context node of the Expression 13:* FigureTrajectory element

The processing of this query type can be described algorithmically as follows:

- *Empty resulting list of trajectories*
- *Empty resulting trajectory*
- *Find all FigureTrajectories of the object/agent object using XPath Expression 19/20*
- *For all FigureTrajectories of the object/agent object*
  - *For all reference regions of FigureTrajectory*
    - *If resulting trajectory is empty*
      - *Compare the reference region with the given region R1*
      - *If two regions are equal with the given threshold*
        *- Add the time point of this reference region to the resulting trajectory*
    - *Else*
      - *Compare the reference region with the given region R2*
      - *If two regions are equal with the given threshold*
        *- Add the time point of this reference region to the resulting trajectory*
        *- Add resulting trajectory to the resulting list of trajectories*
        *- Empty resulting trajectory*

*MovingRegion* elements of objects/agent objects are used to find trajectories of them. Since a *StillRegion* element of an object/agent object represents a region in which this object/agent object appears motionless during a frame interval, *StillRegion* elements cannot be used for trajectory searching.

The *FigureTrajectory* element of a *MovingRegion* element keeps a number of reference regions that show the location of an object/agent object at different time points during a frame interval. XPath Expressions 7 to 12 can be used to find coordinates of these reference regions, and XPath expression 13 can be used to find time points of these reference regions. At the end of the query processing, the resulting list of trajectories keeps the frame intervals of the found trajectories. If the given threshold in the query is less than or equal to the calculated certainty level, regions are assumed to be equal.

*"Events and Temporal Relation to Frame Intervals" Type Queries*

In this query, two events, and a temporal relation (*before*, *during*, *starts*, *finishes*, *meets*, *overlaps*, or *equals*) are given, and the frame intervals during which two events are occurring with the given temporal relation is retrieved. An example to this type of query is *"Find frame intervals in which Ali and Hakan are fighting before Hakan has died"*

Similarly to the *"From Events to Frame Intervals"* type queries, events and their temporal properties are located using XPath Expressions 4, 5, and 6. The temporal properties of events are given with frame intervals. The temporal relationships between events are calculated using the Allen's Temporal Interval Algebra [2], which is used in this query type and shown in Table 8.3.

| Relation | Symbol | Inverse | Meaning |
|---|---|---|---|
| A before B | b | bi | AAA BBB |
| A meets B | m | mi | AAABBB |
| A overlaps B | o | oi | AAA<br>  BBB |
| A during B | d | di |    AAA<br>BBBBBBBB |
| A starts B | s | si | AAA<br>BBBBBB |
| A finishes B | f | fi |      AAA<br>BBBBBB |
| A equal B | e | e | AAA<br>BBB |

**Table 8.3.** Temporal interval relations.

## Database Server

In the prototype system of this study, the Microsoft SQL Server 2000 is used as the database server, and Microsoft ADO (ActiveX Data Objects) library is used to access database tables on the server. ADO enables a data access interface along with OLE DB data access methodology to communicate with all data sources in a variety of environments. With OLE DB, high performance data access from any data source is allowed. Design of the OLE DB includes consumer and provider concepts. Consumer represents the traditional client, and provider places data into a tabular format and returns it to the consumer (client). For developing 1- to n-tier client/server and Web-based, data-driven solutions, ADO is the single data interface. Since we have a web-based database management system, ADO will be a good choice to access data in the database [34].

### 8.5.2 XML and Databases

The first design decision in considering XML and Databases is whether the data or the document has to be stored in the database, i.e., is the solution data centric or document centric. Data centric documents use XML as a data transport mechanism. For data centric documents, it is not important for the application using that data whether it is stored as XML document for some length of time or not. The order of elements in the XML document is not important for that kind of document. Sales orders, flight schedules, scientific data, and stock quotes are examples. On the other hand, document centric documents like books, e-mail, and advertisements have a regular structure, larger grained data and mixed content. The order of elements is almost always significant.

In general, data centric documents are more suitable to be stored in a traditional database, and document centric documents are more suitable to be stored in a native XML database. A Native XML database is characterized as follows:

- A native XML database defines a logical model for an XML document as opposed to the data in that document, and stores and retrieves documents according to that model. At a minimum, the model must include elements, attributes, PCDATA, and document order.
- Native XML databases have an XML document as its fundamental unit of storage.
- The Native XML database is not required to have any particular underlying physical storage model. For example, it can be built on a relational, hierarchical, or object-oriented database, or use a proprietary storage format such as indexed, compressed files [7].

The XML document of the video database system developed is based on a document-centric approach. Using a native XML database for storage is more natural. However, the XML schema support for native XML databases is still developing. For example, the native XML database TAMINO of the Software AG Company supports XML schema in the latest release of TAMINO 3.1. However, this is a very restricted schema support, and implementing the complex description schemas of MPEG-7 standard with TAMINO is difficult.

Our implementation uses Microsoft SQL Server 2000, which supports XML documents in relational database tables. For extraction XML data, SELECT statement is extended with FOR XML construct. For storing XML data, OPENXML row set function, which is analogous to a table or view, is used. However, storing XML data of video entities such as objects and events in separate relational database tables is not suitable for our implementation. The query language used in this implementation is XPath, and the use of multiple tables would require that SQL queries be used. However, in Yukon, the projected new release of SQL Server 2000 will support an XML data type to store pure XML data on columns of database tables [27, 28]. With the availability of Yukon, the model may be extended to use XML data type. For ease of manipulation, we store XML documents on a column, but as a text large object.

## 8.6 Conclusion and Future Work

A web-based video database management system based on the MPEG-7 standard was developed in this research effort. Semantic video entities (objects, agent objects, activities, and events), temporal, spatial, and spatio-temporal properties of these entities are represented using textual representation language of MPEG-7 (DDL). In summary, the following MPEG-7 description tools and their extended versions are used to describe video data:

- Object DS to describe objects,
- AgentObject DS to describe agent objects,
- SemanticBase DS to describe activities,
- Event DS to describe events,
- VideoSegment DS and SegmentSemanticBaseRelation DS to describe temporal properties of events,
- MediaTime DS to describe frame intervals,
- StillRegion DS and MovingRegion DS to describe spatio-temporal properties of objects/agent objects,
- ObjectEventRelation DS to describe role-player pairs of events.

The query model uses XPath language and supports Temporal, spatial, and spatio-temporal queries. The ability to perform fuzzy spatial and spatio-temporal queries is facilitated. Spatial relationships of objects, and object trajectories are defined with certainty levels (membership values). MPEG-7 descriptions of videos are stored in a database table as XML in text large object column. Querying on groups of videos was also facilitated in the implemented system.

This research could be extended to describe more complex visual and audio features of Audio Visual data. While this system uses manual annotation of the video to accurately define the semantic content of multimedia data, existing and new image processing techniques could be used to provide semi-automatic and automatic annotation. For instance, in the automatic approach, feature extraction techniques could be used to extract features of AV data like color, texture, silence, and shape. In semi-automatic approach, after a feature extraction operation, an authorized user performs some manual entries. These are natural extensions to this research and would provide fertile new areas for research.

*Acknowledgement.* The authors would like to thank Dr. Dana Ulery of CISD, Army Research Laboratory, for her guidance and support in this research effort and Ms. Burcin Bostan for her contributions to this chapter.

## References

1. Adali S, Candan K.S., Chen S., Erol K., Subrahmanian V.S. (1996) The Advanced Video Information System: Data structures and Query Processing. Multimedia Systems 4:172–186.
2. Allen J.F. (1983) Maintaining Knowledge about Temporal Intervals. Communications of ACM 26(11):832–843.
3. Ashok M., Biron P.V. (2001) XML Schema Part 2: Datatypes. W3C Recommendation.
4. Beech D., Maloney M., Mendelsohn N., Thompson H.S. (2001) XML Schema Part 1: Structures. W3C Recommendation.
5. Van Beek P., Benitez A.B., Heuer J., Martinez J., Salembier P., Shibata Y., Smith J.R., Walker T. (2001) Text of 15938-5 FCD Information Technology – Multimedia Content Description Interface – Part 5 Multimedia Description. ISO/IEC JTC 1/SC 29/WG 11 MPEG01/M7009, Singapore.
6. Benitez A.B., Chang S.-F. (2000) Validation Experiments on Structural, Conceptual, Collection, and Access Description Schemas for MPEG-7. In: Digest of IEEE 2000 International Conference on Consumer Electronics (ICCE-2000), Los Angeles, CA, pp. 276–277.
7. Bourret R. (2001) XML and Databases.
   Web Site: http://www.rebourret/XML/XMLAndDatabases.htm.
8. Chua T.-S., Chen L., Lee S., Wang J. (2000) Stratification Approach to Modeling Video. Multimedia Tools and Applications Journal, September.
9. Cieplinski L., Kim M., Ohm J.-R., Pickering M., Yamada A. (2001) Text of ISO/IEC 15938-3/FCD Information Technology – Multimedia Content Description Interface – Part 3 Visual. ISO/IEC JTC1/SC29/WG11 MPEG01/M7009, Singapore.
10. Cover R. (2002) The XML Cover Pages, Moving Picture Experts Group: MPEG-7 Standard. Web Site: http://www.xml.coverpages.org.
11. Day N., Martinez J.M. (2001) Introduction to MPEG-7 (V 3.0). ISO/IEC JTC1/SC29/WG11 N4325, Sydney.
12. Day N. (2001) MPEG-7 Projects and Demos. ISO/IEC JTC1/SC29/WG11 N4034, Singapore.
13. Dayen I. (2001) Storing XML in Relational Databases.
    Web Sites: http://www.xml.com/pub/a/2001/06/20/databases.html?page=1

http://www.xml.com/pub/a/2001/06/20/databases.html?page=2
htp://www.xml.com/pub/a/2001/06/20/databases.html?page=3
http://www.xml.com/pub/a/2001/06/20/databases.html?page=4.
14. Donderler M.E. (2002) Data Modeling and Querying for Video Databases. Bilkent Univ., CS Dept.
15. Fallside D.C. (2000) XML Schema Part 0: Primer. W3C Working Draft.
16. Flynn P. (2001) The XML Frequently Asked Questions.
Web Site: http://www.ucc.ie/xml.
17. Hunter J. (1999) MPEG-7 Behind the Scenes. D-Lib Magazine 5(9).
18. Jiang H., Elmagarmid A.K. (1998) WVTDB – A Semantic Content Based Video Text Database System on the World Wide Web. IEEE Trans. on Knowledge and Data Engineering 10(6):947–966.
19. Jiang H., Montesi D., Elmagarmid A.K. (1997) Video Text Database Systems. In: Proc. of the Fourth IEEE Int. Conf. On Multimedia Computing and Systems, Ottawa, Ontario, Canada, IEEE Comp. Soc. Press, Los Alamitos, CA, pp. 344–351.
20. Jiang H., Elmagarmid A.K. (1998) Spatial and Temporal Content-Based Access To Hypervideo Databases. The VLDB Journal 7:226–238.
21. Koprulu M., Cicekli N., Yazici A. (to appear) Spatio-Temporal Querying in Video Databases. Information Sciences: International Journal.
22. Lalmas M., Moutogianni K., Rolleke T., Mory B., Putz W. Searching Multimedia Data Using MPEG-7 Descriptions in a Broadcast Terminal. NecResearchIndex. Web Site: http://citeseer.nj.nec.com/449117.html.
23. Li J.Z., Özsu M.T., Szafron D. (1997) Modeling of Moving Objects in a Video Database. In: Proc. of IEEE International Conference on Multimedia Computing and Systems, Ottawa, Canada, pp. 336–343.
24. Martnez J.M. (2001) Overview of the MPEG-7 Standard (V 5.0). ISO/IEC JTC1/SC29/WG11 N4031.
25. Oomoto E., Tanaka K. (1993) OVID: Design and Implementations of a Video-Object Database System. IEEE Trans. on Knowledge and Data Engineering 5(4)629–643.
26. Pfeiffer S., Srinivasan U. (2000) TV Anytime As An Application Scenario for MPEG-7. In: Proc. of the 2000 ACM workshop on Multimedia, Los Angeles.
27. Sullivan T., Scannel E. (2001) XML Dominates Database File Formats. InfoWorld.
28. Sullivan T., Scannel E. (2002) Tracking Down the XML Reward. Information-Age.
29. Theodoridis Y., Vazirgiannis M., Sellis T. (1996) Spatio-temporal Indexing for Large Multimedia Applications. In: Proc. IEEE Int. Conf. On Multimedia Computing Systems, Japan, IEEE Comp. Soc. Press, Los Alamitos, CA, pp. 441–448.
30. Weiss R., Duda A., Gifford D. (1994) Content-based Access to Algebraic Video. In: Proc. First IEEE Int. Conf. On Multimedia Computing and Systems, Boston, MA, IEEE Computer Society Press, Los Alamitos, CA, pp. 140–153.
31. Zhong D., Chang S.F. (1998) AMOS: An Active System for MPEG-4 Video Object Segmentation. In: Proc. of Int. Conf. on Image Processing.
32. Microsoft MSDN library (2001) Web Site: http://www.msdn.microsoft.com.
33. MSXML Parser 4.0 SDK. Web Site: http://www.msdn.microsoft.com/downloads
34. Microsoft SQL Server 2000 Online Books. Web Site:
http://www.microsoft.com/sql/techinfo/productdoc/2000/books.asp

# 9

# Modeling Spatial Objects Affected by Uncertainty

Eliseo Clementini

University of L'Aquila
Department of Electrical Engineering
loc. Monteluco di Roio
I-67040 Poggio di Roio - L'Aquila, Italy
eliseo@ing.univaq.it

## 9.1 Introduction

The treatment of uncertainty in spatial data is a problem that has been studied from a theoretical point of view (notably, [1]). The various models have never been developed in enough depth to become operational and, till now, the management of uncertainty in current spatial database systems is not supported. The main problem about spatial objects that are stored in spatial databases is that it is impossible to reason about the degree of indetermination of a result, since the data have lost the information about all the uncertainty that added up after having undertaken several transformations. An adequate modeling of geometric uncertainty in spatial data is essential for the assessment of data quality, and, consequently, for helping any process related to dataset acquisition and integration [12, 13, 17].

The main goal of our approach is to record the information about uncertainty together with the data in order to be able to deal with it at each step of the data life cycle. In essence, a new geometric model is needed that overcomes the limits of the current models of spatial databases, which traditionally are a collection of lines (points, polylines and polygons). Also the recent OpenGIS specification for SQL [14] is based on a geometric model that does not support uncertainty. Many models are about the introduction of broad boundaries replacing crisp ones in the case of regions [3, 6, 9, 11, 16]. The advantage of this approach is that it can be implemented on existing database systems at reasonable cost: the new model can be seen as an extension of existing geometric models. The case of lines has received by far less attention in the literature [2]. Based on the geometric models, then more effort is needed to define new operators applying to the new category of objects in order to use them with profit. To this regard, topological relations between regions with a broad boundary have been defined in [5, 6]. Another advantage of our model is that the different semantics of the uncertainty of spatial objects do not influence the representation and management of the spatial objects themselves.

This chapter is organized as follows. In Section 9.2, we propose an ontology of imperfection (taken from [17]) that is useful for understanding how imperfect observations of geographic reality cause uncertainty in spatial data. In Section 9.3, we propose various semantic interpretations of spatial objects with uncertainty. In Section 9.4, we introduce the geometric model for objects with a broad boundary, recalling the definitions for regions with broad boundary and proposing the definition of uncertain lines. Section 9.5 describes the most widely used models for topological relations (the 9-intersection [10] and the CBM [7]) applied to objects with uncertainty. In Section 9.6, we propose specific models for topological relations between uncertain lines and points. Section 9.7 uses the concept of crisping to find additional topological constraints in the case of aggregated objects, such as tessellations and networks. Section 9.8 draws short conclusions.

## 9.2 Ontology of imperfection

Imperfection is an essential aspect of almost all observations of natural phenomena. This is partly because many observational and measurement devices have some inherent level of imprecision and inaccuracy, but also because many representations of the natural world are inherently vague. In the next few paragraphs, we expand upon these ideas and refine our notions of imperfection through the construction of a simple ontology of imperfection (see Figure 9.1).

**Figure 9.1.** Ontology of imperfection.

Researchers use the words 'inaccuracy' and 'imprecision' in different ways. In our ontology, there is a clear distinction between inaccuracy and imprecision, and in fact these concepts are orthogonal. An *inaccurate* proposition is one that lacks correlation with the actual state of affairs, whereas an *imprecise* proposition lacks specificity. Thus, we might propose that the time of writing this sentence is January 1st in the year 1900, at 1530 in the afternoon – fairly precise, but quite inaccurate. On the other hand, we might record the same time by merely observing that the sentence was written in the year 2000, actually accurate but imprecise.

*Vagueness* is that particular kind of imprecision where there are borderline cases for which it is difficult to decide whether they are covered by the concept or not. Many concepts (Bertrand Russell [15] has argued that all of language) contain vagueness. So, for example, 'tallness' is vague, because there will be people for whom is it difficult to decide whether they are tall or not. However, not all imprecise propositions are vague; For example, we could say that we are located in Europe, which is not a vague statement (assuming for the moment a crisp boundary to Europe), but is quite imprecise.

There are two types of errors that are distinguished in our ontology. The distinction here is similar to the distinction between Type I and Type II error in statistical significance testing. In Type I error, a true null hypothesis is incorrectly rejected, while in Type II error, a false null hypothesis fails to be rejected. A Type II error is only an error in the sense that an opportunity to reject the null hypothesis correctly is lost. It is not an error in the sense that an incorrect inference is made, as no conclusion is drawn when the null hypothesis is not rejected.

By analogy with Type I and II errors in statistical significance testing, we distinguish between an *error of commission* (EC) and an *error of omission* (EO). An error of commission (EC) is made when we conclude that a proposition is definitely the case when in fact it is either undetermined or definitely not the case. An error of omission (EO) is made when we do not declare a proposition to be the case when it is the case. For an example of EC, we may be using a remotely sensed image to determine land use at a location, and despite cloud cover, we wrongly assume the location to be a forest because some neighboring pixels might suggest forest. For an example of EO, we might fail to infer that we were in Switzerland because we judged some scanty evidence based on mountain scenery to be insufficient.

In summary, an ontology has been constructed that contains the following components of imperfection in observations and the data resulting from them. Inaccuracy and imprecision are proper and disjoint subconcepts of imperfection, and vagueness is a proper subconcept of imprecision. Also distinguished are the types of inaccuracy referred to as errors of omission and errors of commission. An error of commission occurs when something is declared to be the case when in fact it is not the case. An error of omission occurs when something is not declared to be the case when in fact it is the case.

## 9.3 Spatial objects with uncertainty

In this section, we will discuss the semantic interpretations of various kinds of spatial objects affected by uncertainty. In the course of the discussion we use the concept *uncertainty* to indicate a "cognitive state induced by imperfect observations and states of knowledge".

### 9.3.1 Regions with broad boundary

In the case when the spatial phenomenon under observation is something as simple as a region, then the 3-valued indeterminacy of location in the region has been represented by several authors as a region with broad boundary (see Figure 9.2), also known as the egg-yolk diagram [3, 6, 9, 11].

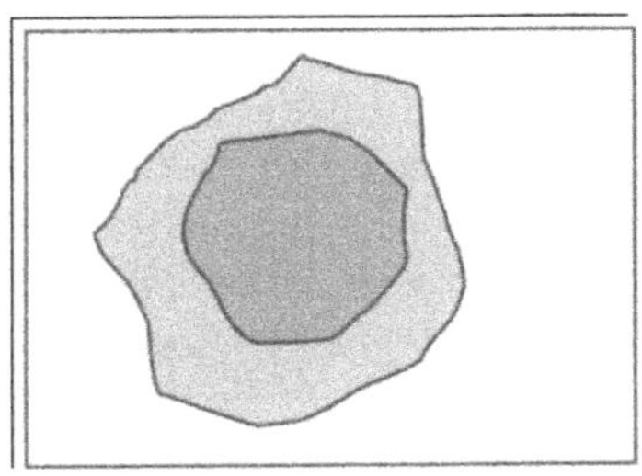

**Figure 9.2.** A region with a broad boundary.

Broad boundaries are a geometric model that can be used to represent uncertainty in many different situations. Therefore, a representation of a spatial phenomenon as a region with broad boundary can have several semantic interpretations (as discussed in [17]):

1. *Incomplete representation of a feature.* In vector databases, if there are missing sides of a polygon or entirely missing polygons, due to omissions in digitization or imperfect data conversion, broad boundaries can be introduced to represent the possible location of the missing lines.
2. *Conflicting representations of a feature.* In the case of cadastral data, or representations of political boundaries (in general, the case of existing data that provide approximations to some real objects), broad boundaries result from the merging of different representation of the same region.
3. *Changing representations of a dynamic spatial phenomenon.* If there are different observations of the same geographic phenomenon taken at different times, the core region may be interpreted as being the collection of locations that are in the region at all times, while the broad boundary may be interpreted as the collection of locations that are in the region at some, but not all, times.

4. *Imprecise observation of a spatial phenomenon.* At a certain resolution (i.e., a partition of the space in which locations indiscernible from the observation are grouped into the same elementary unit of the space), the elementary units impose a granularity of the underlying space. The region with broad boundary represents an imprecise observation of a region at a certain resolution.
5. *Inherently vague representations of real world spatial objects.* Most geographical objects which are not man-made artifacts or conventions fall in this category. They may be represented by fuzzy sets, if a reasonable method of quantifying the membership function for the fuzzy set is available. A region with a broad boundary is interpreted as an approximation in which the broad boundary of the region represents the part of the object where the membership function gradually decreases from 1 to 0. There are two special cases:

   a) *Inherently vague representations of real world spatial objects resulting from scale change.* These are spatial objects that only exist at coarser resolutions (small scale), but are constructed from more finely grained real world objects (large scale). Such objects may be obtained by semantic generalization, aggregating other objects of a different nature that exist at a larger scale: this is the case of urban settlements that are seen as an aggregation of other objects (such as houses, streets, subways, bridges, and so on), or woods that are made up of trees. The broad boundary of such objects represents a peripheral zone where the density of the elements composing the aggregate has an intermediate value.
   b) *Inherently vague representations of real world spatial objects resulting from variation of context.* These correspond to linguistic propositions made up of a geographic location plus a qualitative modifier. The boundary of a region as 'the south of England' is dependent on the context in which the linguistic proposition is placed and its intended use. It might be a different region if the proposition is placed in a historic, architectural, or biological context; finally, it might even depend on different opinions of individuals. Also, the qualitative modifier, like all qualitative spatial terms [8], has a different meaning depending on the granularity with which it is defined: for example, in the two domains for qualitative modifiers {south, center, north} and {south, north}, the proposition the 'south of England' would subsume a rather different underlying region.

### 9.3.2 Uncertain lines

Uncertainty about lines is rather more complicated to model than in the case of regions. In fact, taking a simple line (having the boundary made up of two distinct endpoints), it is not enough to consider a broad boundary instead of a crisp boundary, like in the case of regions. This would take into account a kind of uncertainty only, the one related to the endpoints of the line. We must also consider the uncertainty affecting the interior of the line, which can vary independently of the boundary, and can be modeled by a broad interior. Joining the two contributions, we define the concept of uncertain line, which takes into account both broad boundaries and broad interiors.

To understand the nature of an uncertain line, it is helpful to introduce the concept of a crisping: a crisping of an uncertain line is any line with endpoints and interior that are strictly contained in the uncertain line. Also, un uncertain line can be seen as the set of all possible crispings. Uncertainty affecting an endpoint can be represented by a region (Figure 9.3(a)). Such a region is the set of points such that they can be an endpoint of a crisping. We assume that for each endpoint there is a simple region and call the union of those regions the broad boundary of the uncertain line. This definition of broad boundary of a line has the same semantics of broad boundaries of regions: also for the latter ones, broad boundaries are the set of points that could belong to the boundary of the region's crispings. A problem is that uncertainty on boundary implies some uncertainty on the interior as well, as it is illustrated in Figure 9.3(a). In fact, the crisping of a line starting from an endpoint differs from other crispings also for a piece of the interior.

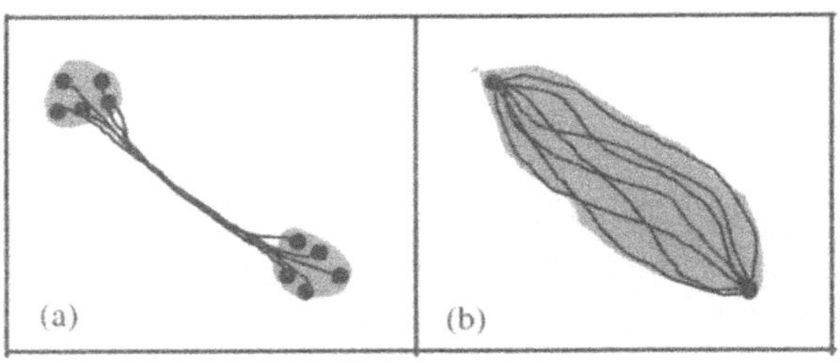


**Figure 9.3.** Broad boundary (a) and broad interior (b) of a simple line.

The second kind is uncertainty affecting the interior, when endpoints are crisp. Such a kind of uncertainty can be represented by the region of all possible crispings connecting the endpoints, as illustrated in Figure 9.3(b). In this region, that we call broad interior, we may find all points that could belong to the interior of a crisping. The concept of broad interior has not a direct correspondence in the case of regions, for which the uncertainty on boundaries takes into account all kinds of uncertainty.

The concept of uncertain line encompasses the two types of uncertainty (see Figure 9.4(a)). We can draw some conclusions about the topology of regions representing broad boundaries and broad interiors. If a point could be part of the boundary of a line's crisping, it means that a neighborhood of that point contains points of the interior of a crisping. This proves that a point of the broad boundary cannot be outside the broad interior. Therefore, the broad boundary is inside the broad interior. A broad interior is always present, even if the uncertainty is only about endpoints. Therefore, the region representing the broad interior is not necessarily a regularly closed set, because in general it may have two-dimensional and one-dimensional parts, as illustrated in Figure 9.4(b). However, it must be connected to be able to contain all possible crispings.

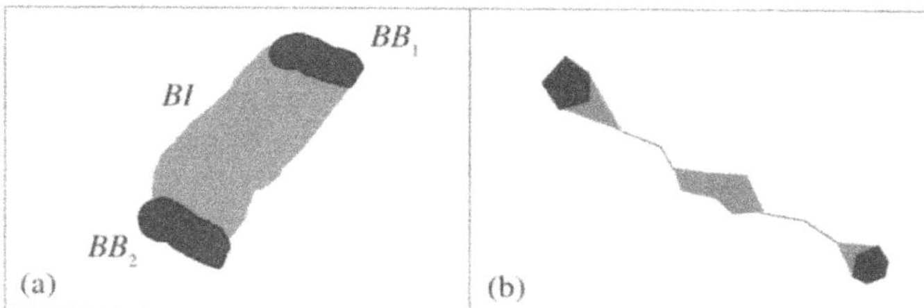


**Figure 9.4.** An uncertain line (a) with its broad interior ($BI$) and broad boundary ($BB_1$ and $BB_2$). An uncertain line whose broad interior has two-dimensional and one-dimensional parts (b).

The following are the possible semantic interpretations of uncertain lines:

1. *Incomplete representation of a line feature.* An example of this case is the situation sometimes arising in vector databases, where in a vector representation of a line feature there might be missing lines or segments of lines, due to omissions in digitization or imperfect data conversion. The broad interior (shaded area in Figure 9.5) can be introduced to represent the possible location of the missing parts. In this case, the phenomena could be represented by crisp concepts, but the observations we have at disposal are not sufficient to be fully specific.

**Figure 9.5.** A broad interior to model an incomplete representation.

2. *Conflicting representations of a line feature.* This might arise in the case of cadastral data, or representations of political boundaries: in general, the case of existing data that provide approximations to some real objects.

There might be both errors of commission and omission depending on which processes took place to obtain the representations. Different logics can be applied during the amalgamation of two resolutions. In the case of ignorance of the origin of the data, the most pessimistic view would need to be adopted (both types of errors in the two resolutions). Figure 9.6 shows an example of this, where a line with broad boundary results from the merging of different positions of the same line. It can be the case of conflicting representations both of the endpoints and of interior of the line. In the case of conflicting representations of the endpoints, the broad boundary can be built by considering the convex hull of various endpoints. In the case of conflicting representations of the interiors, the broad interior is easily built by taking the maximum extent of the lines.

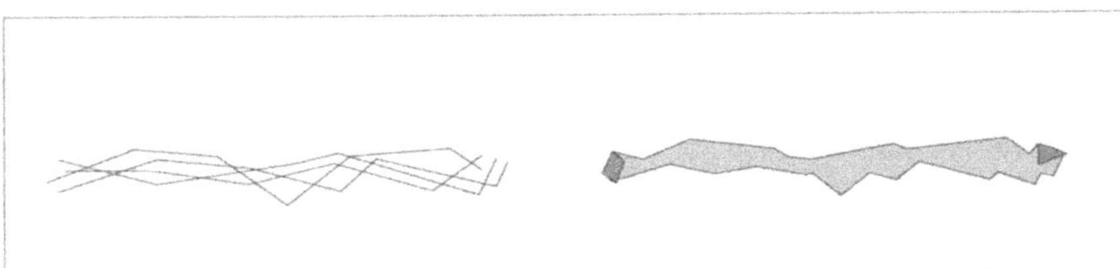

**Figure 9.6.** An uncertain line to model conflicting representations of endpoints and interior.

3. *Changing representations of a dynamic spatial phenomenon.* Variability over time applies when we have different observations of the same geographic phenomenon taken at different times. Given a line that moves its position over times $t_1$ and $t_2$, the method of building the broad boundary and the broad interior of the corresponding uncertain line is the same as in previous case. As another example of this category, a line with a broad interior can represent the set of all possible trajectories of a moving agent starting from point $A$ at time $t_1$ and arriving to point $B$ at time $t_2$ proceeding at a given maximum speed $v_{\max}$ (see Figure 9.7). With $\Delta t = t_2 - t_1$ and indicating with $d$ the distance $\overline{AB}$, the broad interior is bounded by an ellipse of equation:

$$\frac{4x^2}{v_{\max}^2 \cdot \Delta t^2} + \frac{4y^2}{v_{\max}^2 \cdot \Delta t^2 - d^2} = 1.$$

The equation above holds for a reference system having the points $A$ and $B$ on the $x$ axis and equidistant from the origin. It is obtained by considering that the maximum distance that can be travelled by the agent is $v_{\max} \cdot \Delta t$. If the trajectories are straight segments, the moving agent can reach all the points P lying on the ellipse, which, by definition, are the points having constant the sum of distances $\overline{AP} + \overline{PB}$. If the trajectories are curvilinear, they must lie inside the ellipse.

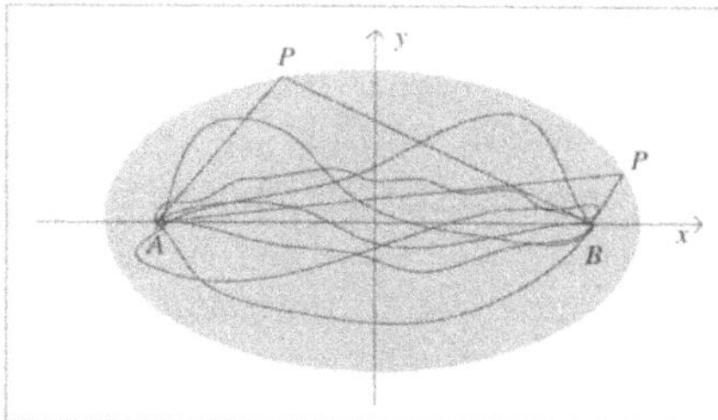


**Figure 9.7.** A line with broad interior to represent possible trajectories of a moving agent.

4. *Inherently vague representations of real world spatial objects.* This case includes line features whose interior is not sharp. Most geographical objects which are not man-made artifacts or conventions fall in this category: for the case of line features, we include in this category rivers, coastal lines, animal paths in the forest, etc. In Figure 9.8, a broad interior is helpful for the representation of a river (and a broad boundary for its endpoint can represent its mouth).

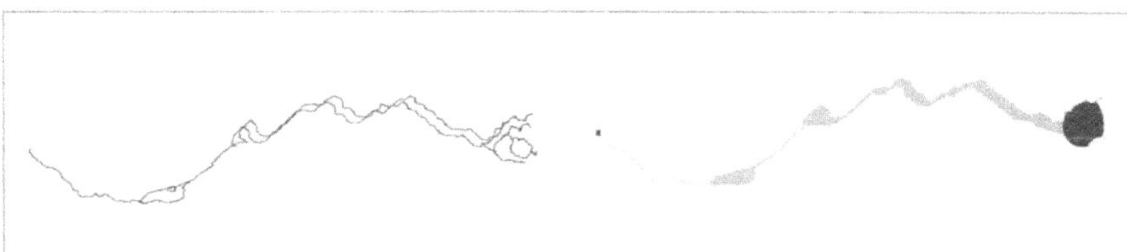

**Figure 9.8.** An uncertain line to model an inherently vague representation of a real world spatial object (for example, a river).

5. *Line features as representations of two-dimensional real world spatial objects.* These are spatial objects that are two-dimensional in reality, but are usually represented as line features, especially at coarser resolutions (small scale). Such lines may be obtained by semantic generalization by a change in dimensionality: for example a street that is represented as a two-dimensional parcel at a large scale and as a line feature at a small scale. These objects have typically a prevalent dimension (roads, highways, railways, utilities lines etc.) and their connection properties are more important than their size and shape: this is why they are usually represented as line features. The proposed model for uncertain lines can be useful in this case for maintaining the two-dimensional representation of the line especially at intermediate scales where the width of the line is still relevant

and, at the same time, for remarking connection properties. To illustrate the concept, let us consider roads that are represented at a certain scale by two-dimensional regions (Figure 9.9): if we apply the models for regions, the topological relations that we are able to distinguish are region-oriented and as such they cannot emphasize the characteristics of lines (connection, crossing, etc.). On the other hand, if roads are represented by uncertain lines, it is possible to represent explicitly the endpoints of the lines and distinguish specific topological relations between two roads (see Section 9.6).

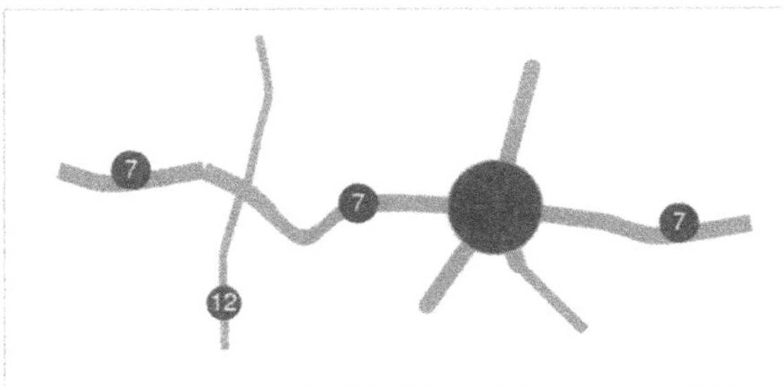


**Figure 9.9.** Uncertain lines as representation of roads: using this model, it is possible to represent connection properties of road 7 with other roads (the roundabout is an endpoint for other roads).

### 9.3.3 Uncertain points

For a point-like spatial phenomenon, the main source of uncertainty is related to the indetermination of its position in space. We call uncertain point the region of all possible positions. Un uncertain point may be represented as an open set that we may assume to be a simply connected region. This definition will allow us to easily reapply the same definitions for topological relations that we had for points, as we will see in Section 9.6.

The semantic interpretations of an uncertain point are the following:

1. *Endpoints of uncertain lines.* Often, the endpoints of uncertain lines can be seen as independent geographic features, like in the previous examples concerning a river, where an endpoint is the mouth of the river, and roads, where endpoints may represent roundabouts and crossroads.
2. *1-dimensional landmarks.* At a small scale, many geographic features are represented by a point (a lighthouse, an archeological site, etc.). The same features at a large scale are represented by regions. At a an intermediate scale, it might be the case that a point-based representation is too imprecise to represent the feature, while a region-based representation looses

the properties of being point-like. A representation based on an uncertain point could be a solution, because it has an extension to take into account imprecision and topological relations for point-like features can be applied to it (see Section 9.6).

3. *Positional uncertainty.* Many point-based geographic features can be known only with some positional uncertainty due to measurement errors (e.g., the peak of a mountain, a wreck on the bottom of sea).

## 9.4 The geometric model of objects with uncertainty

In this section, we give the geometric definitions of various spatial objects affected by uncertainty. We start with regions with a broad boundary that were originally defined in [3, 6]. As preliminary notions, we first define crisp regions and composite regions:

**Definition 1.** A *region* is a regular closed two-dimensional bounded subset of $\mathbb{R}^2$ with a connected interior.

Definition 1 does not exclude the presence of holes. Therefore, the exterior of a region $A$ may be separated in several components: there exist an outer exterior (unbounded set) denoted by $A_0^-$ and $n \geq 0$ inner exteriors (bounded sets) denoted by $A_1^- \dots A_n^-$. Since a region must be a regular closed set, the intersection of the closures of any two different exterior components is empty or equal to a finite set of points. Relaxing the constraint that the interior of a region is connected, we obtain composite regions:

**Definition 2.** A *composite region* is a regular closed two-dimensional bounded subset of $\Re^2$ .

A composite region $A$ may have several interior components denoted by $A_1^\circ \dots A_n^\circ$. Since a composite region must be a regular closed set, the intersection of the closures of any two different interior components is empty or equal to a finite set of points.

Below, we give the definition of regions with a *broad boundary*, which differ from crisp ones with regard to the boundary definition. For a region with a broad boundary, we can define an *inner boundary* and an *outer boundary*, where the inner boundary is surrounded by the outer boundary. The closed annular region comprised between the inner and outer boundary is the broad boundary of the original region.

**Definition 3.** A *region with a broad boundary* $A$ is made up of two regions $A_1$ and $A_2$, with $A_1 \subseteq A_2$, where $\partial A_1$ is the *inner boundary* of $A$ and $\partial A_2$ is the *outer boundary* of $A$.

**Definition 4.** The *broad boundary* $\Delta A$ of a region with a broad boundary $A$ is the closed subset comprised between the inner boundary and the outer boundary of $A$, i.e., $\Delta A = \overline{A_2 - A_1}$, or equivalently $\Delta A = A_2 - A_1^\circ$.

**Definition 5.** Interior, closure, and exterior of a region with a broad boundary $A$ are defined as $A^\circ = A_2 - \Delta A$, $\overline{A} = A^\circ \cup \Delta A$, $A^- = \mathbb{R}^2 - \overline{A}$, respectively.

The interior and exterior of a region with a broad boundary are open sets, while the broad boundary is a closed set. Notice that both the exterior and broad boundary of a region $A$ may have several components, because regions $A_1$ and $A_2$ may have holes. Figure 9.10 illustrates some cases that can arise: case (a) is a region without holes; case (b) is a region with two holes; case (c) is a region where $A_1$ has one hole and $A_2$ has two holes; case (d) is a region where $A_1$ has two holes and $A_2$ has one hole.

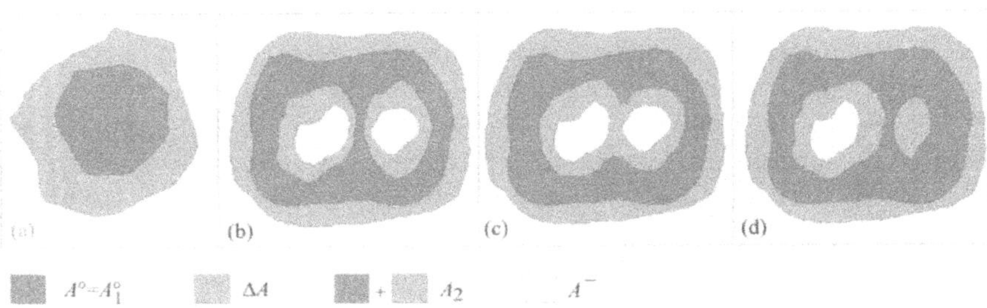


**Figure 9.10.** Regions with a broad boundary.

Finally, we consider the most general case of composite regions with a broad boundary:

**Definition 6.** A *composite region with a broad boundary* $A$ is made up of two composite regions $A_1$ and $A_2$, with $A_1 \subseteq A_2$, where $\partial A_1$ is the *inner boundary* of $A$ and $\partial A_2$ is the *outer boundary* of $A$.

**Definition 7.** The *broad boundary* $\Delta A$ of a composite region with a broad boundary $A$ is the closed subset comprised between the inner boundary and the outer boundary of $A$, i.e., $\Delta A = \overline{A_2 - A_1}$, or equivalently $\Delta A = A_2 - A_1^\circ$.

**Definition 8.** Interior, closure, and exterior of a composite region with a broad boundary $A$ are defined as $A^\circ = A_2 - \Delta A$, $\overline{A} = A^\circ \cup \Delta A$, $A^- = \mathbb{R}^2 - \overline{A}$, respectively.

Figure 9.11 illustrates some configurations of composite regions: case (a) is a region with two components; case (b) is a region where $A_1$ has two components and $A_2$ has one component; case (c) is a region where $A_1$ has one component and $A_2$ has two components.

The initial definitions of lines with uncertainty were given in [5], where authors distinguished two kinds of broad boundaries, modeling either the position of the line or the position of its endpoints. The first kind of broad boundary was an area surrounding the whole line, while the second kind was made up of an area for each endpoint. These definitions are not satisfactory with respect to the topological relations that can be defined over them. A new definition is needed, which takes into account both the uncertainty on endpoints and on the interior of the line. We first define broad interiors and broad boundaries and after that we define the concept of *uncertain line*. A

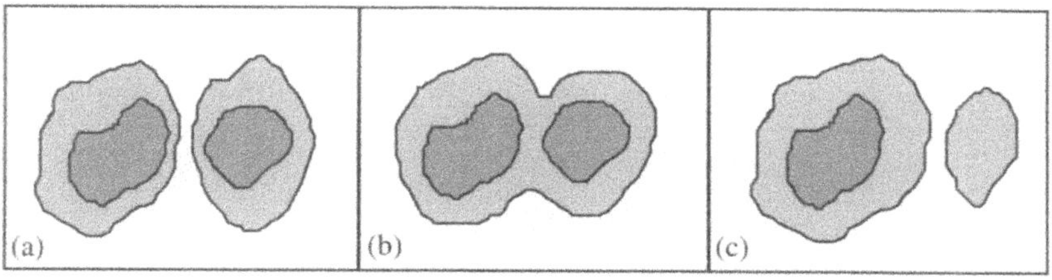


**Figure 9.11.** Composite regions with a broad boundary.

first attempt to give such definitions was given in [2]. As preliminary notions, we recall the definitions of lines without uncertainty.

**Definition 9.** A *simple line* is a closed (non-empty) one-dimensional point-set, $L$, defined as the image of a continuous mapping $f: [0,1] \rightarrow \mathbb{R}^2$, such that $f(t_i) \neq f(t_j)$, $\forall t_i, t_j \in [0,1]$, $t_i \neq t_j$.

The mappings of 0 and 1 through $f$ are the two endpoints of $L$, these two points made up the boundary of a simple line in the plane; i.e., $\partial L = \{f(0),\ f(1)\}$. By relaxing the constraint of no self-intersections in the interior, we get the notion of *line with self-intersections.* Figure 9.12 shows a simple line and four lines with self-intersections. The number of endpoints for a line with self-intersections can be either 2, 1, or 0. In detail, the number of endpoints of L is equal to:

- 2 if $f(0) \notin f((0,1]) \wedge f(1) \notin f([0,1))$; $\partial L = \{f(0),\ f(1)\}$;
- 1 if $f(0) \in f((0,1)) \wedge f(1) \notin f((0,1))$; $\partial L = \{f(1)\}$;
  or if $f(0) \notin f((0,1)) \wedge f(1) \in f((0,1))$; $\partial L = \{f(0)\}$;
- 0 if $(f(0) = f(1)) \vee (f(0) \in f((0,1]) \wedge f(1) \in f([0,1)))$; $\partial L = \emptyset$.

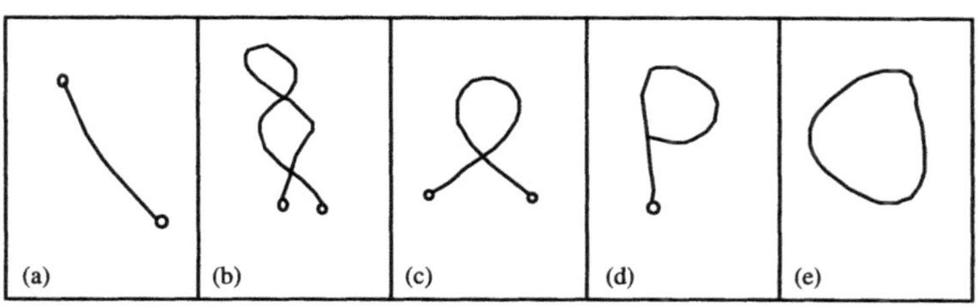


**Figure 9.12.** A simple line (a) and four lines with self-intersections (b-e).

**Definition 10.** Let $f_1, f_2, ...f_n$ be continuous mappings from the interval [0, 1] to the plane. We call *complex line* any closed (non-empty) one-dimensional point-set, $L$, defined as the union of the image of the functions $f_1, f_2, ...f_n$:

$$f_1([0,1]) \cup f_2([0,1]) \cup ... \cup f_n([0,1]) .$$

Figure 9.13 shows a complex line made up of four components. The boundary of $L$, i.e., the set of its endpoints, can be built subtracting from the set of endpoints of single $f_i$, all the endpoints that connect with the lines, that is:

$$\begin{aligned}\partial L = &\{f_1(0), f_1(1), f_2(0), f_2(1), ..., f_n(0), f_n(1)\} - \\ &(\{f_i(0) \mid i \in 1..n,\ f_i(0) \in f_i((0,1])\} \cup \\ &\{f_i(1) \mid i \in 1..n,\ f_i(1) \in f_i([0,1))\} \cup \\ &\{f_i(0) \mid i, \exists j \in 1..n,\ i \neq j,\ f_i(0) \in f_j([0,1])\} \cup \\ &\{f_i(1) \mid i, \exists j \in 1..n,\ i \neq j,\ f_i(1) \in f_j([0,1])\}\ ).\end{aligned}$$

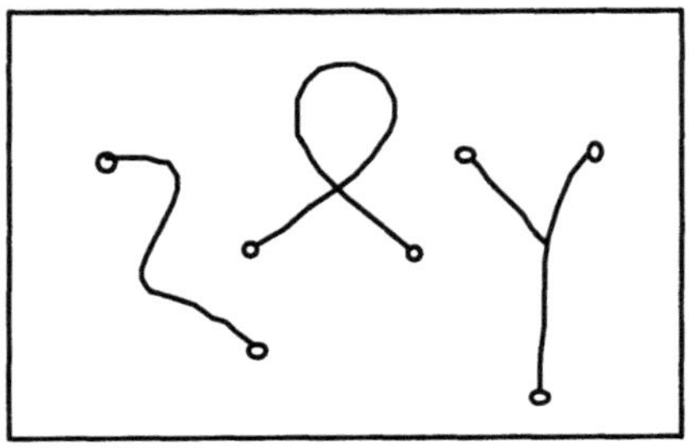

**Figure 9.13.** A complex line.

The closure of a line, $\overline{L}$, is the set of all points of $L$, endpoints included; therefore $\overline{L} = L$. The interior of a line, $L^\circ$, is the set difference between the closure of the line and its boundary: $L^\circ = \overline{L} - \partial L = L - \partial L$. The exterior of a line, $L^-$, is the set difference between the embedding space and the closure of the line: $L^- = (\mathbb{R}^2 - \overline{L}) = (\mathbb{R}^2 - L)$.

Hereafter, we consider the definitions related to lines affected by uncertainty.

**Definition 11.** Given a simple line $L$, whose exact position in $\mathbb{R}^2$ is unknown but can be delimited by a bounded point set, the *broad boundary* of $L$ is the union of two point sets $BB_1(L)$ and $BB_2(L)$ that contain the two endpoints of $L$.

The point sets $BB_1(L)$ and $BB_2(L)$ represent all the possible positions that can be assumed by the two endpoints of the simple line $L$. In theory, these point sets are not required to have any specific topological property, but we assume that they are simple regions (Figure 9.3(a)). Further, $BB_1(L)$ and $BB_2(L)$ could have a nonvoid intersection: in this case, we assume that the two endpoints of $L$ are not coincident, otherwise $L$ would not be a simple line.

**Definition 12.** Given a simple line $L$, whose exact position in $\mathbb{R}^2$ is unknown but can be delimited by a bounded point set, the *broad interior* of $L$ is a point set $BI(L)$ that contains the interior of $L$.

The point set $BI(L)$ represents all the possible positions that can be assumed by the interior of the simple line $L$. The broad interior $BI(L)$ is not necessarily a simple region, since it could contain 1-dimensional parts (Figure 9.4(b)). We assume that $BI(L)$ is a connected set to enforce that it contains all the possible crispings of the line $L$. The broad interior is also neither an open nor a closed set, since we have to exclude the two endpoints of $L$ from its definition.

The two concepts defined so far, broad boundary and broad interior of lines, rarely come separately. For example, a broad boundary implies a broad interior (see Figure 9.3(a)): there is at least a small part of the interior that must connect the endpoint with the main part of the interior, even if we want to consider the latter to be crisp. Therefore, there are parts of space where broad boundary and interior interpenetrate and it is not possible to separate them. We may assume that the broad interior contain the broad boundary of a line.

This brings us to the following definition of uncertain line that encompasses both kinds of uncertainty:

**Definition 13.** Given a simple line $L'$, un *uncertain line* $L$ is made up of a broad boundary $BB_1(L')$ and $BB_2(L')$ and a broad interior $BI(L')$. The *boundary* of $L$, indicated with $\Delta L$, is the union of $BB_1(L')$ and $BB_2(L')$. The *interior* of an uncertain line $L$, indicated by $L^\circ$, is given by the difference $L^\circ = L - \Delta L$.

By the above definition, we enforce that the boundary and interior of an uncertain line are disjoint. This is useful to apply to uncertain lines the models for topological relations that can also be applied to crisp lines, as we will see in Section 9.6.

We extend also the previous definitions to complex lines:

**Definition 14.** Given a complex line $L$, whose exact position in $\mathbb{R}^2$ is unknown but can be delimited by a bounded point set, the *broad boundary* of $L$ is the union of $n$ point sets $BB_1(L)$, $BB_2(L)$, ..., $BB_n(L)$ that contain, respectively, the $n$ endpoints of $L$.

**Definition 15.** Given a complex line $L$, whose exact position in $\mathbb{R}^2$ is unknown but can be delimited by a bounded point set, the *broad interior* of $L$ is a point set $BI(L)$ that contains the interior of $L$.

**Definition 16.** Given a complex line $L'$, un *uncertain complex line* $L$ is made up of a broad boundary $BB_1(L')$, $BB_2(L')$, ..., $BB_2(L')$ and a broad interior $BI(L')$. The *boundary* of $L$, indicated with $\Delta L$,is the union of $BB_1(L')$, $BB_2(L')$, ..., $BB_2(L')$. The *interior* of an uncertain line L, indicated by $L^\circ$, is given by the difference $L^\circ = L - \Delta L$.

We conclude this section by giving the definitions about points affected by uncertainty.

**Definition 17.** A *point* is a (non-empty) zero-dimensional point-set, $P$, consisting of only one element.

The boundary of a point $P$ is empty, while the interior of a point $P$ is equal to the point itself.

As obvious generalization of the previous concept, we introduce the notion of complex point feature as follows.

**Definition 18.** A *complex point* is a (non-empty) zero-dimensional point-set, P, consisting of a finite number of distinct elements.

The boundary of a complex point feature $P$ is empty (i.e., $\partial P = \emptyset$); as a consequence, the interior of a complex point feature $P$ is equal to the union of all the elements in $P$ (i.e., $P^\circ = \bigcup_{i=1}^{n} P_i$).

**Definition 19.** Given a point $P'$, whose exact position in $\mathbb{R}^2$ is unknown but can be delimited by a bounded point set, an *uncertain point* is a point set $P$ that contains $P'$. The boundary of an uncertain point is empty, while the interior of an uncertain point is equal to the uncertain point.

**Definition 20.** Given a complex point $P'$, whose exact position in $\mathbb{R}^2$ is unknown but can be delimited by a bounded point set, an *uncertain complex point* is a point set $P$ that contains $P'$. The boundary of an uncertain complex point is empty, while the interior of an uncertain complex point is equal to the uncertain complex point.

An uncertain point represents the point set of all possible locations of a point whose location is unknown. While there could be no restrictions on the topology of an uncertain point $P$, we assume that an uncertain point is a simply connected open set. Analogously, we assume that an uncertain complex point $P$ is the union of $n$ simply connected open sets, one for each element of the complex point $P'$.

## 9.5 Topological relations between spatial objects with uncertainty

In this section, we discuss models for expressing the topological relations between spatial objects with uncertainty, starting from the main models that apply to crisp objects, namely, the 9-intersection [10] and the Calculus-Based Method (CBM) [7].

The 9-intersection is able to express the binary topological relations between two objects, $A$ and $B$, in $\mathbb{R}^2$, according to the intersection of $A$'s interior, boundary, and exterior with $B$'s interior, boundary, and exterior. The 9-intersection can be concisely represented by the following $3 \times 3$ matrix $M$:

$$M = \begin{pmatrix} A^\circ \cap B^\circ & A^\circ \cap \partial B & A^\circ \cap B^- \\ \partial A \cap B^\circ & \partial A \cap \partial B & \partial A \cap B^- \\ A^- \cap B^\circ & A^- \cap \partial B & A^- \cap B^- \end{pmatrix}.$$

By considering the values empty (0) and nonempty (1), we can distinguish between $2^9 = 512$ binary topological relations. For two simple regions with crisp boundary, only eight of them can be realized, whose names are *disjoint*, *meet*, *overlap*, *coveredBy*, *inside*, *covers*, *contains*, *equal*. For other spatial objects, the number of relations that can be realized is in general bigger.

The 9-intersection model has been extended to simple regions with a broad boundary (non-composite and without holes) in [3, 5] and to composite regions with a broad boundary in [6]. The matrix $M$ needs to be redefined as follows, having a broad boundary in place of a sharp boundary:

$$M = \begin{pmatrix} A^\circ \cap B^\circ & A^\circ \cap \Delta B & A^\circ \cap B^- \\ \Delta A \cap B^\circ & \Delta A \cap \Delta B & \Delta A \cap B^- \\ A^- \cap B^\circ & A^- \cap \Delta B & A^- \cap B^- \end{pmatrix}.$$

In [6], the authors showed that 56 are the realizable matrices for composite regions with a broad boundary, by discussing the geometric criteria that lead to the exclusion of the other configurations.

The CBM is another model for representing the topological relations between geometric objects. As appeared in [7] for the case of simple objects, the CBM was made up of five relations and three boundary operators. For the case of complex objects with crisp boundary, the CBM needs slight modifications to one relation (the *cross*) and to the boundary operators. For regions with a broad boundary, as proved in [6], the definitions are the same as for objects with crisp boundaries. The CBM relations for uncertain lines and points will be discussed in more depth in Section 9.6. In general, given two objects $A$ and $B$, the definitions of the CBM relations are the following:

**Definition 21.** The relation *touch*($A$,$B$) is true, if and only if $(A^\circ \cap B^\circ = \emptyset) \wedge (A \cap B \neq \emptyset)$.

**Definition 22.** The relation *in*($A$,$B$) is true if and only if $(A^\circ \cap B^\circ \neq \emptyset) \wedge (A \cap B = A)$.

**Definition 23.** The relation *cross*($A$,$B$) is true if and only if $(\dim(A^\circ \cap B^\circ) < \max(\dim(A^\circ), \dim(B^\circ))) \wedge (A \cap B \neq A) \wedge (A \cap B \neq B)$.

**Definition 24.** The relation *overlap*($A$,$B$) is true if and only if $(\dim(A^\circ) = \dim(B^\circ) = \dim(A^\circ \cap B^\circ)) \wedge (A \cap B \neq A) \wedge (A \cap B \neq B)$.

**Definition 25.** The relation *disjoint*($A$,$B$) is true if and only if $A \cap B = \emptyset$.

In the above definitions, the function 'dim' returns the dimension of a point-set (0, 1, or 2). In the CBM, the above relations are complemented by the boundary operators, which are able to extract the boundaries from every different kind of spatial objects. The CBM relations have the formal properties of being mutually exclusive and providing a complete coverage of all the realizable geometric configurations [4]. Therefore, the CBM relations can be used to build a set of high level user-oriented operators (touch, in, cross, overlap, disjoint). On the other hand, a low level geometric-oriented operator can be defined with the 9-intersection that is able to test for the detailed topological relation holding among two objects in terms of the corresponding 9-intersection matrix. Since the 9-intersection distinguish in general a high number of different topological relations, each CBM relation corresponds to a cluster of neighboring 9-intersection relations. In [6], strategies of clustering neighboring 9-intersection configurations have been discussed.

## 9.6 Topological relations between uncertain lines and points

In the following, we are going to analyze in more depth the topological relations for uncertain lines. With regard to the 9-intersection, the topological relations between two uncertain lines $L_1$ and $L_2$ are given by different values of the matrix:

$$M = \begin{pmatrix} L_1^\circ \cap L_2^\circ & L_1^\circ \cap \Delta L_2 & L_1^\circ \cap L_2^- \\ \Delta L_1 \cap L_2^\circ & \Delta L_1 \cap \Delta L_2 & \Delta L_1 \cap L_2^- \\ L_1^- \cap L_2^\circ & L_1^- \cap \Delta L_2 & L_1^- \cap L_2^- \end{pmatrix}$$

The 512 values of the matrix can be restricted to those that correspond to a geometric interpretation by considering geometric conditions that must be valid. This approach has been followed for other kinds of objects in [3, 10]. For uncertain lines, we can identify the following set of rules:

1. The exteriors of two uncertain lines must intersect;
2. $L_1$'s boundary must intersect at least one part[1] of $L_2$, and vice versa;
3. $L_1$'s interior must intersect at least one part of $L_2$, and vice versa;
4. If $L_2$'s exterior does not intersect $L_1$'s interior, then $L_1$'s boundary must intersect $L_2$, and vice versa.

Respectively for each of these rules, we can exclude the following patterns[2]:

1. $\begin{pmatrix} \delta & \delta & \delta \\ \delta & \delta & \delta \\ \delta & \delta & 0 \end{pmatrix}$;
2. $\begin{pmatrix} \delta & \delta & \delta \\ 0 & 0 & 0 \\ \delta & \delta & 1 \end{pmatrix}$; $\begin{pmatrix} \delta & 0 & \delta \\ \delta & 0 & \delta \\ \delta & 0 & 1 \end{pmatrix}$;
3. $\begin{pmatrix} 0 & 0 & 0 \\ \delta & \delta & \delta \\ \delta & \delta & 1 \end{pmatrix}$; $\begin{pmatrix} 0 & \delta & \delta \\ 0 & \delta & \delta \\ 0 & \delta & 1 \end{pmatrix}$;
4. $\begin{pmatrix} \delta & \delta & 0 \\ 0 & 0 & \delta \\ \delta & \delta & 1 \end{pmatrix}$; $\begin{pmatrix} \delta & 0 & \delta \\ \delta & 0 & \delta \\ 0 & \delta & 1 \end{pmatrix}$.

By excluding the patterns above, we obtain that the total number of possible topological relations between uncertain lines is 146. In Table 9.1, they are listed by using a linear notation by rows for the values of the 9-intersection matrix.

An example of some topological relations between uncertain lines is given in Figure 9.14. For such two cases, the 9-intersection is exactly the same

[1] The parts of a line are its interior, boundary and exterior.

[2] The symbol $\delta$ indicates any value 0 or 1.

**Table 9.1.** The list of possible 9-intersection values for topological relations between uncertain lines.

| 1 010100001 | 31 111111001 | 61 011011101 | 91 110011011 | 121 111110111 |
|---|---|---|---|---|
| 2 110100001 | 32 010100101 | 62 111011101 | 92 101011011 | 122 001001111 |
| 3 011100001 | 33 110100101 | 63 100111101 | 93 111011011 | 123 101001111 |
| 4 111100001 | 34 011100101 | 64 010111101 | 94 100111011 | 124 011001111 |
| 5 100010001 | 35 111100101 | 65 110111101 | 95 010111011 | 125 111001111 |
| 6 110010001 | 36 100010101 | 66 001111101 | 96 110111011 | 126 100101111 |
| 7 101010001 | 37 010010101 | 67 101111101 | 97 001111011 | 127 010101111 |
| 8 111010001 | 38 110010101 | 68 011111101 | 98 101111011 | 128 110101111 |
| 9 100110001 | 39 001010101 | 69 111111101 | 99 011111011 | 129 001101111 |
| 10 010110001 | 40 101010101 | 70 010100011 | 100 111111011 | 130 101101111 |
| 11 110110001 | 41 011010101 | 71 110100011 | 101 100100111 | 131 011101111 |
| 12 001110001 | 42 111010101 | 72 011100011 | 102 010100111 | 132 111101111 |
| 13 101110001 | 43 100110101 | 73 111100011 | 103 110100111 | 133 100011111 |
| 14 011110001 | 44 010110101 | 74 100010011 | 104 001100111 | 134 010011111 |
| 15 111110001 | 45 110110101 | 75 110010011 | 105 101100111 | 135 110011111 |
| 16 111001001 | 46 001110101 | 76 101010011 | 106 011100111 | 136 001011111 |
| 17 010101001 | 47 101110101 | 77 111010011 | 107 111100111 | 137 101011111 |
| 18 110101001 | 48 011110101 | 78 100110011 | 108 100010111 | 138 011011111 |
| 19 011101001 | 49 111110101 | 79 010110011 | 109 010010111 | 139 111011111 |
| 20 111101001 | 50 011001101 | 80 110110011 | 110 110010111 | 140 100111111 |
| 21 100011001 | 51 111001101 | 81 001110011 | 111 001010111 | 141 010111111 |
| 22 110011001 | 52 010101101 | 82 101110011 | 112 101010111 | 142 110111111 |
| 23 101011001 | 53 110101101 | 83 011110011 | 113 011010111 | 143 001111111 |
| 24 111011001 | 54 011101101 | 84 111110011 | 114 111010111 | 144 101111111 |
| 25 100111001 | 55 111101101 | 85 111001011 | 115 100110111 | 145 011111111 |
| 26 010111001 | 56 100011101 | 86 010101011 | 116 010110111 | 146 111111111 |
| 27 110111001 | 57 010011101 | 87 110101011 | 117 110110111 | |
| 28 001111001 | 58 110011101 | 88 011101011 | 118 001110111 | |
| 29 101111001 | 59 001011101 | 89 111101011 | 119 101110111 | |
| 30 011111001 | 60 101011101 | 90 100011011 | 120 011110111 | |

as for the corresponding one-dimensional crispings. As we pointed out in Section 9.3.2, the model of uncertain lines allow us to emphasize the linear aspects of features, even if they have two-dimensional interiors and boundaries.

With regard to the CBM applied to uncertain lines, let us study first the topological relations by treating separately the uncertainty on the boundary and on the interior. The relations *touch*, *in*, *disjoint* have the same definitions for both kinds of uncertainty:

**Definition 26.** The relation $touch(L_1,L_2)$ is true, if and only if

$$(L_1^\circ \cap L_2^\circ = \emptyset) \wedge (L_1 \cap L_2 \neq \emptyset).$$

**Definition 27.** The relation $in(L_1,L_2)$ is true if and only if

$$(L_1^\circ \cap L_2^\circ \neq \emptyset) \wedge (L_1 \cap L_2 = L_1).$$

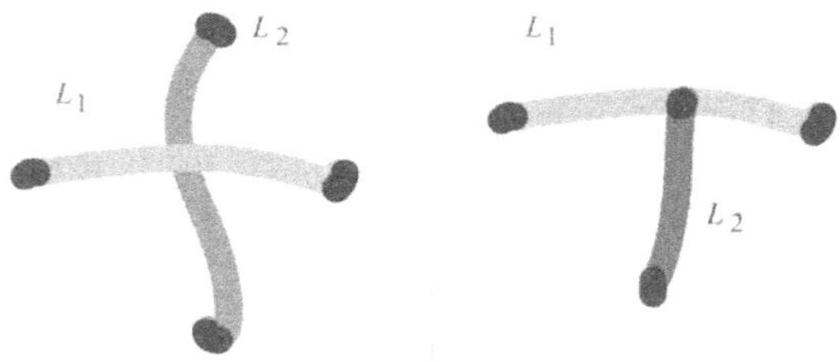


$$\begin{pmatrix} 1 & 0 & 1 \\ 0 & 0 & 1 \\ 1 & 1 & 1 \end{pmatrix} \qquad \begin{pmatrix} 0 & 1 & 1 \\ 0 & 0 & 1 \\ 1 & 1 & 1 \end{pmatrix}$$

**Figure 9.14.** Examples of topological relations between uncertain lines expressed by the 9-intersection.

**Definition 28.** The relation *disjoint*($L_1$,$L_2$) is true if and only if

$$L_1 \cap L_2 = \emptyset.$$

For two lines with a broad boundary and a crisp interior[3], the relations *cross* and *overlap* can be rephrased as follows:

**Definition 29.** The relation *cross*($L_1$,$L_2$) is true if and only if

$$\dim(L_1^\circ \cap L_2^\circ) = 0.$$

**Definition 30.** The relation *overlap*($L_1$,$L_2$) is true if and only if

$$(\dim(L_1^\circ \cap L_2^\circ) = 1) \wedge (L_1 \cap L_2 \neq L_1) \wedge (L_1 \cap L_2 \neq L_2).$$

For two lines with a broad interior and a crisp boundary, the same relations assume the following form:

**Definition 31.** The relation *cross*($L_1$,$L_2$) is true if and only if

$$\dim(L_1^\circ \cap L_2^\circ) \in \{0, 1\}.$$

**Definition 32.**The relation *overlap*($L_1$,$L_2$) is true if and only if

$$(\dim(L_1^\circ \cap L_2^\circ) = 2) \wedge (L_1 \cap L_2 \neq L_1) \wedge (L_1 \cap L_2 \neq L_2).$$

Overall, for relations between uncertain lines and for relations between uncertain lines and other objects such as regions with broad boundary the generic CBM definitions of Section 9.5 are valid. Some configurations are illustrated in Figures 9.15–9.18.

Topological relations involving uncertain points are similar to topological relations involving complex points [4]. With regard to the CBM, the same definitions of Section 9.5 are applicable. Notably, having an uncertain point a two-dimensional interior, the relation *overlap* can also hold among the point/line and point/region groups. See Figure 9.19 for some examples.

[3] This is an ideal case because, as we pointed out in Section 9.3.2, a broad boundary implies always at least a small part of broad interior.

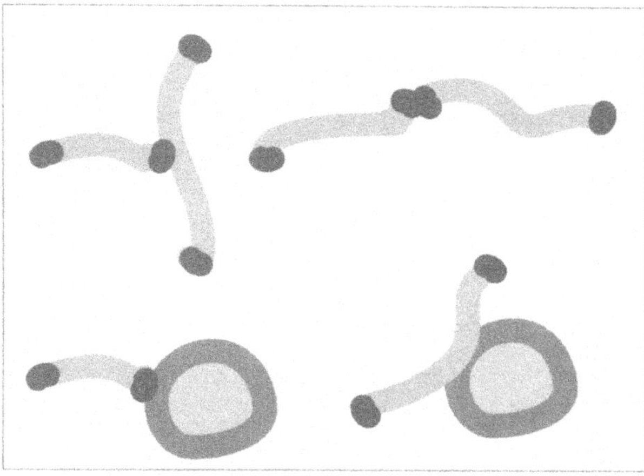

**Figure 9.15.** Examples of the relation *touch*.

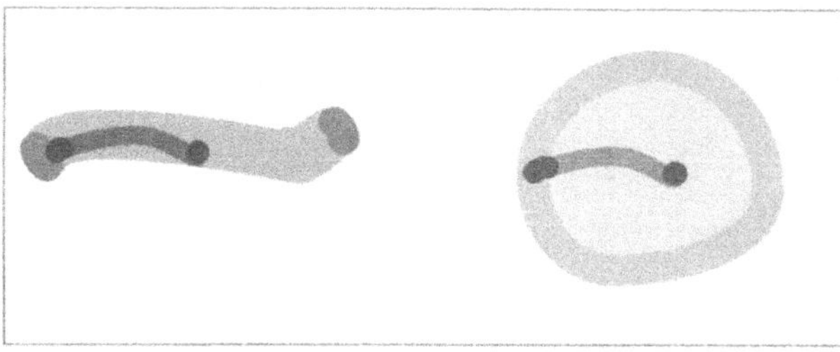

**Figure 9.16.** Examples of the relation *in*.

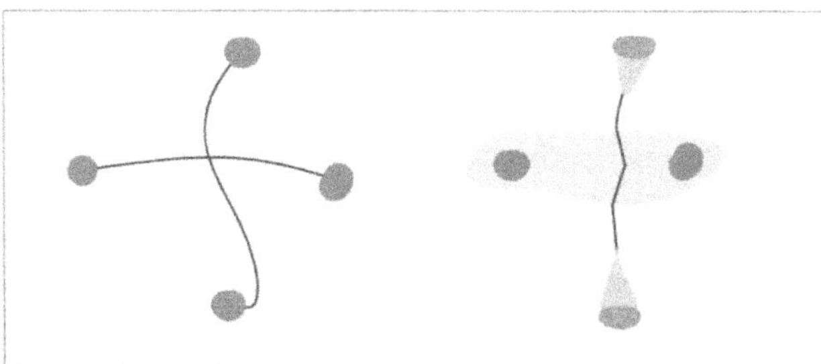

**Figure 9.17.** Examples of the relation *cross*.

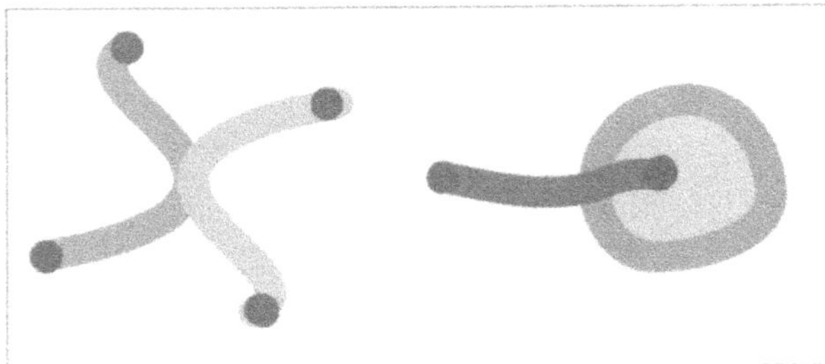

**Figure 9.18.** Examples of the relation *overlap*.

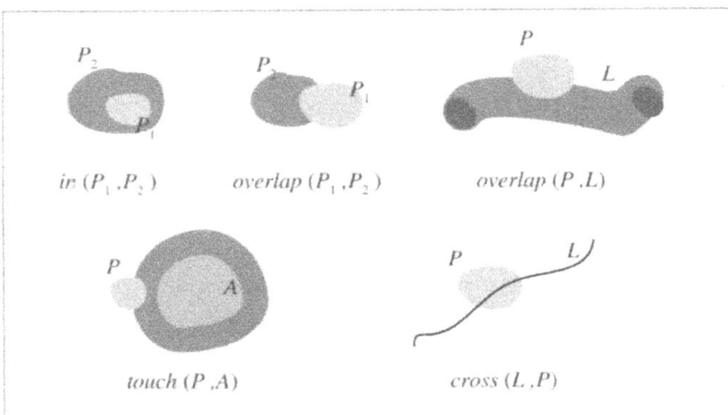


**Figure 9.19.** Examples of CBM relations involving uncertain points.

## 9.7 Topological constraints in aggregated objects

The topological relation between uncertain objects is a weaker version of the topological relation between crisp spatial objects: the higher number of possible topological relations in the case of uncertain objects enables to model different spatial configurations with more gradation than in the case of crisp objects. In some contexts, the knowledge of the topological relation between two objects can be improved by considering additional constraints on the possible topological relations between crispings, as in the case of aggregated objects, such as tessellations and networks.

We consider the concept of *crisping* of an object affected by uncertainty. Starting from a given topological relation between two uncertain objects, we are going to find what are the possible topological relations between the corresponding crispings. A crisping of un uncertain object is any crisp object that is compatible with it. A crisping of a region with a broad boundary is a crisp region with the boundary inside the broad boundary. A crisping of an uncertain line is any simple line with the two endpoints inside the broad boundary and the interior inside the broad interior. A crisping of an uncertain point is any point inside it.

As part of the discussion, let us consider some examples of crispings and the rules on the topological relations. For two lines with a broad boundary (and a crisp interior), the following are the rules on topological relations after a crisping. We indicate with $L$ un uncertain line and with $L'$ a possibile crisping of $L$.

1. If *touch* ($L_1$,$L_2$), then *touch* ($L'_1$, $L'_2$) or *disjoint* ($L'_1$, $L'_2$) or *cross* ($L'_1$, $L'_2$) or *overlap* ($L'_1$, $L'_2$);
2. If *in* ($L_1$,$L_2$), then *in* ($L'_1$, $L'_2$) or *overlap* ($L'_1$, $L'_2$);
3. If *overlap* ($L_1$,$L_2$), then *overlap* ($L'_1$, $L'_2$);
4. If *cross* ($L_1$,$L_2$), then *cross* ($L'_1$, $L'_2$);
5. If *disjoint* ($L_1$,$L_2$), then *disjoint* ($L'_1$, $L'_2$).

The rule for the relation *touch* is illustrated in Figure 9.20. On the possibile crispings for lines having a broad interior, we have the following rules:

1. If *touch* ($L_1$,$L_2$), then *touch* ($L'_1$, $L'_2$) or *disjoint* ($L'_1$, $L'_2$);
2. If *in* ($L_1$,$L_2$), then *touch* ($L'_1$, $L'_2$) or *disjoint* ($L'_1$, $L'_2$) or *cross* ($L'_1$, $L'_2$) or *overlap* ($L'_1$, $L'_2$) or *in* ($L'_1$, $L'_2$);
3. If *overlap* ($L_1$,$L_2$), then *touch* ($L'_1$, $L'_2$) or *disjoint* ($L'_1$, $L'_2$) or *cross* ($L'_1$, $L'_2$) or *overlap* ($L'_1$, $L'_2$) or *in* ($L'_1$, $L'_2$);
4. If *cross* ($L_1$,$L_2$), then *touch* ($L'_1$, $L'_2$) or *disjoint* ($L'_1$, $L'_2$) or *cross* ($L'_1$, $L'_2$) or *overlap* ($L'_1$, $L'_2$) or *in* ($L'_1$, $L'_2$);
5. If *disjoint* ($L_1$,$L_2$), then *disjoint* ($L'_1$, $L'_2$).

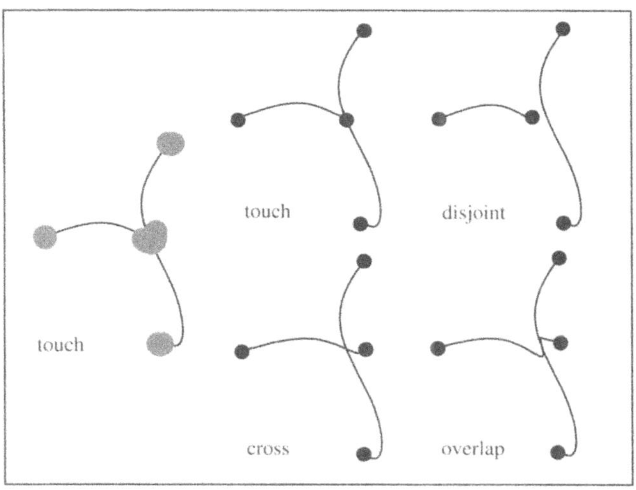


**Figure 9.20.** Possible crispings of a relation *touch* between two lines with broad boundary.

The additional constraints on crispings can be useful to describe the topological relations in aggregated objects affected by uncertainty. Let us consider a tessellation (Figure 9.21(a)) made up of $n$ regions with broad boundary $A_1 \ldots A_n$, such that, $\forall i,j \in 1 \ldots n$, *touch* $(A_i,A_j) \vee$ *disjoint* $(A_i,A_j)$.

In a tessellation, two bordering regions must satisfy the topological constraint that they have a common boundary even if its position is not precisely known. This can be expressed by the following constraint on crispings:

- If *touch* $(A_i,A_j)$, then *touch* $(A'_i, A'_j)$.

Analogously, let us consider a network of lines (Figure 9.21(b)), where the uncertainty is on the position of endpoints. A network is an aggregate of $n$ lines such that, $\forall i, j \in 1 \ldots n$, $touch(L_i, L_j) \vee$ $disjoint(L_i, L_j)$.

Similarly to tessellations, two connected lines must have some common endpoints even without knowing their exact position. This can be expressed by the constraint on crispings:

- If *touch* $(L_i,L_j)$, then *touch* $(L'_i, L'_j)$.

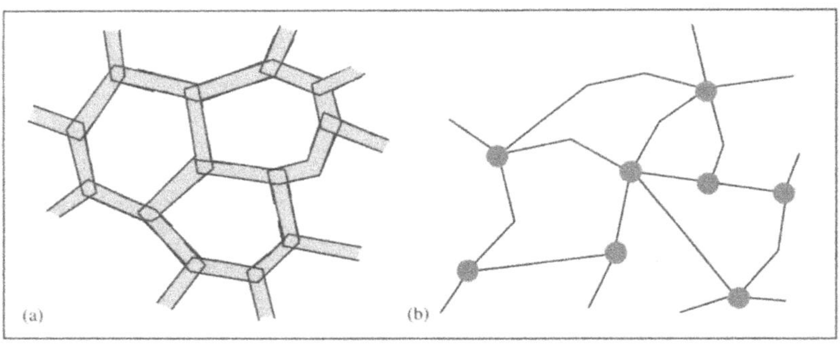


**Figure 9.21.** Aggregated regions and lines with broad boundary (tessellations and networks).

## 9.8 Conclusions

This chapter is a review of data models for spatial objects with a broad boundary and their topological relations and an extension of these models for uncertain lines. The latter ones are lines that incorporate uncertainty description both for boundary and interior. We discussed the possible interpretations of uncertain lines in the geographic context and we developed the models for describing topological relations between uncertain lines. Finally, by using the constraints on possible crispings of objects with uncertainty, we apply the model to aggregated objects, such as tessellations and networks.

*Acknowledgements.* Work supported by M.I.U.R. under project *"Representation and management of spatial and geographic data on the Web".*

## References

1. Burrough P.A., Frank A.U. (Eds.) (1996) Geographic Objects with Indeterminate Boundaries. GISDATA Series, Taylor & Francis, London.
2. Clementini E. (2002) A model for lines with a broad boundary. In: Proc. of the 9th International Conference on Information Processing and Management of Uncertainty in Knowledge-Based Systems (IPMU 2002), Annecy, France.
3. Clementini E., Di Felice P. (1996) An Algebraic Model for Spatial Objects with Indeterminate Boundaries. In: Burrough P.A., Frank A.U. (Eds.) Geographic Objects with Indeterminate Boundaries, Taylor & Francis, London, pp. 155–169.
4. Clementini E., Di Felice P. (1996) A Model for Representing Topological Relationships Between Complex Geometric Features in Spatial Databases. Information Sciences 90(1–4):121–136.
5. Clementini E., Di Felice P. (1997) Approximate Topological Relations. International Journal of Approximate Reasoning 16:173–204.
6. Clementini E., Di Felice P. (2001) A spatial model for complex objects with a broad boundary supporting queries on uncertain data. Data & Knowledge Engineering 37(3):285–305.
7. Clementini E., Di Felice P., van Oosterom P. (1993) A Small Set of Formal Topological Relationships Suitable for End-User Interaction. In: Abel D., Ooi B.C. (Eds.) Advances in Spatial Databases — Third International Symposium, SSD '93, Springer-Verlag, Berlin, pp. 277–295.
8. Cohn A.G. (1995) The challenge of qualitative spatial reasoning. Computing Surveys 27(3):323–327.
9. Cohn A.G., Gotts N.M. (1996) The 'Egg-Yolk' Representation of Regions with Indeterminate Boundaries. In: Burrough P.A., Frank A.U. (Eds.) Geographic Objects with Indeterminate Boundaries, Taylor & Francis, London, pp. 171–187.
10. Egenhofer M.J., Herring J.R. (1991) Categorizing Binary Topological Relationships Between Regions, Lines, and Points in Geographic Databases. Department of Surveying Engineering, University of Maine, Orono, ME.
11. Erwig M., Schneider M. (1997) Vague regions. In: Scholl M., Voisard A. (Eds.) Advances in Spatial Databases — Fifth International Symposium, SSD '97, Springer, Berlin, pp. 298–320.
12. Frank A.U. (1998) Metamodels for Data Quality Description. In: Jeansoulin R., Goodchild M. (Eds.) Data Quality in Geographic Information — From Error to Uncertainty, Editions Hermès, Paris, pp. 15–29.
13. Guptill S., Morrison J. (Eds.) (1995) Elements of spatial data quality. Elsevier Science, Oxford, UK.
14. OpenGIS Consortium (1998) OpenGIS Simple Features Specification for SQL.
15. Russell B. (1923) Vagueness. Australasian Journal of Philosophy and Psychology 1:84–92.

16. Schneider M. (1996) Modelling spatial objects with undetermined boundaries using the Realm/ROSE approach. In: Burrough P.A., Frank A.U. (Eds.) Geographic Objects with Indeterminate Boundaries, Taylor & Francis, London, pp. 141–152.
17. Worboys M., Clementini E. (2001) Integration of imperfect spatial information. Journal of Visual Languages & Computing 12(1):61–80.

# 10

# Fuzzy Minimum Bounding Rectangles

María J. Somodevilla[1] and Fred E. Petry[2]

[1] Universidad de Puebla
Facultad Ciencias Computación BUAP
Puebla, 72000 México
mariasg@solarium.cs.buap.mx

[2] Tulane University
Computer Science and Electrical Engineering Dept.
New Orleans, LA 70118
petry@eecs.tulane.edu

## 10.1 Fuzzy Spatio-Thematic-Temporal Framework for Approximation of Indeterminate Boundaries

We propose to use the spatial, temporal and thematic knowledge about features' relationships as elements to approximate indeterminate boundaries (IB). IB are approximated as a set of locations in space-time with a set of properties characterizing those locations.

Indeterminate boundaries will be defined in a fuzzy framework allowing for imprecise quantitative and qualitative knowledge. Fuzzy set theory provides us with a very suitable theoretical basis to model imprecision since both numerical and symbolic information can be represented together.

Our approach comprises spatial, temporal, and thematic reasoning as shown in figure 10.1. Spatial reasoning is the approximation of the boundary by defining the Fuzzy Minimal Bounding Rectangle ($FMBR$) enclosing the fuzzy region. The core and boundary of the $FMBR$ are then determined based on Euclidean measures. The thematic and temporal reasoning are complementing reasoning spatially with $FMBR$s. Basically the thematic and temporal aspects are treated qualitatively. The qualitative aspects correspond more directly with the human perception of the space, since we are able to describe levels of compatibility or concentration of the attribute of interest. Afterwards, we consider qualitative aspects to semantically enrich the desirable fuzzy region. Also, they allow us to increase or decrease a rigid boundary defined using distance measures since they are concerned with relaxing the metric (spatial) definition of the boundary. In summary, the approach consists of modeling the spatial dimension of geographic features using fuzzy sets to define the spatial extent of the thematic dimension at different times. The framework's steps will be formalized in the next sections.

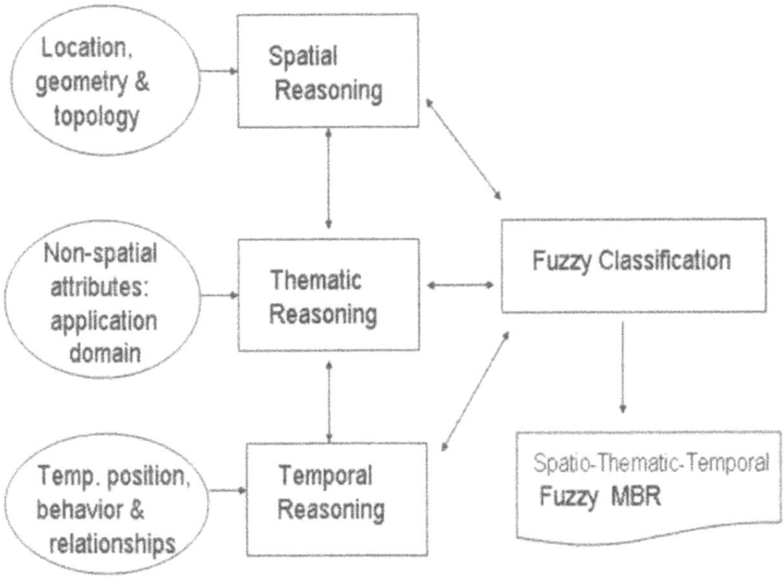


**Figure 10.1.** Fuzzy $MBR$s Framework

## 10.2 Spatial Reasoning

Clementini et al. [3] described how $MBR$ relations can be used as a fast filter to determine whether it is possible for a spatial object to satisfy a given topological relationship. That is, the relationships which hold in the underlying objects also hold for their $MBR$s. To reason spatially, we use $MBR$s as the spatial structure in which the vague boundary can be characterized. We have organized this process in three major steps:

- Definition of the $MBR$ of the entire fuzzy region ($FMBR$). This $FMBR$ encloses all the points of the map space where our feature of interest is located. Geographic features are a direct representation of the geographic entities rather than geometric elements such as a point, line or polygon. A feature is then defined as an entity with common attributes and relationships.Then, the $FMBR$ represents the generalization of the underlying irregular polygon delimiting the fuzzy region. Also, the $FMBR$ can be considered as the circumscribed rectangle (CR) of the underlying fuzzy polygon.
- Iterative generation of the inner bounding rectangles until we have the inscribed rectangle (IR) of the underlying object. So, the IR is the maximum inner rectangle inside the object and it corresponds to the core of

the fuzzy region. Distances between the IR and the $FMBR$ are used to represent the fuzzy boundary.
- A spatial membership function based on Euclidean distance will be used to determine the degree of belonging of a feature to the fuzzy set. Thus, features inside the IR or core will have degree of membership equal to 1. This degree will be gradually decreased while we move away from the core. Points located outside of the $FMBR$ will have a membership degree of 0.

An $FMBR$ is a natural representation for many commonly occurring spatial situations. The problems of identifying a spatial boundary have been under considerable attention for the GIS area [2]. For example consider photo-interpreters who are trying to label a forest in an image. There is clearly a region (core) which agree is the heart of the forest and merits the specific labeling. However, as the forest thins out into meadows all around, there is no sharp boundary delimiting the forest area. Rather the density of the trees decreases gradually until there is just open meadow land. It is just such a situation that we are trying to model by means of $FMBR$.

A graphical representation of the $FMBR$, as described above, is shown in figure 10.2. The underlying vague region $\widetilde{A}$ is approximated by the $FMBR(\widetilde{A})$. This first approximation is also called the circumscribed rectangle (CR) of the fuzzy region. In other words, the $FMBR$ or CR corresponds to the minimal rectangle with edges parallel to the x and y axes that optimally encloses the vague region $\widetilde{A}$.

$\alpha MBR$-cuts allow us to make finer distinctions inside the fuzzy region since $\alpha MBR$s are individual crisp regions inside the $FMBR$. Thus, we can think of a fuzzy structured region as an aggregation of crisp $\alpha$-level regions. $\alpha MBR$s start to be defined from the edge of the $FMBR(\widetilde{A})$ to the core of $\widetilde{A}$. The more external the $\alpha MBR$-cut, the lower the degree of membership in the fuzzy set representing $\widetilde{A}$ as locations which are closer to the core will have higher membership degrees. The shadowed rectangle labeled as *Core* corresponds to the inscribed rectangle. Since the *IR* is totally inside $\widetilde{A}$ we assume that the points in the core belong to the fuzzy region with a degree of membership of 1. In the next subsections, we formally define and analyze the $FMBR$s in a metric space context.

### 10.2.1 Definition of the $FMBR$

**Definition 1:** Given a universe, $U$, of objects, a fuzzy set $\widetilde{A} \in F(U)$ is a mapping, denoted $f_{\widetilde{A}}$ from $U$ to the interval $[0,1]$, where $f_{\widetilde{A}}(x)$ is the membership value of $x$ in $\widetilde{A}$ for any $x \in U$, and $F$ stands for fuzzy powerset of $U$.

**Definition 2:** A map space $S$ is a bounded subset of $\mathbb{R}^2$.

An $FMBR$ is a fuzzy region and so can be represented as a fuzzy set which is delimited by an $MBR$, with edges parallel to the x and y axes, in the universe (map space).

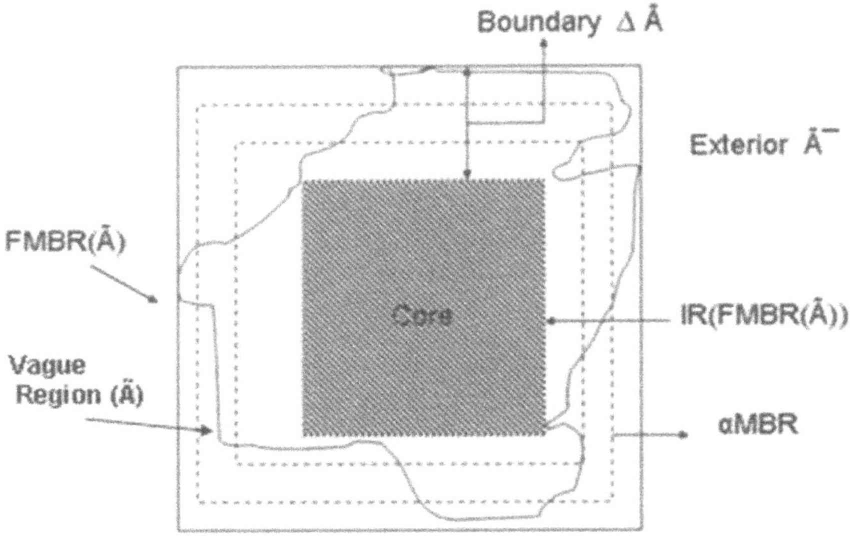


**Figure 10.2.** Fuzzy Regions as $MBR$s

**Definition 3:** Let $\widetilde{A} \in F(U)$ be a fuzzy region and $FMBR(\widetilde{A})$ its approximation, then $FMBR(\widetilde{A}) \subset U$, where:

- $$FMBR(\widetilde{A}) = CR(\widetilde{A}) = core(FMBR(\widetilde{A})) \bigcup \triangle\widetilde{A}(FMBR(\widetilde{A})) = \left\{u \in U \mid 0 < f_{FMBR(\widetilde{A})}(u) \leq 1\right\}$$
- $$core\left(FMBR(\widetilde{A})\right) = IR(FMBR(\widetilde{A})) = \left\{u \in U \mid f_{FMBR(\widetilde{A})}(u) = sup_{\omega \in U} f_{FMBR(\widetilde{A})}(w)\right\} = \left\{u \in U \mid f_{FMBR(\widetilde{A})}(u) = 1\right\}$$
  (since $sup_{\omega \in U} f_{FMBR(\widetilde{A})}(w) = 1$)
- Let the *fuzzy boundary*
$$\triangle\widetilde{A} = \left\{u \in U \mid 0 < f_{FMBR(\widetilde{A})}(u) < sup_{\omega \in U} f_{FMBR(\widetilde{A})}(w)\right\}$$

Definition 1 defines the universe of spatial objects and a membership function to denote the membership of the objects in the fuzzy set. In Definition 2, we defined a map as a spatial subset of the universe $U = \mathbb{R}^2$ from Definition 1. A $FMBR$ is a fuzzy region in a map according Definition 3.

We have defined a fuzzy region as a region delimited by a fuzzy boundary. The external edge of this boundary is the $MBR$ of the underlying region, which is itself an approximation structure, and so can be considered as a fuzzy representation. Structurally, as defined in Definition 3, a $FMBR$ is the union of its core and its boundary. Those elements with a degree of membership of 1 belong to the core, that means we admit with certainty (degree of membership=1) that the features in the core belong to the fuzzy region. The remaining elements are then in the boundary with a membership greater than zero and less than 1. This fuzzy definition of boundary allows us to differentiate between features strongly identified with the region and those weakly identified.

Above we discussed the boundary definition of one fuzzy region. Then, the boundary between two different fuzzy regions can also be defined as below:

**Definition 4:** Let $FMBR(\widetilde{A})$ and $FMBR(\widetilde{B})$ represent two fuzzy regions. Let

$$I = sup_{u \in U} min\left\{f_{FMBR(\widetilde{A})}(u), f_{(FMBR(\widetilde{B}))}(u)\right\}$$

Then, the

$$boundary(FMBR(\widetilde{A}), FMBR(\widetilde{B})) = \left\{u \in U \mid 0 < min\left\{f_{FMBR(\widetilde{A})}(u), f_{(FMBR(\widetilde{B}))}(u)\right\} < I\right\}$$

Since we work with fuzzy regions, the core can be located at any position inside the fuzzy region, that is, not necessarily at center of the $FMBR$. The next section makes these distinctions about the different positions that the core of the fuzzy region may adopt.

### 10.2.2 Morphology of the $FMBR$

The morphology of the $FMBR$ in this context refers to a concentration of an attribute towards a certain position within the fuzzy region. The area with the highest concentration of the attribute is considered the core of the region. The following identified cases of the core correspond to the number of intersections between the $FMBR(\widetilde{A})$, and $IR(FMBR(\widetilde{A}))$ edges of the fuzzy region [10]. These cases, as shown in figure 10.3, are the following:

a) Central core: touch no sides $\#\mathrm{I}(\triangle^{IR(FMBR(\widetilde{A}))}, \triangle^{FMBR(\widetilde{A})}) = 0$
b) Corner core: touch 2 sides $\#\mathrm{I}(\triangle^{IR(FMBR(\widetilde{A}))}, \triangle^{FMBR(\widetilde{A})}) = 2$
c) Side core: touch 1 side $\#\mathrm{I}(\triangle^{IR(FMBR(\widetilde{A}))}, \triangle^{FMBR(\widetilde{A})}) = 1$

where $\#\mathrm{I}$ corresponds to the number of intersections between the parts of the objects indicated in the parenthesis, and $\triangle$ represents the indeterminate boundary, see figure 10.2 (to distinguish from the crisp boundary $\delta$, as represented in the 9-intersection model [5]).

In figure 10.3, each case of a fuzzy region is represented along its $\alpha$-cuts that indicate the smooth variation of the attribute. So we have nested $\alpha$-cut level sets $\{F^{\alpha_1} \subseteq F^{\alpha_2} \subseteq \cdots \subseteq F^{\alpha_n}\}$. It follows that, for any *point* $x \in F^{\alpha_i} - F^{\alpha_{i+1}}$, its membership in F is given by $\alpha_i - \alpha_{i+1}$.

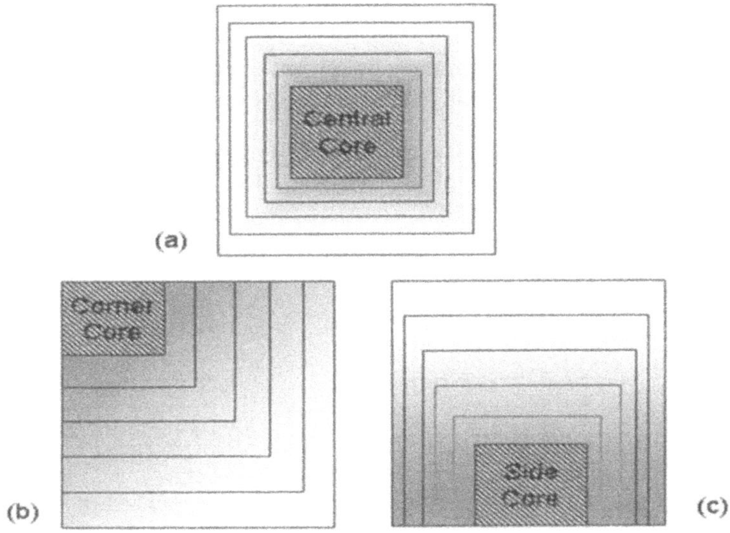


**Figure 10.3.** *FMBR*s Morphology

### 10.2.3 Generation of the *FMBR*'s $\alpha$-cuts

The external edge of the *FMBR* or CR is calculated using the maximum and the minimum points out of the set of points comprising the fuzzy region. The maximum point ($p_{max}$) is the one with the maximum values for its coordinates $x$ and $y$. On the other hand the minimum, ($p_{min}$) is the one whose $x$ and $y$ coordinates have the minimum values. Then, $p_{max}$ is the upper right corner and $p_{min}$ is the lower left corner of the *FMBR*. The *FMBR* is now taken as a reference to generate its inner rectangles. These inner rectangles are the $\alpha$-cuts of the *FMBR*, i.e., crisp regions inside the *FMBR*, where the membership to the fuzzy region can be sharply defined.

We need an even distribution of the $\alpha$-cuts to model the smooth distribution of the attributes. In doing so, we follow the next steps:

- Calculation of the *FMBR*'s diagonal by the Euclidean distance between $p_{max}$ and $p_{min}$.

$$FMBR_{diagonal} = \sqrt{\left((x_2 - x_1)^2 + (y_2 - y_1)^2\right)}$$

where $p_{max} = (x_2, y_2)$ and $p_{min} = (x_1, y_1)$

- Calculation of the ratio of generation of the $\alpha$-cuts.

$$r_\alpha = \frac{FMBR_{diagonal}}{k}$$

where $k$ is a constant. The size of $k$ determines the separation between the $\alpha$-cuts. The greater $k$, the smaller generation's ratio of the inner rectangles. As a consequence, the greater the number of $\alpha MBR$s, the better the approximation of the fuzzy region.

- Calculation of the $ith$ inner rectangle.

The calculation of the inner $\alpha MBR$s basically consists of decreasing the original size of the $FMBR$. Depending on the case of the core we specify the corresponding equations.

i. Central core
Equation (10.1) is the generic form of generation of the inner central core rectangles.

$$FMBR\alpha_i : (X_{2\alpha_i} = X_{2\alpha_{i-1}} - r_\alpha,\ Y_{2\alpha_i} = Y_{2\alpha_{i-1}} - r_\alpha),$$
$$(X_{1\alpha_i} = X_{1\alpha_{i-1}} + r_\alpha,\ Y_{1\alpha_i} = Y_{1\alpha_{i-1}} + r_\alpha) \quad (10.1)$$

This denotes that $FMBR\alpha_i$ is defined by the points $(X_{2\alpha_i} = X_{2\alpha_{i-1}} - r_\alpha,\ Y_{2\alpha_i} = Y_{2\alpha_{i-1}} - r_\alpha)$ and $(X_{1\alpha_i} = X_{1\alpha_{i-1}} + r_\alpha,\ Y_{1\alpha_i} = Y_{1\alpha_{i-1}} + r_\alpha)$ where $x$ and $y$, coordinates of the upper corner ($p_{max}$), are decreasing by $r_\alpha$, and the lower corner coordinates of ($p_{min}$) are increasing by $r_\alpha$ as well to generate the next inner $\alpha MBR$. Since the side and corner cores do not generate concentric $\alpha MBR$s, as in the central case, we give separate expressions for each of these cases.

ii. Corner core
This case of the core may present itself in four distinct positions corresponding to the four corners: Top-Left (TL), Top-Right (TR), Bottom-Left (BL), and Bottom-Right (BR), as shown in figure 10.4. The set of equations (10.2) correspond to the generation of the next $ith$ inner $\alpha MBR$ for each case.

$$
\begin{aligned}
TL:\ &(X_{2\alpha_i} = X_{2\alpha_{i-1}} - r_\alpha,\ Y_{2\alpha_i} = Y_{2\alpha_{i-1}}),\\
&\qquad (X_{1\alpha_i} = X_{1\alpha_{i-1}},\ Y_{1\alpha_i} = Y_{1\alpha_{i-1}} + r_\alpha)\\
TR:\ &(X_{2\alpha_i} = X_{2\alpha_{i-1}},\ Y_{2\alpha_i} = Y_{2\alpha_{i-1}}),\\
&\qquad (X_{1\alpha_i} = X_{1\alpha_{i-1}} + r_\alpha,\ Y_{1\alpha_i} = Y_{1\alpha_{i-1}} + r_\alpha)\\
BL:\ &(X_{2\alpha_i} = X_{2\alpha_{i-1}} - r_\alpha,\ Y_{2\alpha_i} = Y_{2\alpha_{i-1}} - r_\alpha),\\
&\qquad (X_{1\alpha_i} = X_{1\alpha_{i-1}},\ Y_{1\alpha_i} = Y_{1\alpha_{i-1}})\\
BR:\ &(X_{2\alpha_i} = X_{2\alpha_{i-1}},\ Y_{2\alpha_i} = Y_{2\alpha_{i-1}} - r_\alpha),\\
&\qquad (X_{1\alpha_i} = X_{1\alpha_{i-1}} + r_\alpha,\ Y_{1\alpha_i} = Y_{1\alpha_{i-1}})
\end{aligned}
\tag{10.2}
$$

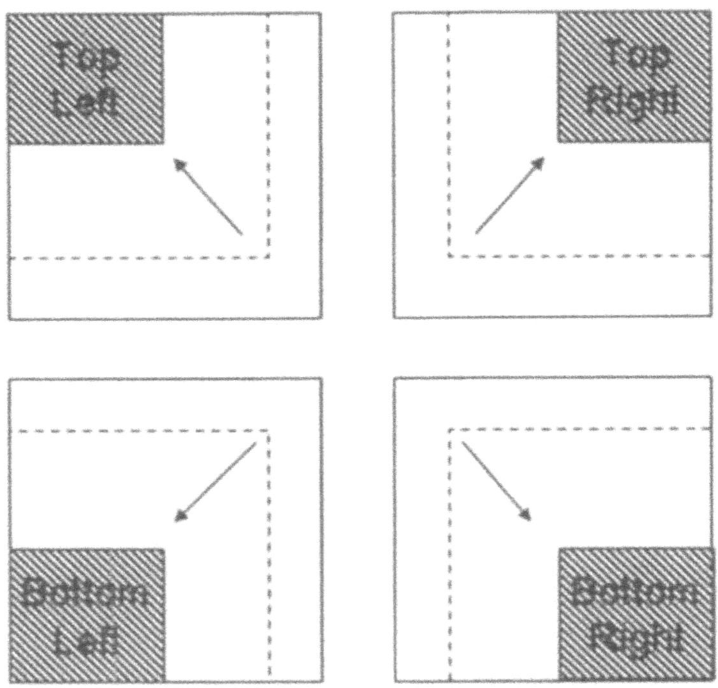


**Figure 10.4.** Corners core positions

iii. Side core
The side core has four cases too, one for each side of the $FMBR$, see figure 10.5, they are: Top (T), Bottom (B), Left (L), and Right (R). The generation of its inner rectangles is made according to the set of equations described by (10.3).

$$
\begin{aligned}
T: & (X_{2\alpha_i} = X_{2\alpha_{i-1}} - r_\alpha,\ Y_{2\alpha_i} = Y_{2\alpha_{i-1}}) \\
& \qquad (X_{1\alpha_i} = X_{1\alpha_{i-1}} + r_\alpha,\ Y_{1\alpha_i} = Y_{1\alpha_{i-1}} + r_\alpha) \\
B: & (X_{2\alpha_i} = X_{2\alpha_{i-1}} - r_\alpha,\ Y_{2\alpha_i} = Y_{2\alpha_{i-1}} - r_\alpha) \\
& \qquad (X_{1\alpha_i} = X_{1\alpha_{i-1}} + r_\alpha,\ Y_{1\alpha_i} = Y_{1\alpha_{i-1}}) \\
L: & (X_{2\alpha_i} = X_{2\alpha_{i-1}} - r_\alpha,\ Y_{2\alpha_i} = Y_{2\alpha_{i-1}} - r_\alpha) \\
& \qquad (X_{1\alpha_i} = X_{1\alpha_{i-1}},\ Y_{1\alpha_i} = Y_{1\alpha_{i-1}} + r_\alpha) \\
R: & (X_{2\alpha_i} = X_{2\alpha_{i-1}},\ Y_{2\alpha_i} = Y_{2\alpha_{i-1}} - r_\alpha) \\
& \qquad (X_{1\alpha_i} = X_{1\alpha_{i-1}} + r_\alpha,\ Y_{1\alpha_i} = Y_{1\alpha_{i-1}} + r_\alpha)
\end{aligned} \tag{10.3}
$$

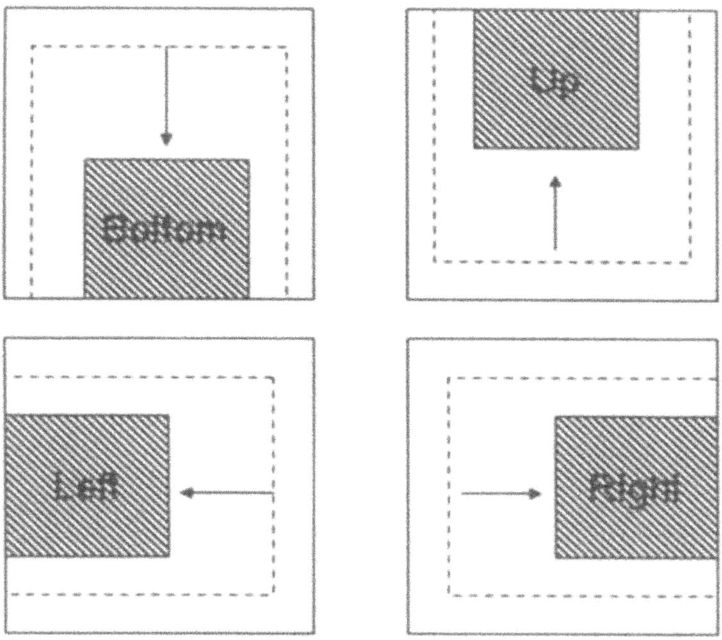


**Figure 10.5.** Sides core positions

- Checking of the inscribed rectangle condition.

  **Definition 5:** A rectilinear figure is said to be inscribed in a rectilinear figure when the respective angles of the inscribed figure lie on the respective sides of the figure in which it is inscribed [7].

Since a rectangle is a rectilinear figure, the inscribed $\alpha MBR$ is one which satisfies the condition (10.4).

$$(x_{min\alpha_i MBR} \leq x_{inscribedR} \leq x_{max\alpha_i MBR}) \wedge \\ (y_{min\alpha_i MBR} \leq y_{inscribedR} \leq y_{max\alpha_i MBR}) \quad (10.4)$$

where $x_{\alpha_i}$ and $y_{\alpha_i}$ are the coordinates of the last generated $\alpha MBR$, *max* correspond to the upper right and *min* to the lower left corners of the and $x_{inscribedR}$ and $y_{inscribedR}$ are the coordinates of the points that belong to the inscribed rectangle of the fuzzy region.
This condition needs to be evaluated after each inner rectangle is generated. When the condition is fulfilled the process of generation of $\alpha MBR$s is completed. The last $\alpha MBR$ generated is the core of the $FMBR$.

### 10.2.4 Membership function of $FMBR$

The property of the intrinsic smoothness of fuzzy regions can be modeled by the concept of continuity and results in special continuous membership functions for fuzzy regions. Since the distribution of the attributes' values within a region may be continuous, we use (10.5) as a membership function. The membership of the fuzzy region $\mu(\alpha_i MBR)$ can be interpreted according the statement "$\alpha_i MBR$ is the true representation of $FMBR$" [4]. Then, a degree of membership $\mu(\alpha_i MBR)$ can be attached to each $\alpha_i MBR$. It is defined as $\mu(\alpha_i MBR) = \alpha_i MBR - \alpha_{i-1} MBR$ for $1 < i < n+1$ with $\alpha_1 = 0$ and $\alpha_n = 1$.

It is easy to see that $\sum_{i=2}^{n} \mu(\alpha_i MBR) = \alpha_n MBR - \alpha_1 MBR = 1 - 0 = 1$. Consider a region $\widetilde{A}$ defined as $FMBR(\widetilde{A}) = IR(FMBR(\widetilde{A})) \cup \Delta\widetilde{A}$,

$$\mu_{FMBR(\widetilde{A})} = \begin{cases} 1 & \text{if } (x,y) \in IR(FMBR(\widetilde{A})) \\ \dfrac{\lambda - d((\alpha MBR), edgeIR(FMBR(\widetilde{A}))}{\beta} & \text{if } (x,y) \in \Delta\widetilde{A} \\ 0 & \text{if } (x,y) \in \widetilde{A}^- \end{cases} \quad (10.5)$$

where:

- $\lambda$ is the maximum distance between an edge of $FMBR(\widetilde{A})$ and the edge of $IR(FMBR(\widetilde{A}))$,
- $d((\alpha MBR), edgeIR(FMBR(\widetilde{A})))$ is the maximum distance between an edge of the $\alpha MBR$ and the edge of $IR(FMBR(\widetilde{A}))$,
- $\beta$ is the maximum Euclidean distance between the centroid of $IR(FMBR(\widetilde{A}))$ and an edge of $IR(FMBR(\widetilde{A}))$, see figure 10.6 for an example.

The values in $0 < \mu_{FMBR(\widetilde{A})} < 1$ interval generated by the function (10.5) correspond to the $\alpha MBR$-cuts. $\alpha MBR$-cuts are the values delimiting the crisp regions in which the indeterminate boundary is going to be approximated. All the features within a region on the edge of an $\alpha MBR$ or between two consecutive $\alpha$-cuts will have the same membership for the fuzzy set according to the fuzzy membership function (10.5). Therefore, to calculate a feature membership, we need first, to determine the $\alpha MBR$-cut that the feature belongs to. If a feature is between two $\alpha MBR$s, its membership is the difference of the two $\alpha MBR$s memberships.

The constant $\lambda$ bounds the size of the indeterminate boundary, and therefore $d((\alpha MBR), edgeIR(FMBR(\widetilde{A}))) \leq \lambda$. In other words, while the distance $d$ increases to $\lambda$, $\mu_{FMBR(\tilde{A})}$ decreases to 0. This function exploits the geometric concept of Euclidean distance as for those functions defined by Usery in [12].

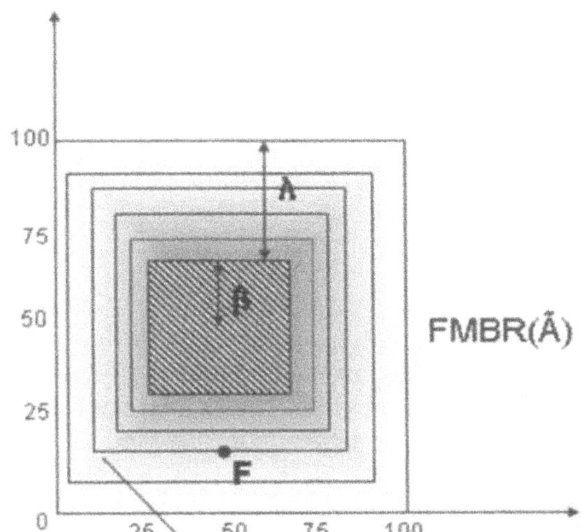


**Figure 10.6.** $\alpha MBR$s membership calculation

**Example:**

Let $\widetilde{A}$ be a fuzzy region, and its $FMBR$ upper corner $= (100, 100)$, and its lower corner $= (0, 0)$. Let $k = 28$, then

$$r_\alpha = \frac{141.42}{28} \approx 5$$

By increasing and decreasing the upper and lower corners of the $FMBR$ respectively we will have the next $\alpha MBR$s:

$FMBR = [(95, 95), (5, 5)]$
$FMBR = [(90, 90), (10, 10)]$
$FMBR = [(85, 85), (15, 15)]$
$FMBR = [(80, 80), (20, 20)]$

The $FMBR = [(75, 75), (25, 25)]$ is the inscribed rectangle, that is, the core of the fuzzy region. The centroid of the $IR(FMBR(\widetilde{A}))$ is $(50, 50)$.

Next, we show calculation of the parameters of $\lambda$ and $\beta$ for figure 10.6.

$$\lambda = x_{edge(core)} - x_{min(FMBR)} = 50 - 25 = 25$$

Thus, 25 is the IR width, the maximum distance from the edge of the core to the edge of the $FMBR$.

$$\beta = 25$$

is the half of the core edge $IR(FMBR(\widetilde{A}))$.

Let $F$ be a feature located at the $\alpha MBR = 15$ of the indeterminate boundary of $FMBR(\widetilde{A})$, as shown in figure 10.6. Then, the $F$ membership in $\widetilde{A}$ is:

$$\mu_{FMBR(\widetilde{A})} = \frac{25 - 15}{25} = 0.4$$

### 10.2.5 Qualitative Spatial Reasoning

Quantitative spatial reasoning, as described previously, delivers precise results but is often too rigid [8]. The reason is that quantitative statements like *"All houses that are less than 200m from lakes"* and *"All houses that are within 500m from a road"* may eventually result in an empty map, as they restrict the search space too dramatically by excluding any areas, for example, which are 490 from lakes or 510 for a road. Such an area, however, might be the best choice available and therefore perfectly acceptable.

This problem can be solved by using qualitative spatial statements as an alternative to the quantitative ones, when the situation demands it. The alternative qualitative statement may be *"All houses that are very close to lakes"* and *"All houses that are quite close to the roads"*.

To achieve this goal, we interpret spatial relations among objects as restrictions on linguistic variables which represent spatial information about the objects. Instead of directly using a degree of membership, we can describe the phenomena using natural language.

### 10.2.6 Topological Relationships in the *FMBR*

Membership in a $FMBR$ provides us with a measure about how close a feature is with respect to the core. Using geometric relationships we are able to know the relationship between the element and its set, that is, whether a feature is still in its boundary or does not belong to the fuzzy region at all.

The next issue is related to the relationships between fuzzy regions, for instance, those features that are in the common boundary between a mountain and a valley. For cases like this, we need to determine whether or not features in the boundaries of two regions share the same geographic locations. The relations between objects with indeterminate boundaries are called approximate topological relations, since they constitute an approximation of classical topological relations between objects with crisp boundaries.

As in the geometric analysis, we use $FMBR$s and the degree of membership of the fuzzy region calculated in section 10.2.4 to approximate the topological relationships between fuzzy regions. The remaining question is how to employ the $\alpha MBR$s for determining these topological relationships. Consider the example in figure 10.7.

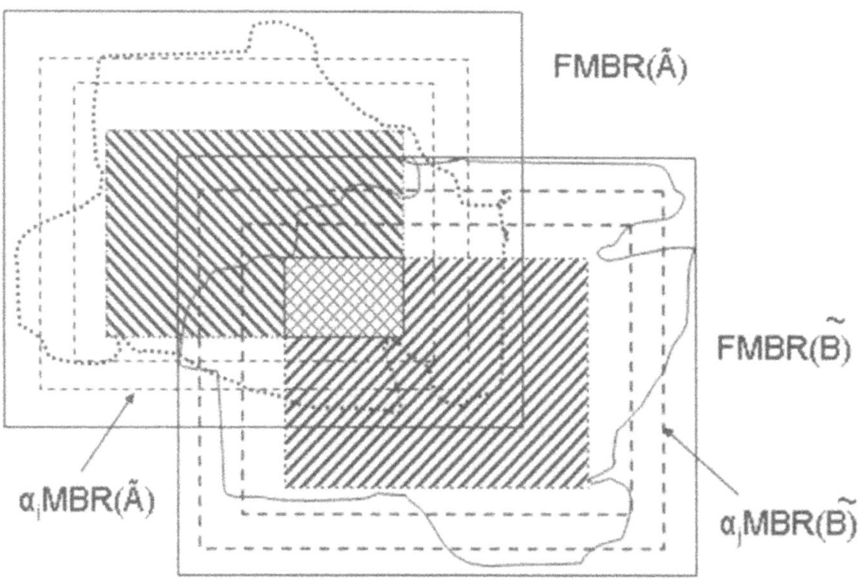


**Figure 10.7.** A relationship between two $FMBR$s

Let $\tau(A, B)$, stand for one of the eight topological relationships (i.e. disjoint, meet, overlap, coveredby, inside, covers, contains, and equal) between the fuzzy regions $A$ and $B$. In figure 10.7, $\tau$ represents the overlap operation. Thus, the topological relations between the $FMBR$s of $A$ and $B$ can be determined by equation 10.6.

$$\tau(FMBR(\widetilde{A}), FMBR(\widetilde{B})) = \sum_{i=1}^{n}\sum_{j=1}^{n} \mu(\alpha_i MBR(\widetilde{A}))\mu(\alpha_j MBR(\widetilde{B}))\tau(\alpha_i MBR(\widetilde{A}), \alpha_j MBR(\widetilde{B})) \quad (10.6)$$

where $\tau(\alpha_i MBR(\widetilde{A}), \alpha_j MBR(\widetilde{B})) \in \{0,1\}$ checks the validity of the predicate $\tau$ between two simple crisp $\alpha$-level regions $\alpha_i MBR(\widetilde{A})$ and $\alpha_j MBR(\widetilde{B})$. Equation (10.7) is a generalization of equation (10.6) to be applied when a point feature lies between two $\alpha MBR$s.

$$\tau(FMBR(\widetilde{A}), FMBR(\widetilde{B})) = \sum_{i=2}^{n}\sum_{j=2}^{n} (\alpha_i MBR(\widetilde{A}) - \alpha_{i-1} MBR(\widetilde{A}))(\alpha_j MBR(\widetilde{B}) - \alpha_{j-1} MBR(\widetilde{B}))\, \tau(\alpha_i MBR(\widetilde{A}), \alpha_j MBR(\widetilde{B})) \quad (10.7)$$

The value of $\tau(\alpha_i MBR(\widetilde{A}), \alpha_j MBR(\widetilde{B}))$ denotes the membership value for $\alpha_i MBR(\widetilde{A})\tau(\alpha_i MBR(\widetilde{A}), \alpha_j MBR(\widetilde{B}))\alpha_j MBR(\widetilde{B})$, that is, how the topological relationship holds in a particular $\alpha MBR$-cut. Equation (10.7) is applicable to any of the eight basic topological relations given that an $\alpha$-cut level region is a crisp region and one of the relations must be satisfied.

Since $\alpha MBR$-cuts level regions are crisp regions that can be represented as $MBR$s, topological relationships are determined for existing GIS algorithms. The values $\tau(\alpha_i MBR(\widetilde{A}), \alpha_j MBR(\widetilde{B}))$ are computed for all pairs $(\alpha_i MBR(\widetilde{A}), \alpha_j MBR(\widetilde{B}))$.

We only present formulations for approximately determing three out of the eight binary topological relations between fuzzy regions in the following paragraphs. The remaining five topological relations can be easily determined using formulations similar to equations (10.8), (10.9), (10.10), as shown below.

*Disjoint.* Let $\tau_{disjoint}(\alpha_i MBR(\widetilde{A}), \alpha_j MBR(\widetilde{B}))$ denote the topological relation *disjoint* between two $\alpha MBR$ level regions $FMBR(\widetilde{A})$ and $FMBR(\widetilde{B})$. Based on (10.7), the binary topological relationship *disjoint* can be determined using equation (10.8).

$$\tau_{disjoint}(FMBR(\widetilde{A}), FMBR(\widetilde{B})) = \sum_{i=1}^{n}\sum_{j=1}^{n} (\alpha_i MBR(\widetilde{A}) - \alpha_{i-1} MBR(\widetilde{A}))(\alpha_j MBR(\widetilde{B}) - \alpha_{j-1} MBR(\widetilde{B}))\, \tau_{disjoint}(\alpha_i MBR(\widetilde{A}), \alpha_j MBR(\widetilde{B})) \quad (10.8)$$

*Equal.* The binary topological relationship *equal* between two fuzzy regions should be treated differently from the other seven topological relations. Two fuzzy regions are said to be equal if and only if all corresponding $\alpha MBR$-cuts of the two fuzzy regions are equal. This relation can be determined using equation (10.9).

$$\tau_{equal}(FMBR(\widetilde{A}), FMBR(\widetilde{B})) = \sum_{i=1}^{n}\sum_{j=1}^{n}(\alpha_i MBR(\widetilde{A}) - \alpha_{i-1}MBR(\widetilde{A}))\tau_{equal}(\alpha_i MBR(\widetilde{A}), \alpha_j MBR(\widetilde{B})) \tag{10.9}$$

*Overlap.* Denote $\tau_{overlap}(\alpha_i MBR(\widetilde{A}), \alpha_j MBR(\widetilde{B}))$ as the topological relation *disjoint* between two $\alpha MBR$ level regions $FMBR(\widetilde{A})$ and $FMBR(\widetilde{B})$. The binary topological relationship *overlap* can be determined using equation 10.10.

$$\tau_{overlap}(FMBR(\widetilde{A}), FMBR(\widetilde{B})) = \sum_{i=1}^{n}\sum_{j=1}^{n}(\alpha_i MBR(\widetilde{A}) - \alpha_{i-1}MBR(\widetilde{A}))(\alpha_j MBR(\widetilde{B}) - \alpha_{j-1}MBR(\widetilde{B})) \tau_{overlap}(\alpha_i MBR(\widetilde{A}), \alpha_j MBR(\widetilde{B})) \tag{10.10}$$

Since the result of the operation $\tau(\alpha_i MBR(\widetilde{A}), \alpha_j MBR(\widetilde{B}))$ has to be 0 (false) or 1 (true) and the summation of the memberships for all the $\alpha$-cuts is 1, then the topological relation for the whole fuzzy region, $\tau(FMBR(\widetilde{A}), FMBR(\widetilde{B}))$ has a value between 0 and 1.

It remains to show that $0 \leq \tau(FMBR(\widetilde{A}), FMBR(\widetilde{B})) \leq 1$ actually holds, that is, $\tau$ is a fuzzy predicate. Since $\alpha_i MBR - \alpha_{i-1}MBR > 0$ for all $2 \leq i \leq n$ and since $\tau(\alpha_i MBR(\widetilde{A}), \alpha_j MBR(\widetilde{B})) \geq 0$ for all $2 \leq i, j \leq n$, $\tau(FMBR(\widetilde{A}), FMBR(\widetilde{B})) \geq 0$ holds. We can show the other inequality by determining an upper bound for $\tau(FMBR(\widetilde{A}), FMBR(\widetilde{B}))$:

$$\begin{aligned}\tau(FMBR(\widetilde{A}), FMBR(\widetilde{B})) &= \sum_{i=2}^{n}\sum_{j=2}^{n}(\alpha_i MBR(\widetilde{A}) - \alpha_{i-1}MBR(\widetilde{A})) \\ &\quad (\alpha_j MBR(\widetilde{B}) - \alpha_{j-1}MBR(\widetilde{B}))\tau(\alpha_i MBR(\widetilde{A}), \alpha_j MBR(\widetilde{B})) \\ &\leq \sum_{i=2}^{n}\sum_{j=2}^{n}(\alpha_i MBR(\widetilde{A}) - \alpha_{i-1}MBR(\widetilde{A}))(\alpha_j MBR(\widetilde{B}) - \alpha_{j-1}MBR(\widetilde{B}))\end{aligned}$$

(since, $\tau(\alpha_i MBR(\widetilde{A}), \alpha_j MBR(\widetilde{B})) \leq 1$)

$$
\begin{aligned}
=&(\alpha_2 MBR(\widetilde{A}) - \alpha_1 MBR(\widetilde{A}))(\alpha_2 MBR(\widetilde{B}) - \alpha_1 MBR(\widetilde{B}))\\
&+\cdots+(\alpha_2 MBR(\widetilde{A}) - \alpha_1 MBR(\widetilde{A}))(\alpha_n MBR(\widetilde{B}) - \alpha_{n-1} MBR(\widetilde{B}))\\
&+\cdots+(\alpha_n MBR(\widetilde{A}) - \alpha_{n-1} MBR(\widetilde{A}))(\alpha_2 MBR(\widetilde{B}) - \alpha_1 MBR(\widetilde{B}))\\
&+\cdots+(\alpha_n MBR(\widetilde{A}) - \alpha_{n-1} MBR(\widetilde{A}))(\alpha_n MBR(\widetilde{B}) - \alpha_{n-1} MBR(\widetilde{B}))\\
=&(\alpha_2 MBR(\widetilde{A}) - \alpha_1 MBR(\widetilde{A}))((\alpha_2 MBR(\widetilde{B}) - \alpha_1 MBR(\widetilde{B}))\\
&+\cdots+(\alpha_n MBR(\widetilde{B}) - \alpha_{n-1} MBR(\widetilde{B})))\\
&+\cdots+(\alpha_n MBR(\widetilde{A}) - \alpha_{n-1} MBR(\widetilde{A}))((\alpha_2 MBR(\widetilde{B}) - \alpha_1 MBR(\widetilde{B}))\\
&+\cdots+(\alpha_n MBR(\widetilde{B}) - \alpha_{n-1} MBR(\widetilde{B})))\\
=&(\alpha_2 MBR(\widetilde{A}) - \alpha_1 MBR(\widetilde{A})) + ... + (\alpha_n MBR(\widetilde{A}) - \alpha_{n-1} MBR(\widetilde{A}))
\end{aligned}
$$

(since $\sum_{i=2}^{n}(\alpha_i - \alpha_{i-1}) = 1$)
$= 1$

Hence, $\tau(FMBR(\widetilde{A}), FMBR(\widetilde{B})) \leq 1$ holds.

**Example:**

Let $\widetilde{A}$ and $\widetilde{B}$ be two fuzzy regions. The $FMBR(\widetilde{A})$ *contains* $FMBR(\widetilde{B})$ to some degree as shown in figure 10.8. We apply equation (10.7) using the $\alpha_i MBR$ membership generated by the function (10.5). For simplicity reasons, we assume that in both regions, $\widetilde{A}$ and $\widetilde{B}$, $\alpha_1 = 0$, $\alpha_2 = 0.25$, $\alpha_3 = 0.50$, $\alpha_4 = 0.75$, and $\alpha_5 = 1$. Then, we substitute these values in the equation (10.7) to obtain the solution:

$$
\sum_{i=2}^{n}\sum_{j=2}^{n}(\alpha_i MBR(\widetilde{A}) - \alpha_{i-1} MBR(\widetilde{A}))(\alpha_j MBR(\widetilde{B}) - \alpha_{j-1} MBR(\widetilde{B}))
$$

$$
\tau_{contains}(\alpha_i MBR(\widetilde{A}), \alpha_j MBR(\widetilde{B}))
$$

thus

$$
\tau_{contains}(FMBR(\widetilde{A}), FMBR(\widetilde{B})) = 0.875
$$

The evaluation of the *contains* predicate for $\alpha_2 MBR(\widetilde{A})$ results in 0 for the $\alpha_1 MBR(\widetilde{B})$ and $\alpha_2 MBR(\widetilde{B})$ which are the $\alpha MBR$s of $FMBR(\widetilde{B})$ that are outside of the $FMBR(\widetilde{A})$. Therefore, we can say that $FMBR(\widetilde{A})$ *contains* the $FMBR(\widetilde{B})$ with a degree of belief 0.875.

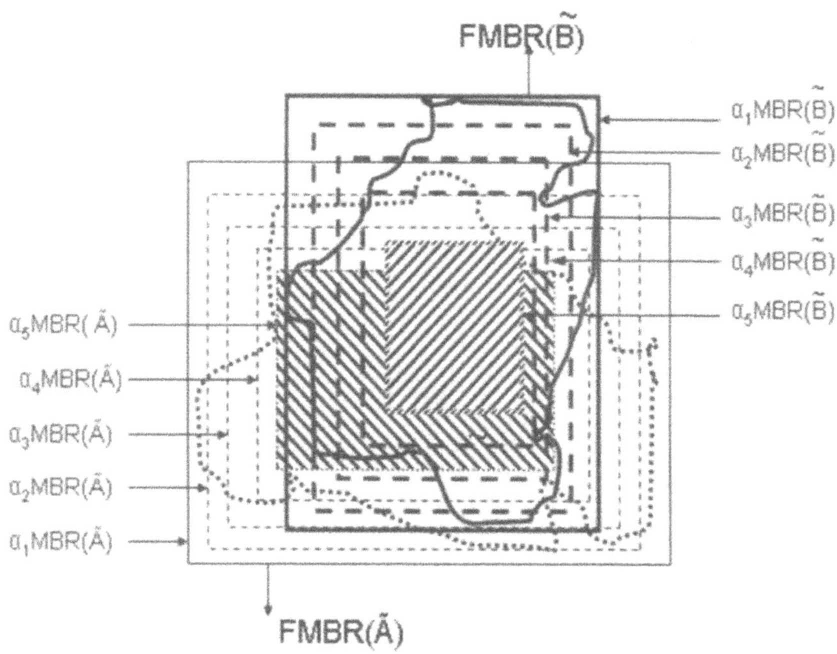


**Figure 10.8.** Contains relationship between $FMBR(\widetilde{A})$ and $FMBR(\widetilde{B})$

## 10.3 Thematic Reasoning

The thematic data associated with geographic entities and characterizing them depends on the application domain. Fuzzy set methodologies can be used to provide a more realistic representation of thematic dimension. Specifically, the application domain defines a set of $j$ lexical values characterizing a phenomenon, for example, for the application domain: *land-moisture*, a set of lexical values may have be defined as: [*dry, moderate, wet, water*]. To each individual entity, related to that phenomenon, is attached some degree of membership, one for each lexical value.

The data model, as shown in figure 10.9, can be viewed as a hierarchy of data [11]. At the highest level of abstraction there is a space-time set, which is called *map*. The map is composed of *thematic layers*. Each thematic layer refers to the theme characterizing geographic entities, for instance, *vegetation, elevation, temperature,* etc. The thematic layers are comprised of *lexical layers.* Each lexical layer accommodates the degree of membership of each *individual space-time location* in the lexical value characterizing a theme. A space-time location is the smallest spatio-temporal unit of interest for the corresponding layer in the application domain. Geographic entities of interest are represented by space-time *zones* and identified uniquely. Therefore, each entity or zone is represented by a set of spatio-temporal locations approximated by its $FMBR$. For each thematic space that may have a hierarchical structure, the themes are conceived of as regions having topological relationships in a theme space. The overlay operation is defined as the generation of a new layer and the assignment of the attribute values to its individual locations resulting from the combination of the corresponding values in two or more existing thematic layers. It is important to notice that a geographic area may change form or attributes over time and the proposed model may represent this information.

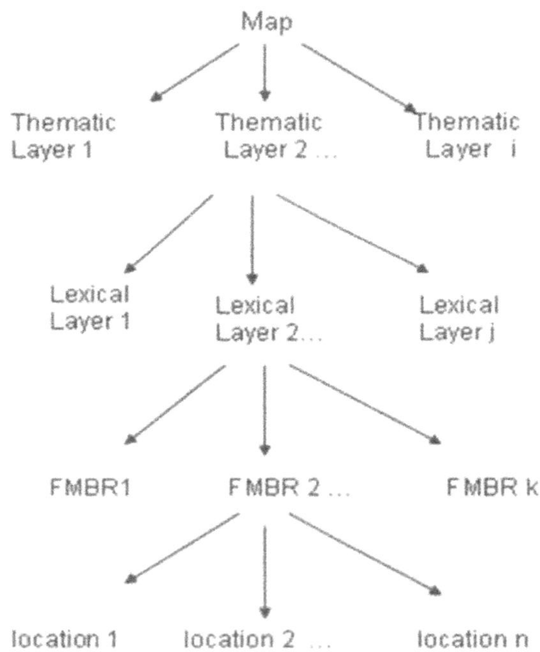


**Figure 10.9.** Hierarchy of data

An example of a map might be Mexico in 1980–1990 decade (figure 10.10). Examples of thematic layers are states layer, road network layer, hydrology layer, etc. The hydrology layer consists of the following lexical sub-layers: river, stream, lake, lagoon, pond, marsh, etc. An individual location $(X, Y, T, t)$ that belongs to the $FMBR[(x_1, y_1), (x_2, y_2)]$ corresponds to the location $(X, Y)$ in the $ith$ $FMBR$ of the lexical layer of the thematic layer $T$ at time $t$. For instance, an individual location, which is assigned a degree of membership equal to 0.7, means that this location belongs to a lake with a degree of membership

0.7. All individual locations in a lake cube with a degree of membership value in the interval $[0.7, 1]$ and $T = summer$, compose a zone, which represents the lakes areas in Mexico at summer time.

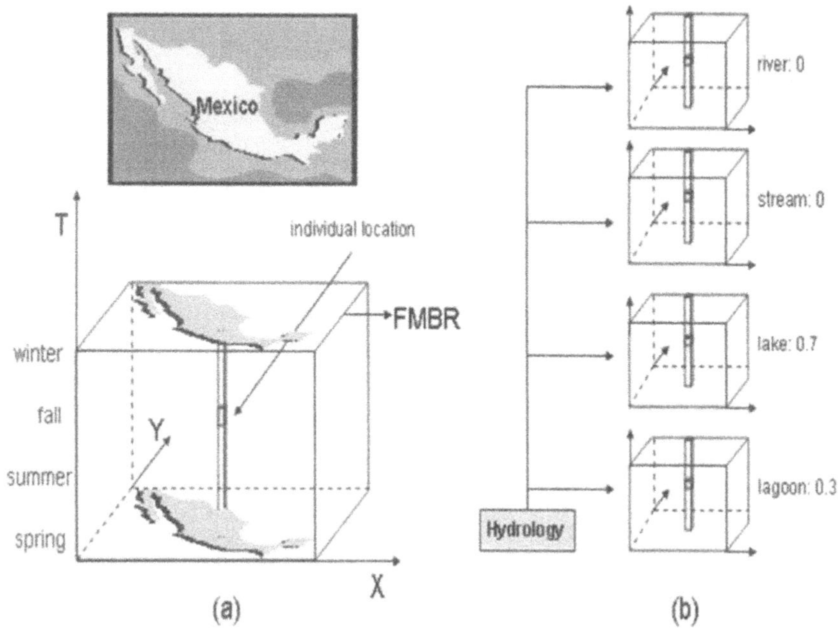


**Figure 10.10.** (a) Map of Mexico (b) The hydrology layer and its lexical layers with the degree of membership of the locations

## 10.4 Temporal Reasoning

The need to evaluate past situations and previous values of the variables (which in many cases are fuzzy) and principally, to reason about them, has led us to incorporate temporal reasoning. The use of the conventional fuzzy rules would not permit the direct treatment of the knowledge, since use of average values would not reflect sharp variations of a variable in a cycle, or it would take a long time to detect a gentle and constant change in a variable.

Due to the dynamic structure of the geographical entities and phenomena, we use an integrated fuzzy spatio-thematic-temporal framework to model them. Trying to capture the dynamic nature of the geographic phenomena, we formulate a statement involving spatio, thematic, and temporal variables. This model is similar to the one used in [1] and [9].

Two basic temporal entities in the statements are used: fuzzy instant and fuzzy intervals. Statements like "*X is A for T*" are allowed, with $T$ being a fuzzy instant or fuzzy interval.The formulation for the statement is:

$$V \; in \; FMBR \; is \; A \; \langle in \; Q \; of \rangle \; T \tag{10.11}$$

where

- $V$ is a linguistic variable;
- $A$ is a linguistic value of $V$;
- $T$ is a temporal reference or entity;
- $Q$ is a fuzzy quantifier.

The temporal entities $T$ may represent both instant as well as fuzzy temporal intervals, being in both cases, membership functions defined on a discrete set of values $\tau = \{\tau_0, \tau_1, \ldots, \tau_k, \ldots, \tau_{now}\}$, where each $\tau_k$ represents a precise instant of time, $\tau_0$ represents the origin, and $\tau_{now}$ the current or possibly a future time point. Syntactic constructions $\langle in \; Q \; of \rangle$ may exhibit well differentiated semantics when $T$ represents a fuzzy temporal interval. In figure 10.11 some definitions of the membership functions $\mu_Q$ associated to the temporal persistence quantifiers ("In") used in the application are represented.

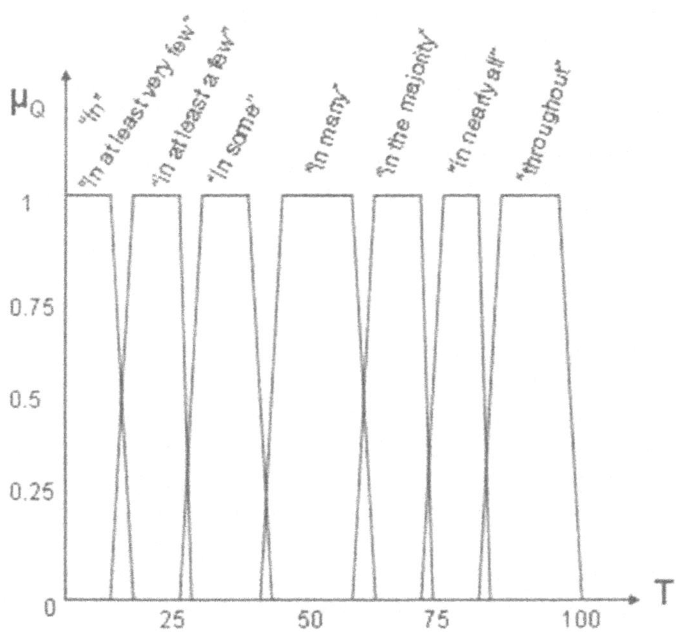


**Figure 10.11.** Membership functions of the temporal quantifiers

The calculation of the degree of fulfillment (DOF) of the statement like (10.11) is accomplished by taking into account all the points belonging to the support of $T$, in the following manner:

- *Non-persistence: "V in FMBR is A in T"*

$$DOF = max(min(\mu_{FMBR}(X(\tau_k)), \mu_A(X(\tau_k)), \mu_T(X(\tau_k)))) \quad (10.12)$$

- *Persistence: "V in FMBR is A throughout T"*

$$DOF = min(max(\mu_{FMBR}(X(\tau_k)), \mu_A(X(\tau_k)), 1 - \mu_T(X(\tau_k)))) \quad (10.13)$$

- *Intermediate Case: "V in FMBR is A in Q of T"*

$$DOF = \frac{\sum_{\tau_k \in SUPP_T} (min(\mu_{FMBR}(X(\tau_k)), \mu_A(X(\tau_k)), \mu_T(X(\tau_k))))}{\sum_{\tau_k \in SUPP_T} \mu_T(\tau_k)}$$

where[3]

- $\mu_{FMBR}$ is the membership function associated to the $FMBR$;
- $\mu_A$ is the membership function associated to the value $A$ of the statement;
- $X(\tau_k)$ is the value observed for the variable $X$ in the instant $\tau_k$;
- $\mu_T$ is the membership function of the temporal reference;
- $\mu_Q$ is the membership function associated with the linguistic quantifier $Q$.

Figure 10.12 shows an example of the calculation of the DOF for the statement *"lake very-close to forest is high-level throughout spring"*. The previous statement means that the level of the lake, which is very close to the forest should be high throughout the spring. The sub-statement *"lake very-close to forest"* is a qualitative form of the sub-statement *"V in FMBR"* in statement (10.11). The membership of the lake in the fuzzy set that topologically relates the lake with the forest corresponds to the linguistic variable "*very-close*". In other words, the membership function based on distance that determines the degree of membership of the $FMBR$ has been labeled with a linguistic variable.

The process is as follows: first, the membership degree of the variable *lake* for the linguistic label *very-close* for each instant $\tau_k$ in the support of the fuzzy set corresponding to "spring" is calculated. The time line has been divided in equally spaced intervals, $\triangle$. Second, the membership degree of the variable *level* for the linguistic modifier *high* is determined for each instant $\tau_k$. Thus, six membership degrees are obtained, one for each temporal instant belonging to the support of $T$. Next, and given that the statement is of a persistent type (*"throughout"*), expression (10.13) is applied. In the example, this calculation leads to a result of 0.6, which corresponds to the value measured at the temporal instant $\tau_{now}$ $6\triangle$.

[3] The support $SUPP_A$ of a membership function $\mu_A$ defined in an universe of discourse $U$ is defined as $SUPP_A = \{u \in U \mid \mu_A(u) > 0\}$

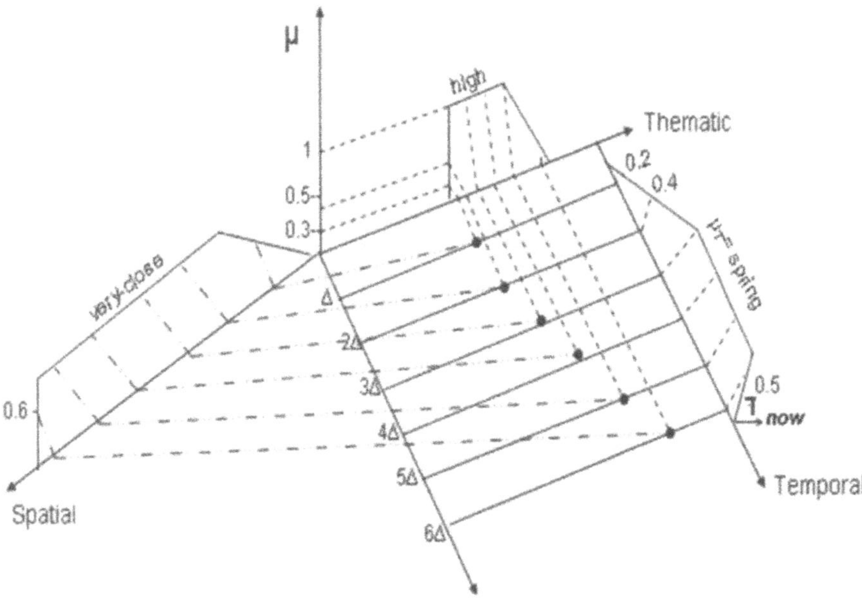


**Figure 10.12.** Calculation of the DOF for the statement *"lake very-close to forest is high-level throughout spring"*

## 10.5 Querying a Fuzzy MBR

A spatial database stores objects that have spatial characteristics which describe them. The spatial relationships among the objects are also important, and they are often required when querying the database. The main extensions that are needed for spatial databases are models that can interpret spatial characteristics. The basic extensions needed are two-dimensional geometric concepts, such as points, lines and polygons, in order to specify the spatial characteristics of the object. In addition, special indexing and storage structures are crucial to improve performance.

Spatial operations are required to operate on an object's spatial characteristics, for instance, to compute the distance between two objects, or to check whether two spatial objects overlap. The following categories represent the typical types of spatial queries:

- Feature-based query

  Range query: It finds the objects of a particular type that are inside a given spatial area or within a particular distance from a given location, for example —select all rivers that pass through a city?— or —are there any train depots within region x?—

Spatial joins or overlays: These join objects of two types based on some spatial condition, such as the objects intersecting or overlapping spatially, for instance —find all cities that fall on major highways?—

- Location-based queries

Nearest neighbor query: It finds an object of a particular type that is closest to a given location, for example —what is the nearest waterway to train depot x?—

A Fuzzy $MBR$s database, besides the extensions needed to implement a spatial database has to manage the uncertainty associated with the borders of the geographic phenomena to be able to answer queries like: —find all locations that are about 500 m from a road— or —find all locations that are very close to the roads—.

### 10.5.1 Indexing and storage structures

For the type of spatial queries mentioned to be answered efficiently, spatial techniques for spatial indexing are needed. One of the best known techniques is the use of Grid Files. We propose to use Grid Files for storing and indexing vague regions represented by $FMBR$s.

The grid file access method [6] retrieves records by at most two disk access and efficiently handles range queries. This is done by using a grid directory consisting of grid cells. All records in one grid cell are stored in the same bucket. However, several grid cells can share a bucket as long as the union of these grid cells forms a rectangle in the space of records. This guarantees that the records stored in the same bucket be close each other. Although the regions of the buckets are piecewise disjoint, together they span the space of records. To guarantee that data items are always found with no more than two disk access for exact match queries, the grid itself is kept in main memory, represented by n-dimensional arrays called *scales*, where n is the number of search keys.

Figure 10.13 shows a grid file which has a bucket capacity of sixteen data points. The center of the figure shows the grid directory with 1-dimensional scales on the x- and y-axes. To answer a query, we first use the scales to locate the cell containing the search point. If the appropriate grid cell is not in the main memory, one disk access is necessary, and the loaded cell contains the reference to the page of where to find the possibly matching data.

Since we are working with $FMBR$s to represent vague regions, we have decided to store points belonging to the same $\alpha MBR$ in the same bucket to enhance querying performance. Thus, we might have many buckets as $\alpha MBR$s. In figure 10.13, the two buckets above the grid directory and the one at the bottom right corner store just one $\alpha MBR$. On the other side, the bucket at the bottom left corner stores two different $\alpha MBR$s. This does not violate

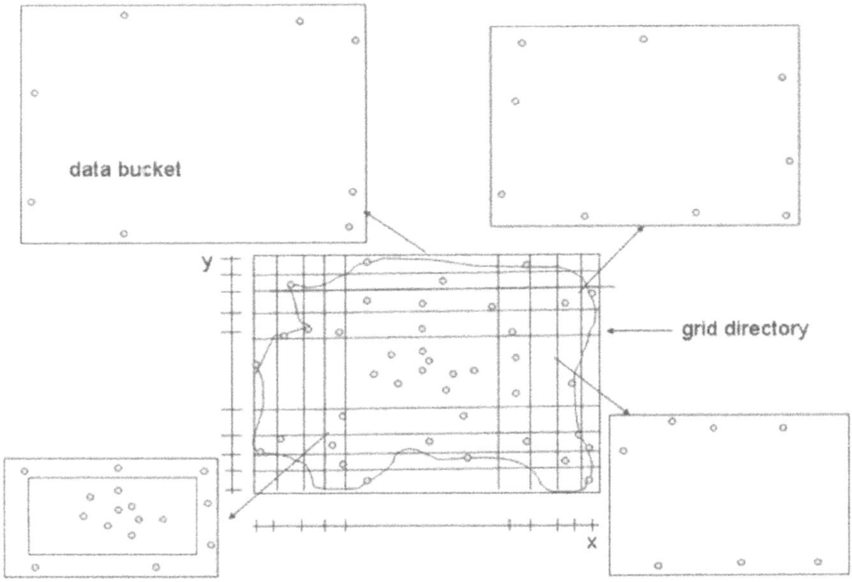


**Figure 10.13.** $FMBR$s Grid File

the organization principle of the grid files because the points of a particular $\alpha MBR$ forms a rectangle. In addition, a relationship of closeness with respect to the core of the vague region is fulfilled. Moreover, points within two consecutive $\alpha MBR$s are close enough among them and do not decrease the querying process performance, thus, they can be merged in the same bucket.

### 10.5.2 Querying FMBRs by location

To answer queries by location is straightforward. Once we know the coordinates of a feature we can know the relevant cell by looking in the grid file. Considering again the query, —what is the nearest waterway ($W_i$) to the train depot $X$?—, the solution of the query is as follow:

- Obtain the coordinates of the train depot $X$.
- Use the coordinates to locate the bucket containing $X$.
- Look for other data points representing waterways. The searching starts in the same bucket where is $X$ stored. If more than one waterway were found, the one having the least distance to $X$ is selected for the answer. If no waterways were found in the same bucket as $X$, the searching continues at the next inner $\alpha FMBR$, otherwise we need to look for waterways at the next outer $\alpha FMBR$ and so on. Figure 10.14 shows an example of this kind of query.

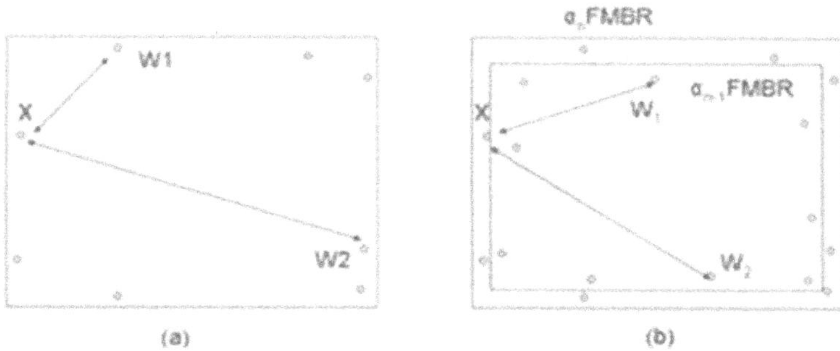


**Figure 10.14.** Searching of the nearest waterway to the train depot $X$

In figure 10.14 the $\alpha FMBR$ that contains $X$ does not contain any waterways, so the searching continues in the next $\alpha FMBR$. This $\alpha FMBR$ contains $W_1$ and $W_2$ waterways. Since the distance between the train depot $X$ and $W_1$ is the minimum, $W_1$ is the waterway that satisfies the answer.

### 10.5.3 Querying FMBRs by features

In order to query the database by features we need to keep a look up table mapping features and their respective locations. To answer the previous query —are there any waterway in the region $X$?—, we need then:

- To get the coordinates of the region $X$ from the table. These coordinates should correspond to the corners of the $FMBR$ approximating $X$.
- Each $\alpha FMBR$ of the entire $FMBR$ needs to be checked for any waterway lying in it. Since features within an $\alpha FMBR$ have a degree of membership for the fuzzy region depending on the distance to the $FMBR$'s core, different degrees of membership will be assigned to the waterways belonging to the region $X$. An example of the evaluation of this query is shown in figure 10.15.

In figure 10.15, first the $\alpha FMBR$ enclosing the region $X$ is located. Then search for the waterways starts at the $\alpha FMBR$ closest to the core of the region (inner $\alpha FMBR$). Thus, $W_1$ is the waterway that satisfies the query best, since its membership in the fuzzy region $X$ is higher than the membership of $W_2$.

## 10.6 Summary

We have defined a formal framework that is capable of describing the different aspects of vague spatial regions. It is an extension of a determinate region

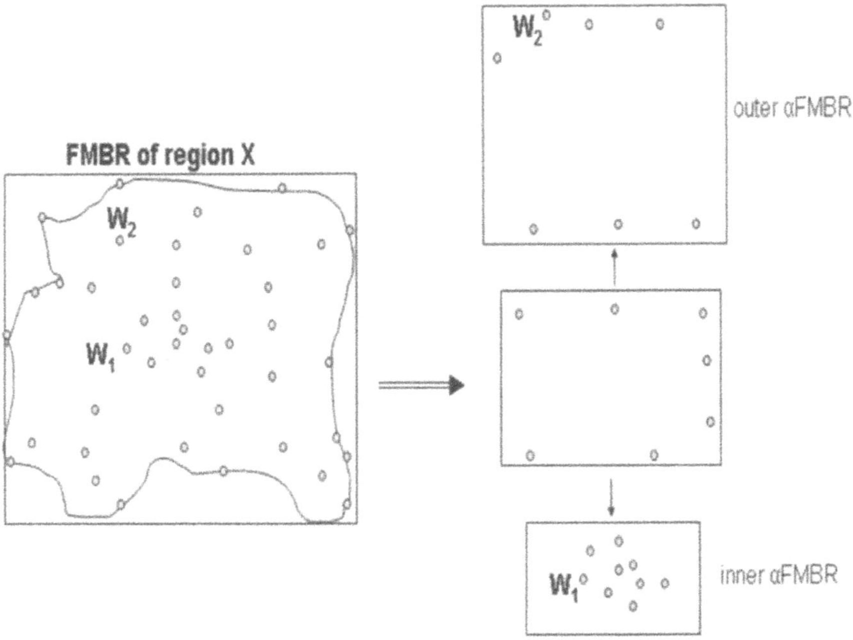


**Figure 10.15.** Finding any waterway in the region $X$

model which facilitates the treatment of vague regions by applying fuzzy sets concepts.

A vague region is modeled by using a $FMBR$. We start the process taking the $MBR$ as the first approximation of the vague region, that is, the external edge. Beyond the $MBR$ edge, the membership of the fuzzy region is equal to zero. By consecutive generation of $\alpha MBR$-cuts, we obtain higher degrees of membership until we have an $\alpha MBR$-cut with a degree of membership of 1, that is, the core of the fuzzy region. Besides the representation of individual vague regions, we are able to represent the topological relationships among two fuzzy regions modeled each as an $FMBR$.

We incorporate the thematic and temporal dimensions to complement the $FMBR$ spatial reasoning, in order to model a complete view of the geographic world. Thematic aspects were modeled by the definition of spatial ontologies, where maps are at the higher level and $FMBR$s containing single areas constitute the bottom level. Instants and temporal intervals are used to specify the time when a thematic fuzzy region exists in its $FMBR$ position.

Finally, to query an $FMBR$, a representation based on a grid file indexing mechanism has been proposed. This strategy allow us to query the $FMBR$s by location and features.

## References

1. Bugarin A., Carinene P., Felix P., Barro S. (1999) Fuzziness in Petri Nets. In: Studies in Fuzziness and Soft Computing, Vol. 22, pp. 174–202.
2. Burrough, P. (1996) Natural Objects with Inderminate Boundaries. In: Geographic Objects with Indeterminate Boundaries, pp. 3–28.
3. Clementini E., Sharma J., Egenhofer M. (1994) Modeling Topological Spatial : Strategies for Query Processing Relationships. International Journal of Computer and Graphics 18:815–822.
4. Dubois D., Jaulent M. (1987) A general approach to parameter evaluation in fuzzy digital pictures. Pattern Recognition Letters 6:251–259.
5. Egenhofer M., Herring J. (1991) Categorizing binary topological relationships between regions, lines, and points in geographic databases. Technical Report, Univ. of Maine.
6. Elmasri R., Navathe S. (2000) Fundamentals of Database Systems. 3th. Edition, Addison Wesley.
7. Euclid (1956) The Thirteen Books of Euclid's Elements. Addison and Wesley.
8. Guesgen H., Albrecht J. (2000) Imprecise reasoning in geographic information systems. Fuzzy Sets and Systems 113:121–131.
9. Mucientes M., Iglesias R., Regueiro C., Bugarin A., Carinena P., Barro S. (2001) Fuzzy Temporal Rules for Mobile Robot Guidance in Dynamics Enviroments. IEEE Transactions on Systems, MAN, and Cybernetics-Part C: Applications and Reviews, 31(3):391–398.
10. Somodevilla M., Petry F. (2003) Approximation of Topological Relations on Fuzzy Regions: An approach using Minimal Bounding Rectangles. In: Proc. of the NAFIPS03 Conference, Chicago, Il, pp. 1–23.
11. Stefanakis E. (2001) A Unified Framework for Fuzzy Spatio-Temporal Representation and Reasoning. In: Proc. of the 20th International Cartographic Conference, pp. 345–360.
12. Usery E. (1996) A Conceptual Framework and Fuzzy Set Implementation for Geographic Features. In: Geographic Objects with Indeterminate Boundaries, pp. 71–85.

# 11

# Fuzzy Spatial Data Types and Predicates: Their Definition and Integration into Query Languages

Markus Schneider

University of Florida
Department of Computer & Information Science & Engineering
E450 CSE Building
PO Box 116120
Gainesville, FL 32611-6120
mschneid@cise.ufl.edu

## 11.1 Introduction

Representing, storing, quering, and manipulating spatial information is important for many non-standard database applications. Specialized systems like geographical information systems (GIS) and spatial database systems to a certain extent provide the needed technology to support these applications. So far, spatial data modeling has implicitly assumed that the extent and hence the borders of spatial phenomena are precisely determined, homogeneous, and universally recognized. From this perspective, spatial phenomena are typically represented by sharply described *points* (with exactly known coordinates), *lines* (linking a series of exactly known points), and *regions* (bounded by exactly defined lines which are called *boundaries*). Special data types called *spatial data types* (see [23] for a survey) have been designed for modeling these spatial data. We speak of *spatial objects* as instances of these data types. The properties of the space at the points, along the lines, or within the regions are given by attributes whose values are assumed to be constant over the total extent of the objects. Well known examples are especially man-made spatial objects representing engineered artifacts like highways, houses, or bridges and some predominantly immaterial spatial objects exerting social control like countries, districts, and land parcels with their political, administrative, and cadastral boundaries. We will denote this kind of entities as *crisp* or *determinate* spatial objects.

Increasingly, researchers are beginning to realize that the current mapping of spatial phenomena of the real world to exclusively crisp spatial objects is an insufficient abstraction process for many spatial applications and that the feature of *spatial vagueness* or *spatial indeterminacy* is inherent to many geographic data [4]. Moreover, there is a general consensus that applications based on this kind of indeterminate spatial data are not covered by current GIS and spatial database systems. In this chapter we focus on a special kind of spatial vagueness called *spatial fuzziness*. Fuzziness captures the property of many spatial objects in reality which do not have sharp boundaries or whose boundaries cannot be precisely determined. Examples are natural, social, or cultural phenomena like land features with continuously changing properties (such as population density, soil quality, vegetation, pollution, temperature, air pressure), oceans, deserts, English speaking areas, or mountains and valleys. The transition between a valley and a mountain usually cannot be exactly ascertained so that the two spatial objects "valley" and "mountain" cannot be precisely separated and defined in a crisp way. We will designate this kind of entities as *fuzzy* spatial objects.

Topological relationships characterize the relative locations of two spatial objects, for example, whether they overlap, meet, or are disjoint. In spatial databases and GIS they are important for formulating spatial selections and spatial joins. The definition of topological predicates on fuzzy spatial objects in general and fuzzy regions in particular is currently an open problem. For two fuzzy regions $A$ and $B$ we would like to be able to pose and answer queries like

- Do regions $A$ and $B$ overlap *a little bit*?
- Determine all pairs of regions that *nearly completely* overlap.
- Does region $A$ *somewhat* contain region $B$?
- Which regions lie *quite* inside $B$?

In this chapter we present a formal object model for *fuzzy points*, *fuzzy lines*, and *fuzzy regions* in the two-dimensional Euclidean space, an effort which is to lead to a *fuzzy spatial algebra*. We propose *fuzzy set theory* and *fuzzy topology* as appropriate conceptual tools for modeling indeterminate spatial data. Fuzzy set theory is an extension and generalization of classical set theory; the approach of fuzzy sets replaces the crisp boundary of a classical set by a gradual transition zone and permits partial and multiple set membership. For fuzzy regions, different views give a better understanding of their nature and also demonstrate how these objects can be represented as (collections of) crisp regions. Consequently, the current exact object models for crisp spatial objects can be considered as simplified special cases of a richer class of models for general spatial objects.

Based on the data type definitions, we consider topological predicates on fuzzy spatial objects and show their integration into a *fuzzy spatial query language*. *Fuzzy modifiers* may be used in the query language to emphasize nuances of topological predicates and thus to modify their semantics.

Section 11.2 explains different aspects of spatial vagueness and presents related work. Section 11.3 introduces some basic definitions of fuzzy set theory and fuzzy topology as far as they are needed in this paper. Sections 11.4, 11.5 and 11.6 formally define fuzzy points, fuzzy lines, and fuzzy regions, respectively. Since the definition for fuzzy regions does not expose their geometric structure, Section 11.7 provides several structured views of fuzzy regions based on collections of crisp complex regions. Section 11.8 introduces topological predicates on complex crisp regions as they are needed in Section 11.9 for defining topological predicates on fuzzy complex regions. Section 11.10 demonstrates how querying with fuzzy topological predicates can be done. Section 11.11 draws some conclusions and gives a prospect of future research.

## 11.2 Related Work

In the past, there has been a strong tendency to force vague spatial reality into crisp objects. In practice, however, there is no apparent reason for the whole boundary of a region to be determined. There are a lot of geographical application examples illustrating that the boundaries of spatial objects can be partially or totally indeterminate or blurred. For instance, boundaries of geological, soil, and vegetation units [1, 3, 19, 20] are often sharp in some places and vague in others; many human concepts like "the Indian Ocean" are implicitly vague.

In the real world, there are essentially two categories of indeterminate boundaries: sharp boundaries whose position and shape are unknown or cannot be measured precisely, and boundaries which are not well-defined or which are useless (e.g., between a mountain and a valley) and where essentially the topological relationship between spatial objects is of interest. According to these two categories, mainly two kinds of spatial vagueness can be identified: uncertainty and fuzziness. *Spatial uncertainty* is traditionally equated with randomness and chance occurrence and relates either to a lack of knowledge about the position and shape of an object with an existing, real boundary (*positional* uncertainty) or to the inability of measuring such an object precisely (*measurement* uncertainty). *Spatial fuzziness* is an intrinsic feature of an object itself and describes the vagueness of an object which certainly has an extent but which inherently cannot or does not have a precisely definable boundary.

The subject of modeling spatial vagueness has so far been predominantly treated by geographers but rather neglected by computer scientists. At least three alternatives are proposed as general design methods: (1) *exact models* [7, 8, 15, 22] which transfer type systems and concepts for spatial objects with sharp boundaries to objects with unclear boundaries and which model both uncertainty and fuzziness but in a restricted way, (2) *probabilistic models* [2, 3, 16, 24] which are based on probability theory and predominantly model positional and measurement uncertainty, and (3) *fuzzy models* [1, 3, 10, 11, 19, 20, 26, 27, 28, 31] which are all based on fuzzy set theory and predominantly model fuzziness.

The exact object model approach profits from existing definitions, techniques, data structures, algorithms, etc. which need not be redeveloped but only modified and extended. Except for [15], the approaches are based on some kind of *zone* concept. Vague boundaries of a region are modeled as zones expressing the minimal and maximal possible extent of a region. *Vague regions* [15] are a generalization of these models. A vague region is defined as a pair of disjoint, crisp regions. The first region called the *kernel* describes the area which definitely and always belongs to the vague region. The second region called the *boundary* describes the *area* for which we cannot say with any certainty whether it or parts of it belong to the vague region or not. *Maybe* it is the case, *maybe* it is not. Or we could say that this is *unknown*. Vague regions are based on a *three-valued logic*, and boundaries need not necessarily be one-dimensional structures but can be regions.

Probability theory is able to represent uncertainty and defines the probability of an entity in a set by a statistically defined probability function. It deals with the *expectation* of a future event, based on something known now. Examples are the uncertainty about the spatial extent of regions defined by some property such as temperature, or the water level of a lake.

Fuzzy set theory deals only with fuzziness, that is, it describes the *admission of the possibility* (given by a so-called *membership function*) that an individual is a member of a set or that a given statement is true. Hence, the vagueness represented by fuzziness is not the uncertainty of expectation. It is the vagueness resulting from the imprecision of meaning of a concept. Examples of fuzzy spatial objects include mountains, valleys, biotopes, oceans, and many other geographic features which cannot be rigorously bounded by a sharp line.

The potential importance of fuzzy sets for geographical applications is demonstrated in [3, 20, 26] where also examples of application-specific membership functions are given. The benefits of fuzzy set theory for approximate spatial reasoning and fuzzy query languages is shown in [10, 11, 19, 27]. [28] models fuzzy objects by means of the relational data model.

Our definitions of fuzzy topological relationships are based on the so-called 9-*intersection model* [12] from which a complete collection of mutually exclusive topological relationships can be derived for each combination of spatial types. The model is based on the nine possible intersections of *boundary* ($\partial A$), *interior* ($A^\circ$), and *exterior* ($A^-$) of a spatial object $A$ with the corresponding components of another object. Each intersection is tested for the topologically invariant criteria of emptiness and non-emptiness. $2^9 = 512$ different configurations are possible from which only a limited subset makes sense depending on the combination of spatial objects just considered.

A restriction of the 9-intersection model with respect to regions is that regions must be homeomorphic to the closed disc, that is, they must be connected and are not allowed to have holes. These regions are usually called *simple regions*. For two simple regions, eight meaningful configurations have been identified which lead to the eight predicates of the set $T_{sr}$ = {*disjoint*, *meet*, *overlap*, *equal*, *inside*, *contains*, *covers*, and *coveredBy*}. Each predicate is uniquely determined so that all predicates are mutually exclusive and complete with regard to the topologically invariant criteria of emptiness and non-emptiness.

Topological predicates on crisp *complex regions* have so far not been defined. In [6] the so-called TRCR (Topological Relationships for Composite Regions) model only allows sets of disjoint simple regions without holes. In [13] only topological relationships of simple regions with holes are considered; multi-part regions are not permitted. Topological predicates on fuzzy spatial objects, let them be simple or complex, have so far not been defined.

## 11.3 Fuzzy Sets and Fuzzy Topology

Crisp regions have been formally defined on the basis of point sets and point set topology (e.g., [15, 17, 23]) which mainly rest on the set operations of union, intersection, and difference. In a straightforward way, for fuzzy spatial objects we will now use extensions of these two concepts. They are provided by fuzzy set theory and fuzzy topology. Fuzzy set theory [30] is an extension and generalization of Boolean set theory. Let $X$ be a classical (crisp) set of objects, called the *universe (of discourse)*. All elements of $X$ are mapped to the real interval [0,1] indicating the *degree of membership* of these elements in the set in question. Hence, fuzzy set theory permits an element to have partial membership. Larger values designate higher grades of set membership. Let $X$ again be the universe. Then $\mu_{\tilde{A}} : X \to [0,1]$ is called the *membership function* of $\tilde{A}$, and the set $\tilde{A} = \{(x, \mu_{\tilde{A}}(x)) \mid x \in X\}$ is called a *fuzzy set* in $X$.

There are many ways of extending the set inclusion as well as the basic crisp set operations to fuzzy sets. We will comply with the definitions in [30]. Let $\tilde{A}$ and $\tilde{B}$ be fuzzy sets in $X$. Then

(i) $\neg\tilde{A} = \{(x, \mu_{\neg\tilde{A}}(x)) \mid x \in X, \mu_{\neg\tilde{A}}(x) = 1 - \mu_{\tilde{A}}(x)\}$
(ii) $\tilde{A} \subseteq \tilde{B} \Leftrightarrow \forall\, x \in X : \mu_{\tilde{A}}(x) \leq \mu_{\tilde{B}}(x)$
(iii) $\tilde{A} \cap \tilde{B} = \{(x, \mu_{\tilde{A}\cap\tilde{B}}(x)) \mid x \in X \ \wedge\ \mu_{\tilde{A}\cap\tilde{B}}(x) = \min(\mu_{\tilde{A}}(x), \mu_{\tilde{B}}(x))\}$
(iv) $\tilde{A} \cup \tilde{B} = \{(x, \mu_{\tilde{A}\cup\tilde{B}}(x)) \mid x \in X \ \wedge\ \mu_{\tilde{A}\cup\tilde{B}}(x) = \max(\mu_{\tilde{A}}(x), \mu_{\tilde{B}}(x))\}$
(v) $\tilde{A} - \tilde{B} = \tilde{A} \cap \neg\tilde{B}$

A [*strict*] *α-cut* or [*strict*] *α-level set* of a fuzzy set $\tilde{A}$ for a specified value $\alpha$ is the crisp set $A_\alpha\ [A^*_\alpha] = \{x \in X \mid \mu_{\tilde{A}}(x) \geq [>]\, \alpha\ \wedge\ 0 \leq \alpha \leq [<]\, 1\}$. The strict $\alpha$-cut for $\alpha = 0$ is called *support* of $\tilde{A}$, i.e., $supp(\tilde{A}) = A^*_0$. For a fuzzy set $\tilde{A}$ and $\alpha, \beta \in [0,1]$ holds (i) $X = A_0$ and (ii) $\alpha < \beta \Rightarrow A_\alpha \supseteq A_\beta$. The set of all levels $\alpha \in [0,1]$ that represent distinct $\alpha$-cuts of a given fuzzy set $\tilde{A}$ is called the *level set* $\Lambda_{\tilde{A}} = \{\alpha \in [0,1] \mid \exists\ x \in X : \mu_{\tilde{A}}(x) = \alpha\}$ of $\tilde{A}$.

Fuzzy (point set) topology [5] is a straightforward extension and generalization of ordinary point set topology and allows one to distinguish specific topological structures of a fuzzy set like its closure or interior.

A *fuzzy topology* on a universe $X$ is a family $\tilde{T}$ of fuzzy sets in $X$ satisfying the following conditions:

(i) $X \in \tilde{T}, \emptyset \in \tilde{T}$
(ii) $\tilde{A} \in \tilde{T}, \tilde{B} \in \tilde{T} \Rightarrow \tilde{A} \cap \tilde{B} \in \tilde{T}$
(iii) $\tilde{S} \subseteq \tilde{T} \Rightarrow \bigcup_{\tilde{A} \in \tilde{S}} \tilde{A} \in \tilde{T}$

The pair $(X, \tilde{T})$ is said to be a *fuzzy topological space.* The elements of $\tilde{T}$ are called *open fuzzy sets.* The family $\tilde{T}'$ of all *closed fuzzy sets* in a fuzzy topological space $(X, \tilde{T})$ is given by $\tilde{T}' = \{\neg\tilde{A} \mid \tilde{A} \in \tilde{T}\}$. The *closure* [*interior*] of a fuzzy set $\tilde{A}$ in a fuzzy topological space $(X, \tilde{T})$ is the smallest closed [largest open] fuzzy set containing $\tilde{A}$ [contained in $\tilde{A}$].

## 11.4 Fuzzy Points

Due to our assumption of the point set paradigm, an understanding of the nature of a point, or more precisely a *fuzzy point*, is necessary. There are at least two meaningful definitions for a fuzzy point.

The first definition views a fuzzy point as a point in two-dimensional Euclidean space with a membership value greater than 0, since 0 documents the non-existence of a point. A *fuzzy point* $\tilde{p}$ at $(a, b)$ in $\mathbb{R}^2$, written $\tilde{p}(a, b)$, is a fuzzy singleton in $\mathbb{R}^2$ defined by

$$\mu_{\tilde{p}(a,b)}(x, y) = \begin{cases} m & \text{if } (x, y) = (a, b) \\ 0 & \text{otherwise} \end{cases}$$

with $0 < m \leq 1$. Point $\tilde{p}$ is said to have support $(a, b)$ and value $m$. Let $P_f$ be the set of all fuzzy points. $P_f$ is, of course, a proper superset of $P_c$, the set of all crisp points in $\mathbb{R}^2$. For $\tilde{p} = p = (a, b) \in P_c$, we obtain $\mu_{\tilde{p}(a,b)}(x, y) = \chi_p(x, y) = 1$, if $(x, y) = (a, b)$, and 0 otherwise.

The second definition uses a membership function that returns the *degree of proximity* of a point to a reference point $\tilde{p}$. That is, we consider that the point $(x, y)$ is "approximately $(a, b)$" or "about $(a, b)$" to the degree $\mu_{\tilde{p}(a,b)}(x, y)$. A fuzzy point $\tilde{p}(a, b)$ is then generally defined by

(i) $\mu_{\tilde{p}(a,b)}$ is upper semicontinuous[1]
(ii) $\mu_{\tilde{p}(a,b)}(x, y) = 1$ if and only if $(x, y) = (a, b)$
(iii) $\forall\, 0 \leq \alpha \leq 1 : \tilde{p}_\alpha$ is a convex[2] subset of $\mathbb{R}^2$

Next, we define three *geometric primitives* on fuzzy points which are valid for both definitions of fuzzy points. Let $\tilde{p}(a, b), \tilde{q}(c, d) \in P_f$ with $a, b, c, d \in \mathbb{R}$. Then

(i) $\tilde{p}(a, b) = \tilde{q}(c, d) \;:\Leftrightarrow\; a = c \,\wedge\, b = d \,\wedge\, \mu_{\tilde{p}(a,b)} = \mu_{\tilde{q}(c,d)}$

---

1 A function $f : X \to \mathbb{R}$ is *upper semicontinuous* $:\Leftrightarrow \forall\, r \in \mathbb{R} : \{x \mid f(x) < r\}$ is open. (The notation ":$\Leftrightarrow$" means "is defined as being equivalent".)

2 A set $X \subseteq \mathbb{R}^2$ is called *convex* $:\Leftrightarrow \forall\, p, q \in \mathbb{R}^2\ \forall\, \lambda \in \mathbb{R}^+$ with $0 < \lambda < 1 : r = \lambda\, p + (1 - \lambda)\, q \in X$ ($p$, $q$, and $r$ are here regarded as vectors)

(ii) $\tilde{p}(a,b) \neq \tilde{q}(c,d) \;:\Leftrightarrow\; \neg(\tilde{p}(a,b) = \tilde{q}(c,d))$
(iii) $\tilde{p}(a,b)$ and $\tilde{q}(c,d)$ are *disjoint* $:\Leftrightarrow\; supp(\tilde{p}(a,b)) \cap supp(\tilde{q}(c,d)) = \emptyset$

In contrast to crisp points, for fuzzy points we also have a predicate for disjointedness. We are now able to define an object of the fuzzy spatial data type *fpoint* as a set of disjoint fuzzy points:

$$fpoint = \{Q \subseteq P_f \mid \forall\, \tilde{p}, \tilde{q} \in Q : \tilde{p}(a,b) \text{ and } \tilde{q}(c,d) \text{ are } disjoint \;\wedge\; Q \text{ is finite}\}$$

## 11.5 Fuzzy Lines

In this section we define a data type for *fuzzy lines*. For that, we first introduce a *simple* fuzzy line as a continuous curve with smooth transitions of membership grades between neighboring points of the line. We assume a total order on $\mathbb{R}^2$ which is given by the lexicographic order "<" on the coordinates (first $x$, then $y$). The membership function of a *simple fuzzy line* $\tilde{l}$ is then defined by

$$\mu_{\tilde{l}} : f_{\tilde{l}} \to [0,1] \text{ with } f_{\tilde{l}} : [0,1] \to \mathbb{R}^2 \text{ such that}$$

(i) $\mu_{\tilde{l}}$ is continuous
(ii) $f_{\tilde{l}}$ is continuous
(iii) $\forall\, a, b \in ]0,1[ : a \neq b \;\Rightarrow\; f_{\tilde{l}}(a) \neq f_{\tilde{l}}(b)$
(iv) $\forall\, a \in \{0,1\}\; \forall\, b \in ]0,1[ : f_{\tilde{l}}(a) \neq f_{\tilde{l}}(b)$
(v) $f_{\tilde{l}}(0) < f_{\tilde{l}}(1) \;\vee\; (f_{\tilde{l}}(0) = f_{\tilde{l}}(1) \;\wedge\; \forall a \in ]0,1[) : f_{\tilde{l}}(0) < f_{\tilde{l}}(a))$

Function $f_{\tilde{l}}$ on its own models a continuous, simple *crisp* line (a *simple curve*). The points $f_{\tilde{l}}(0)$ and $f_{\tilde{l}}(1)$ are called the *end points* of $f$. The definition allows loops ($f_{\tilde{l}}(0) = f_{\tilde{l}}(1)$) but prohibits equality of interior points and thus self-intersections (condition (iii)) and equality of an interior with an end point (condition (iv)). The last condition ensures uniqueness of representation, i.e., in a *closed simple line* $f_{\tilde{l}}(0)$ must be the leftmost point.

Let $S$ be the set of fuzzy simple lines, and let $T \subseteq S$. An *S-complex $C$ over $T$* is a finite set $C = \{\tilde{l}_1, \ldots, \tilde{l}_n\} \subseteq T$ such that[3]

(i) $\forall\, 1 \leq i < j \leq n : f_{\tilde{l}_i}(]0,1[) \cap f_{\tilde{l}_j}(]0,1[) = \emptyset$
(ii) $\forall\, 1 \leq i < j \leq n : \{f_{\tilde{l}_i}(0), f_{\tilde{l}_i}(1)\} \cap f_{\tilde{l}_j}(]0,1[) = \emptyset$
(iii) $\forall\, 1 \leq i \leq n\; \exists\, 1 \leq j \leq n, j \neq i : \{f_{\tilde{l}_i}(0), f_{\tilde{l}_i}(1)\} \cap \{f_{\tilde{l}_j}(0), f_{\tilde{l}_j}(1)\} \neq \emptyset$
(iv) For all $1 \leq i,j \leq n$ and for all $a, k \in \{0,1\}$ let $V^a_{\tilde{l}_i} = \{(j,k) \mid f_{\tilde{l}_i}(a) = f_{\tilde{l}_j}(k)\}$. Then we require: $\forall\, 1 \leq i \leq n\; \forall\, a \in \{0,1\} : (|V^a_{\tilde{l}_i}| = 1) \vee (|V^a_{\tilde{l}_i}| > 2)$
(v) $\forall\, 1 \leq i \leq n\; \forall\, a \in \{0,1\}\; \forall\, (j,k) \in V^a_{\tilde{l}_i} : \mu_{\tilde{l}_i}(f_{\tilde{l}_i}(a)) = \mu_{\tilde{l}_j}(f_{\tilde{l}_j}(k))$

[3] The application of a function $f$ to a set $X$ of values is defined as $f(X) = \{f(x) \mid x \in X\}$.

Condition (i) requires that the elements of an $S$-complex do not intersect or overlap within their interior. Moreover, they may not be touched within their interior by an endpoint of another element (condition (ii)). Condition (iii) ensures the property of connectivity of an $S$-complex; isolated fuzzy simple lines are disallowed. Condition (iv) expresses that each endpoint of an element of $C$ must belong to exactly one or more than two incident elements of $C$ (note that always $(i, a) \in V_{\tilde{l}_i}^a$). This condition supports the requirement of maximal elements and hence achieves minimality of representation. Condition (v) requires that the membership values of more than two elements of $C$ with a common end point must have the same membership value; otherwise we get a contradiction saying that a point of an $S$-complex has more than one different membership value.

All conditions together define an $S$-complex over $T$ as a *connected planar fuzzy graph* with a unique representation. The corresponding point set of $C$ is $points(C) = \bigcup_{\tilde{l} \in C} f_{\tilde{l}}([0,1])$. The set of all $S$-complexes over $T$ is denoted by $SC(T)$. The disjointedness of any two $S$-complexes $C_1, C_2 \in SC(T)$ is defined as follows:

$$C_1 \text{ and } C_2 \text{ are } \mathit{disjoint} :\Leftrightarrow \mathit{points}(C_1) \cap \mathit{points}(C_2) = \emptyset$$

A fuzzy spatial data type for fuzzy lines called *fline* can now be defined in two equivalent ways. The "structured view" is based on $S$-complexes:

$$\mathit{fline} = \{D \subseteq SC(S) \mid \forall\, C_1, C_2 \in D : C_1 \text{ and } C_2 \text{ are } \mathit{disjoint} \wedge D \text{ is finite}\}$$

Let $1_r = \mathbb{R}^2 \times \{1\}$. The "flat view" emphasizing the point set paradigm is:

$$\mathit{fline} = \{\tilde{Q} \subseteq 1_r \mid \exists\, D \subseteq SC(S) : \bigcup_{C \in D} \mathit{points}(C) = \mathit{supp}(\tilde{Q})\}$$

## 11.6 Fuzzy Regions

The aim of this section is to develop and formalize the concept of a *fuzzy region*. Section 11.6.1 informally discusses some intrinsic features of fuzzy regions and compares them to classical crisp regions. Section 11.6.2 provides their formal definition.

### 11.6.1 What are Fuzzy Regions?

The question what a *crisp region* is has been treated in many publications. A very general definition defines a crisp region as a set of disjoint, connected areal components, called *faces*, possibly with disjoint *holes* [15, 18, 23] in the Euclidean space $\mathbb{R}^2$. This model has the nice property that it is closed under (appropriately defined) geometric union, intersection, and difference operations. It allows crisp regions to contain holes and islands within holes to any finite level.

By analogy with the generalization of crisp sets to fuzzy sets, we strive for a generalization of crisp regions to fuzzy regions on the basis of the point set paradigm and fuzzy concepts. At the same time we would like to transfer the structural definition of crisp regions (i.e., the component view) to fuzzy regions. Thus, the structure of a fuzzy region is supposed to be the same as for a crisp region but with the exception and generalization which amounts to a relaxation and hence greater flexibility of the strict belonging or non-belonging principle of a point in space to a specific region and which enables a partial membership of a point in a region. This is just what the term "fuzzy" means here.

There are at least three possible, related interpretations for a point in a fuzzy region. First, this situation may be interpreted as the *degree of belonging* to which that point is *inside* or *part of* some areal feature. Consider the transition between a mountain and a valley and the problem to decide which points have to be assigned to the valley and which points to the mountain. Obviously, there is no strict boundary between them, and it seems to be more appropriate to model the transition by partial and multiple membership. Second, this situation may indicate the *degree of compatibility* of the individual point with the attribute or concept represented by the fuzzy region. An example are "warm areas" where we must decide for each point whether and to which grade it corresponds to the concept "warm". Third, this situation may be viewed as the *degree of concentration* of some attribute associated with the fuzzy region at the particular point. An example is air pollution where we can assume the highest concentration at power stations, for instance, and lower concentrations with increasing distance from them. All these related interpretations give evidence of *fuzziness.*

When dealing with crisp regions, the user usually does not employ point sets as a method to conceptualize space. The user rather thinks in terms of sharply determined *boundaries* enclosing and grouping areas with *equal* properties or attributes and separating different regions with different properties from each other; he or she has purely *qualitative* concepts in mind. This view changes when fuzzy regions come into play. Besides the qualitative aspect, in particular the *quantitative* aspect becomes important, and boundaries in most cases disappear (between a valley and a mountain there is no strict boundary!). The distribution of attribute values within a region and transitions between different regions may be *smooth* or *continuous.* This feature just characterizes fuzzy regions. There are a lot of spatial phenomena showing a smooth behavior. Application examples are air pollution (Figure 11.1), temperature zones, magnetic fields, storm intensity, and sun insolation.

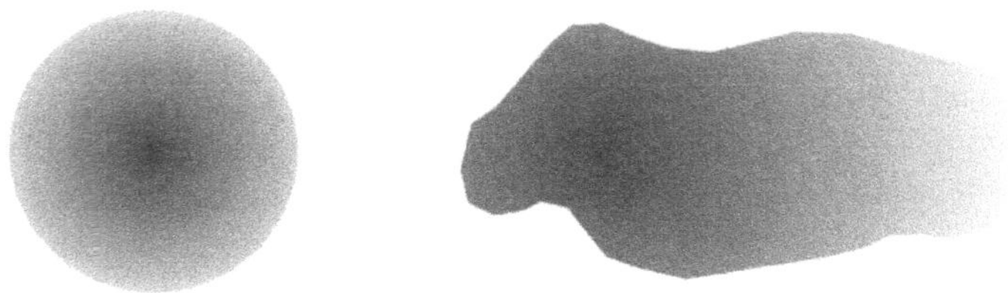

**Figure 11.1.** This figure demonstrates a possible visualization of a fuzzy region which could model the expansion of air pollution caused by a power station. The left image shows a radial expansion where the degree of pollution concentrates in the center (darker locations) and decreases with increasing distance from the power station (brighter locations). The right image has the same theme but this time we imagine that the power station is surrounded by high mountains to the north, the south, and the west. Hence, the pollution cannot escape in these directions and finds its way out of the valley in eastern direction. In both cases we can recognize the smooth transitions to the exterior.

### 11.6.2 Formal Definition of Fuzzy Regions

Since our objective is to model two-dimensional fuzzy areal objects for spatial applications, we consider a fuzzy topology $\tilde{T}$ on the Euclidean space (plane) $\mathbb{R}^2$. In this spatial context we denote the elements of $\tilde{T}$ as *fuzzy point sets.* The membership function for a fuzzy point set $\tilde{A}$ in the plane is then described by $\mu_{\tilde{A}} : \mathbb{R}^2 \to [0, 1]$.

From an application point of view, there are two observations that prevent a definition of a fuzzy region simply as a fuzzy point set. We will discuss them now in more detail and at the same time elaborate properties of fuzzy regions.

**Avoiding Geometric Anomalies: Regularization.** The first observation refers to a necessary *regularization* of fuzzy point sets. The first reason for this measure is that fuzzy (as well as crisp) regions that actually appear in spatial applications in most cases cannot be just modeled as arbitrary point sets but have to be represented as point sets that do not have "geometric anomalies" and that are in a certain sense *regular*. Geometric anomalies relate to isolated or dangling line or point features and missing lines and points in the form of cuts and punctures. Spatial phenomena with such degeneracies never appear as entities in reality. The second reason is that, from a data type point of view, we are interested in fuzzy spatial data types that satisfy closure properties for (appropriately defined) geometric union, intersection, and difference.

We are, of course, confronted with the same problem in the crisp case where the problem can be avoided by the concept of *regularity* [15, 23, 25]. It turns out to be useful to appropriately transfer this concept to the fuzzy case. Let $\tilde{A}$ be a fuzzy set of a fuzzy topological space $(\mathbb{R}^2, \tilde{T})$. Then

$$\tilde{A} \text{ is called a } \textit{regular open fuzzy set} \text{ if } \tilde{A} = int_{\tilde{T}}(cl_{\tilde{T}}(\tilde{A}))$$

Whereas crisp regions are usually modeled as *regular closed crisp sets*, we will use *regular open fuzzy sets* due to their vagueness and their usual lack of boundaries. Regular open fuzzy sets avoid the aforementioned geometric anomalies, too. Since application examples show that fuzzy regions can also be partially bounded, we admit *partial boundaries* with a crisp or fuzzy character. For that purpose we define the following fuzzy set (The notation ":=" means "is defined as".):

$$frontier_{\tilde{T}}(\tilde{A}) := \{((x,y), \mu_{\tilde{A}}(x,y)) \mid (x,y) \in supp(\tilde{A}) - supp(int_{\tilde{T}}(\tilde{A}))\}$$

A fuzzy set $\tilde{A}$ is now called a *spatially regular fuzzy set* iff

(i) $int_{\tilde{T}}(\tilde{A})$ is a regular open fuzzy set
(ii) $frontier_{\tilde{T}}(\tilde{A}) \subseteq frontier_{\tilde{T}}(cl_{\tilde{T}}(int_{\tilde{T}}(\tilde{A})))$
(iii) $frontier_{\tilde{T}}(\tilde{A})$ is a partition of $n$ connected boundary parts (fuzzy sets)

We can conclude that $frontier_{\tilde{T}}(\tilde{A}) = \emptyset$ if $\tilde{A}$ is regular open. We will base our definition of fuzzy regions on spatially regular fuzzy sets and define a *regularization* function $reg_f$ which associates the interior of a fuzzy set $\tilde{A}$ with its corresponding regular open fuzzy set and which restricts the partial boundary of $\tilde{A}$ (if it exists at all) to a part of the boundary of the corresponding regular closed fuzzy set of $\tilde{A}$:

$$reg_f(\tilde{A}) := int_{\tilde{T}}(cl_{\tilde{T}}(\tilde{A})) \cup (frontier_{\tilde{T}}(\tilde{A}) \cap frontier_{\tilde{T}}(cl_{\tilde{T}}(int_{\tilde{T}}(\tilde{A}))))$$

The different components of the regularization process work as follows: the *interior* operator $int_{\tilde{T}}$ eliminates dangling point and line features since their interior is empty. The *closure* operator $cl_{\tilde{T}}$ removes cuts and punctures by appropriately adding points. Furthermore, the *closure* operator introduces a fuzzy boundary (similar to a crisp boundary in the ordinary point-set topological sense) separating the points of a closed set from its exterior. The operator $frontier_{\tilde{T}}$ supports the restriction of the boundary.

The following statements about set operations on regular open fuzzy sets are given informally and without proof. The intersection of two regular open fuzzy sets is regular open. The union, difference, and complement of two regular open fuzzy sets are not necessarily regular open since they can produce anomalies. Correspondingly, this also holds for spatially regular fuzzy sets. Hence, we introduce *regularized set operations* on spatially regular fuzzy sets that preserve regularity. Let $\tilde{A}$, $\tilde{B}$ be spatially regular fuzzy sets of a fuzzy topological space $(\mathbb{R}^2, \tilde{T})$, and let $a \dot{-} b = a - b$ for $a \geq b$ and $a \dot{-} b = 0$ otherwise $(a, b \in \mathbb{R}_0^+)$. Then

(i) $\tilde{A} \cup_r \tilde{B} := reg_f(\tilde{A} \cup \tilde{B})$
(ii) $\tilde{A} \cap_r \tilde{B} := reg_f(\tilde{A} \cap \tilde{B})$
(iii) $\tilde{A} -_r \tilde{B} := reg_f(\{((x,y), \mu_{\tilde{A}-_r\tilde{B}}(x,y) \mid (x,y) \in \tilde{A} \;\; \wedge$
$\mu_{\tilde{A}-_r\tilde{B}}(x,y) = \mu_{\tilde{A}}(x,y) \dot{-} \mu_{\tilde{B}}(x,y)\})$

(iv) $\neg_r \tilde{A} := reg_f(\neg \tilde{A})$

Note that we have changed the meaning of difference (i.e., $\tilde{A} -_r \tilde{B} \neq \tilde{A} \cap_r \neg \tilde{B}$) since the right side of the inequality does not seem to make great sense in the spatial context. Regular open fuzzy sets, spatially regular fuzzy sets, and regularized set operations express a natural formalization of the desired dimension-preserving property of set operations. In the crisp case this is taken for granted but mostly never fulfilled by spatial type systems, geometric algorithms, spatial database systems, and GIS.

Whereas the subspace *RCCS* of *r*egular *c*losed *c*risp *s*ets together with the crisp regular set operations "$\cup$" and "$\cap$" and the set-theoretic order relation "$\subseteq$" forms a Boolean lattice [15], this is not the case for *SRFS* denoting the subspace of *s*patially *r*egular *f*uzzy *s*ets. Here we obtain the (unproven but obvious) statement that *SRFS* together with the regularized set operations "$\cup_r$" and "$\cap_r$" and the fuzzy set-theoretic order relation "$\subseteq$" is a pseudo-complemented distributive lattice.

This implies that (i) (*SRFS*, $\subseteq$) is a partially ordered set (reflexivity, antisymmetry, transitivity), (ii) every pair $\tilde{A}$, $\tilde{B}$ of elements of *SRFS* has a least upper bound $\tilde{A} \cup_r \tilde{B}$ and a greatest lower bound $\tilde{A} \cap_r \tilde{B}$, (iii) (*SRFS*, $\subseteq$) has a maximal element $1_r := \{((x, y), \mu(x, y)) \mid (x, y) \in \mathbb{R}^2 \;\; \wedge \;\; \mu(x, y) = 1\}$ (identity of "$\cap_r$") and a minimal element $0_r := \{((x, y), \mu(x, y)) \mid (x, y) \in \mathbb{R}^2 \;\; \wedge \;\; \mu(x, y) = 0\}$ (identity of "$\cup_r$"), and (iv) algebraic laws like idempotence, commutativity, associativity, absorption, and distributivity hold for "$\cup_r$" and "$\cap_r$".

(*SRFS*, $\subseteq$) is not a complementary lattice. Although the algebraic laws of involution and dualization hold, this is not true for the laws of complementarity. If we take the standard fuzzy set operations presented in Section 11.3 as a basis, the law of excluded middle $\tilde{A} \cup_r \neg \tilde{A} = 1_r$ and the law of contradiction $\tilde{A} \cap_r \neg \tilde{A} = 0_r$ do *not* hold in general. This fact explains the term "pseudo-complemented" from above and is no weakness of the model but only an indication of fuzziness.

**Modeling Smooth Attribute Changes: Predominantly Continuous Membership Functions.** The second observation is that according to the application cases shown in Section 11.6.1 the mapping $\mu_{\tilde{A}}$ itself may not be arbitrary but must take into account the intrinsic smoothness of fuzzy regions. This property can be modeled by the well known mathematical concept of *continuity* and results in special continuous membership functions for fuzzy regions. We say that a function $f$ contains a *continuity gap* at a point $x_0$ of its domain if $f$ is semicontinuous but not continuous at $x_0$. Function $f$ is called *predominantly continuous* if $f$ is continuous and has at most a *finite* number of continuity gaps.

**Defining Fuzzy Regions.** The type *fregion* for fuzzy regions can now be defined in the following way:

$$fregion = \{\tilde{R} \in SRFS \mid \mu_{\tilde{R}} \text{ is predominantly continuous}\}$$

## 11.7 Structured Views of Fuzzy Regions

The formal definition of a fuzzy region given in Section 11.6.2 is conceptually somehow "structureless" in the sense that only "flat" point sets are considered and no structural information is revealed. In the following four subsections some "semantically richer" characterizations of fuzzy regions are presented which enable a better understanding of fuzzy regions. On the one hand they subdivide fuzzy regions into fuzzy components and on the other hand they describe them as collections of crisp regions. Moreover, they give hints for a possible implementation.

### 11.7.1 Fuzzy Regions as Multi-Component Objects

The first structured view considers a fuzzy region as a set of *fuzzy components*. For a definition we need a notion of *connectedness* for fuzzy regions. A *separation* of a fuzzy region $\tilde{Y}$ is a pair $\tilde{A}, \tilde{B}$ of fuzzy subregions satisfying the following four conditions:

(i) $\tilde{A} \neq \emptyset, \tilde{B} \neq \emptyset$
(ii) $\tilde{Y} = \tilde{A} \cup_r \tilde{B}$
(iii) $\tilde{A} \cap int_{\tilde{T}}(\tilde{B}) = \emptyset \;\wedge\; int_{\tilde{T}}(\tilde{A}) \cap \tilde{B} = \emptyset$
(iv) $|\tilde{A} \cap_r \tilde{B}|$ is finite

If a separation of $\tilde{Y}$ into $\tilde{A}$ and $\tilde{B}$ exists, then $\tilde{Y}$ is said to be *separated*, and we call $\tilde{A}$ and $\tilde{B}$ to be *disjoint*. Otherwise $\tilde{Y}$ is said to be *connected*. Note that condition (iii) of the definition uses the usual fuzzy intersection operation and not the one on spatially regular fuzzy sets since the latter requires two fuzzy regions as operands. The property of disjointedness (condition (iv)) requires that the two fuzzy subregions $\tilde{A}$ and $\tilde{B}$ may at most share a finite number of boundary points; this makes sense since otherwise they could be simply merged into one fuzzy subregion. We now continue this separation process and decompose a fuzzy region $\tilde{Y}$ into its maximal set of pairwise disjoint fuzzy components $\tilde{Y} = \{\tilde{A}_1, \ldots, \tilde{A}_n\}$ (in the spatial context this decomposition is always finite) so that we obtain with $I = \{1, \ldots, n\}$:

(i) $\forall\, i \in I : \tilde{A}_i \neq \emptyset$
(ii) $\tilde{Y} = \bigcup_{r\, i \in I} \tilde{A}_i$
(iii) $\forall\, i, j \in I, i \neq j : \tilde{A}_i \cap int_{\tilde{T}}(\tilde{A}_j) = \emptyset \;\wedge\; int_{\tilde{T}}(\tilde{A}_i) \cap \tilde{A}_j = \emptyset$
(iv) $\forall\, i, j \in I, i \neq j : |\tilde{A}_i \cap_r \tilde{A}_j|$ is finite
(v) $\forall\, i \in I : (\tilde{A}_i$ is connected $\wedge\; \nexists\, \tilde{B} \supset \tilde{A}_i : \tilde{B}$ is connected)

We call each fuzzy component $\tilde{A}_i$ a *fuzzy face*. Hence, we obtain:

A *fuzzy region* is a set of pairwise disjoint *fuzzy faces*.

A question arises whether also *fuzzy holes* can be identified from the point set view of a fuzzy region. This question has to be negated. Let us briefly consider the crisp case. If $A$ is a crisp region, its faces can have holes which belong to the complement (exterior) of $A$, i.e., to $\mathbb{R}^2 - A$, and are "enclosed" by $A$. Unfortunately, ordinary point set topology offers no method to extract holes from a (regular closed) point set as separate components; they are simply part of the complement. Note that this does not mean that regions with holes cannot be modeled. Some research work in [13, 23, 29], for example, shows that this is possible by selecting a constructive approach. Roughly speaking, the idea is to assume that the holes of $A$ are already given as regions and to subtract these holes from a "generalized region A*" being isomorphic to a closed disc and being the union of $A$ and the holes. But since this a pure set operation, afterwards $A$ "forgets" how it was produced and cannot reconstruct its past. Similarly to the crisp case, holes cannot be identified from a (spatially regular) fuzzy point set, since fuzzy topology also offers no concept of holes.

Moreover, we are here faced with the problem of the nature of a fuzzy hole. By analogy with the crisp case, we could say that the fuzzy holes of a fuzzy region $\tilde{A}$ exclusively contain all points that are enclosed by any fuzzy face of $\tilde{A}$ and that have membership grade 0 in $\tilde{A}$. But then, a fuzzy hole is crisp and a subset of the set

$$H = \{((x,y),1) \mid (x,y) \in supp(\neg\tilde{A})\}$$

This model of a fuzzy hole is unsatisfactory in the sense that it only deals with those points enclosed by $\tilde{A}$ that definitely do not belong to $\tilde{A}$. It does not take into account the complement of those points of $\tilde{A}$ belonging only partially to $\tilde{A}$, i.e., the model does not consider the set

$$\underline{\tilde{A}} = \{((x,y),m) \mid (x,y) \in supp(\tilde{A}) \ \wedge \ m = 1 - \mu_{\tilde{A}}(x,y)\}$$

called the *anti-fuzzy region* of $\tilde{A}$.

One could argue that the points of $\underline{\tilde{A}}$ also belong to the fuzzy holes. And indeed, we will take this view. The consequence is that for a fuzzy face there exists exactly one fuzzy hole.

### 11.7.2 Fuzzy Regions as Three-Part Crisp Regions

The second structured view leads to a simplification of an originally smooth fuzzy region to a core-boundary region and thus to a change from a quantitative to a qualitative perspective. It distinguishes between the kernel, the boundary, and the exterior as the three parts of a fuzzy region. For a fuzzy region $\tilde{A}$, these parts are defined as crisp regions (regular closed sets)[4]:

[4] Correspondingly, for a crisp set $A$, the regularization function $reg_c$ is defined as $reg_c(A) = cl_T(int_T(A))$ where $T$ is a topology for a universe $X$ and $cl_T$ and $int_T$ are the *closure* and *interior* operators on a topological space $(X,T)$.

$$kernel(\tilde{A}) = reg_c(\{(x,y) \in \mathbb{R}^2 \mid \mu_{\tilde{A}}(x,y) = 1\})$$
$$boundary(\tilde{A}) = reg_c(\{(x,y) \in \mathbb{R}^2 \mid 0 < \mu_{\tilde{A}}(x,y) < 1\})$$
$$exterior(\tilde{A}) = reg_c(\{(x,y) \in \mathbb{R}^2 \mid \mu_{\tilde{A}}(x,y) = 0\})$$

The *kernel* identifies the part that definitely belongs to $\tilde{A}$. The *exterior* determines the part that definitely does not belong to $\tilde{A}$. The indeterminate character of $\tilde{A}$ is summarized in the *boundary* of $\tilde{A}$ in a unified and simplified manner. Kernel and boundary can be adjacent with a common border, and kernel and/or boundary can be empty. This view corresponds exactly to the already described concept of *vague regions* with its three-valued logic [15].

All in all, this view presents only a very coarse and restricted description of fuzzy regions since it differentiates only between three parts. The original gradation in the membership values of the points of the boundary gets lost. The benefit of this view lies in the implementation since efficient representation methods and algorithms for crisp regions can be used.

### 11.7.3 Fuzzy Regions as Collections of Crisp α-Level Regions

The third structured view attempts to diminish the drawbacks of the three-part view of fuzzy regions and to avoid the great information loss in this representation. It describes a fuzzy region in terms of nested $\alpha$-level sets. Let $\tilde{A}$ be a fuzzy region. Then we represent a region $A_a$ for an $\alpha \in [0,1]$ as

$$A_a = reg_c(\{(x,y) \in \mathbb{R}^2 \mid \mu_{\tilde{A}}(x,y) \geq \alpha\})$$

We call $A_a$ an *α-level region.* Clearly, $A_a$ is a crisp region whose boundary is defined by all points with membership value greater than or equal to $\alpha$. Note that $A_a$ can have holes. The kernel of $\tilde{A}$, as it has been defined in Section 11.7.2, is then equal to $A_{1.0}$. A property of the $\alpha$-level regions of a fuzzy region is that they are nested, i.e., if we select membership values $1 = \alpha_1 > \alpha_2 > \cdots > \alpha_n > \alpha_{n+1} = 0$ for some $n \in \mathbb{N}$, then

$$A_{\alpha_1} \subseteq A_{\alpha_2} \subseteq \cdots \subseteq A_{\alpha_n} \subseteq A_{\alpha_{n+1}}$$

We here describe the finite, discrete case that enables us to model and implement finite-valued and interval-based regions. If $\Lambda_{\tilde{A}}$ is infinite, then there are obviously infinitely many $\alpha$-level regions which can only be finitely represented within this view if we make a finite selection of values. In the discrete case, if $|\Lambda_{\tilde{A}}| = n+1$ and we take all these occurring membership values of a fuzzy region, we can replace "$\subseteq$" by "$\subset$" in the inclusion relationships above. This follows from the fact that for any $p \in A_{\alpha_i} - A_{\alpha_{i-1}}$ with $i \in \{2, \ldots, n+1\}$, $\mu_{\tilde{A}}(p) = \alpha_i$. For the continuous case, we get $\mu_{\tilde{A}}(p) \in [\alpha_i, \alpha_{i-1})$ which leads to interval-based regions. As a result, we obtain:

A *fuzzy region* is a (possibly infinite) set of $\alpha$-level regions, i.e., $\tilde{A} = \{A_{\alpha_i} \mid 1 \leq i \leq |\Lambda_{\tilde{A}}|\}$ with $\alpha_i > \alpha_{i+1} \Rightarrow A_{\alpha_i} \subseteq A_{\alpha_{i+1}}$ for $1 \leq i \leq |\Lambda_{\tilde{A}}| - 1$.

From the implementation perspective, one of the advantages of using (a finite collection of) $\alpha$-level sets to describe fuzzy regions is that existing geometric data structures and geometric algorithms known from Computational Geometry [21] can be applied.

### 11.7.4 Fuzzy Regions as $\alpha$-Partitions

The fourth structured view is partially motivated by the previous one and describes a fuzzy region as a partition. A partition in the spatial context, called a *spatial partition* [14], is a subdivision of the plane into pairwise disjoint (crisp) regions (called *blocks*) where each block is associated with an attribute and where adjacent blocks are not allowed to be labeled with the same attribute. It differs from the set-theoretic notion of a partition in the sense that it, of course, relates to space and that it incorporates a treatment of common boundary points which at the same time may belong to two adjacent blocks.

From an application point of view, different blocks of a spatial partition are often marked differently, i.e., different *labels* of some set $L$ are assigned to different blocks. Thus, in a certain way, $L$ determines the type of a partition. This leads to spatial partitions of type $L$ that are functions $\pi : \mathbb{R}^2 \rightarrow L$. In most cases, partitions are defined only partially, i.e., there are blocks (frequently called the *exterior* of a partition) which have no explicitly assigned labels. To complete $\pi$ to a total function, we assume a label $\bot_L$ (called *undefined*) for each label type $L$ and require that the exterior of a partition is labeled by $\bot_L$.

Like for crisp regions, we also desire regularity for the blocks of a spatial partition. We require the interiors of blocks to be regular open sets. Since points on the boundary cannot be uniquely assigned to either adjacent block, we cannot simply map them to single $L$-values. Instead, boundary points are mapped to the set of values given by the labels of all adjacent blocks. This leads to the definition of a *spatial mapping* of type $L$ as a total mapping $\pi : \mathbb{R}^2 \rightarrow L \cup 2^L$. The *range* of a spatial mapping $\pi$ yields the set of labels actually used in $\pi$ and is denoted by $range(\pi)$. The blocks of a spatial mapping $\pi$ are point sets that are mapped to the same labels. The block for a single label $l$ (or a set $S$ of labels) is given[5] by $f^{-1}(l)$ ($f^{-1}(S)$). The common label of a block $b$ of $\pi$ is denoted by $\pi[b]$, i.e., $\pi(b) = \{l\} \Rightarrow \pi[b] = l$. Obviously, the cardinality of block labels identifies different parts of a partition. A *region* of $\pi$ is any block of $\pi$ that is mapped to a single element of $L$, and a *border* of $\pi$ is given by a block that is mapped to a set of $L$-values, or formally for a spatial mapping $\pi$ of type $L$:

(i) $\rho(\pi) = \pi^{-1}(range(\pi) \cap L))$ (*regions*)
(ii) $\beta(\pi) = \pi^{-1}(range(\pi) \cap 2^L))$ (*borders*)

[5] We use the following definition of function inverse: for $f : X \rightarrow Y$ and $\forall\, y \in Y$ : $f^{-1}(y) := \{x \in X \mid f(x) = y\}$. Note that $f^{-1}$ applied to a set yields a set of sets.

Now we can finally define a spatial partition by topologically constraining regions to regular open sets and by semantically constraining boundary labels to those of adjacent regions.

A *spatial partition* of type $L$ is a spatial mapping $\pi$ of type $L$ with

(i) $\forall\, r \in \rho(\pi) : r$ is a regular open set (i.e., $r = int_T(cl_T(r))$)
(ii) $\forall\, b \in \beta(\pi) : \pi[b] = \{\pi[r] \mid r \in \rho(\pi) \ \wedge \ b \subseteq cl_T(r)\}$

The set of all spatial partitions of type $L$ is denoted by $[L]$, i.e., $[L] \subseteq \mathbb{R}^2 \to L \cup 2^L$.

Using the representation based on $\alpha$-level regions defined in the preceding subsection, we are now able to define a fuzzy region as a spatial partition. In our case $L = \Lambda_{\tilde{A}}$, i.e., the labels are formed by all possible membership values $\alpha$. We have now to determine the different blocks for regions and borders. The regions of $\tilde{A}$ are given[6] by the set $\{int_T(A_{\alpha_i} -_c A_{\alpha_{i-1}}) \mid i \in \{2, \ldots, n+1\}\}$, and the borders of $\tilde{A}$ are represented by the set $\{bound_T(A_{\alpha_i} -_c A_{\alpha_{i-1}}) \mid i \in \{2, \ldots, n+1\}\}$. The object $A_{\alpha_i} -_c A_{\alpha_{i-1}}$ is a region possibly with holes. Each region is uniquely associated with an $\alpha \in \Lambda_{\tilde{A}}$, and each border has all $\alpha$-labels of adjacent regions.

A *fuzzy region* $\tilde{A}$ is a spatial partition of type $\Lambda_{\tilde{A}}$ (i.e., $\tilde{A} \in [\Lambda_{\tilde{A}}]$), called an *$\alpha$-partition*.

If $\Lambda_{\tilde{A}}$ is infinite, we get an infinite spatial partition.

## 11.8 Topological Predicates on Complex Crisp Regions

In this section we extend the eight topological predicates on simple crisp regions to complex crisp regions. The extended predicates serve as a basis for topological predicates on complex fuzzy regions modeled as collections of crisp $\alpha$-level regions (Section 11.9).

In the following we use as a syntactical simplification the notation $(P_1|P_2|\ldots|P_n)(F,G)$ for the term $P_1(F,G) \ \vee \ P_2(F,G) \ \vee \ \ldots \ \vee \ P_n(F,G)$ where $P_i : \mathit{region} \times \mathit{region} \to \mathbb{B}$ is a topological predicate for each $1 \leq i \leq n$.

### 11.8.1 Topological Predicates on Simple Regions with Holes

As a first step to a general definition of topological predicates for complex crisp regions we consider such predicates for simple regions with holes and base their definition on the topological predicates for simple regions as they have been derived from the 9-intersection model (Section 11.2).

[6] In the following, the operation "$-_c$" denotes the regular difference operation on regular closed sets. The operation $bound_T$ applied to a regular closed set yields its point-set topological boundary.

Let $F$ and $G$ be two simple regions with holes, that is, $F = F_0 - \bigcup_{i=1}^{n} F_i$ and $G = G_0 - \bigcup_{j=1}^{m} G_j$. We consider $F$ and $G$ to be disjoint if they have nothing in common, that is, either $F_0$ and $G_0$ are disjoint, and thus implicitly also their corresponding holes due to the definition of $F$ and $G$, or $F_0$ (or $G_0$, respectively) and implicitly its holes are completely inside a hole $G_j$ of $G$ ($F_i$ of $F$, respectively) (Figure 11.2a). Formally, we can then define the predicate $disjoint_{srh}$ as

$$\begin{aligned} disjoint_{srh}(F,G) \;:=\; & disjoint(F_0,G_0) \lor \\ & (\exists 1 \le i \le n : inside(G_0,F_i)) \lor \\ & (\exists 1 \le j \le m : inside(F_0,G_j)) \end{aligned}$$

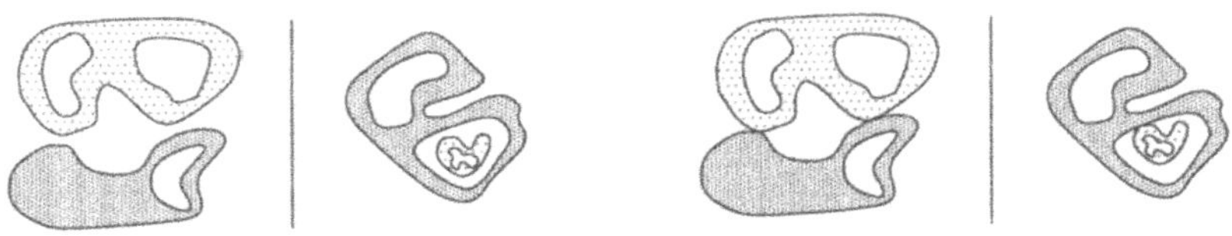

(a) (b)

**Figure 11.2.** Examples for the predicates $disjoint_{srh}(F,G)$ (a) and $meet_{srh}(F,G)$ (b).

The predicate $meet_{srh}$ is defined as follows (Figure 11.2b):

$$\begin{aligned} meet_{srh}(F,G) \;:=\; & meet(F_0,G_0) \lor \\ & (\exists 1 \le i \le n : coveredBy(G_0,F_i)) \lor \\ & (\exists 1 \le j \le m : coveredBy(F_0,G_j)) \end{aligned}$$

We consider $F$ to be inside $G$ if $F_0$ is inside $G_0$ and if each hole $G_j$ of $G$ is either disjoint from $F_0$ or inside a hole $F_i$ of $F$. (Figure 11.3a). The definition for the predicate $inside_{srh}$ is:

$$\begin{aligned} inside_{srh}(F,G) \;:=\; & inside(F_0,G_0) \land \\ & (\forall 1 \le j \le m : disjoint(F_0,G_j) \lor \\ & \quad (inside(G_j,F_0) \;\land\; \exists 1 \le i \le n : inside(G_j,F_i))) \end{aligned}$$

We do not have to take into account the topological relationships between the $F_i$'s and $G_0$ in our definition, because $inside(F_0,G_0) \Rightarrow inside(F_i,G_0)$ due to $F_i \subset F_0$ for $1 \le i \le n$. The predicate $contains_{srh}$ is symmetric to the predicate $inside_{srh}$, that is, $contains_{srh}(F,G) := inside_{srh}(G,F)$.

We consider $F$ and $G$ to be equal if $F_0$ and $G_0$ are equal, if $F$ and $G$ have the same number of holes, and if each hole $F_i$ of $F$ coincides with a hole $G_j$ of $G$ and vice-versa, that is,

$$\begin{aligned} equal_{srh}(F,G) \;:=\; & equal(F_0,G_0) \;\land\; n = m \;\land \\ & \exists \pi : \{1,\ldots,n\} \to \{1,\ldots,n\}, \pi \text{ bijective}, \\ & \forall 1 \le i \le n : equal(F_i, G_{\pi(i)}) \end{aligned}$$

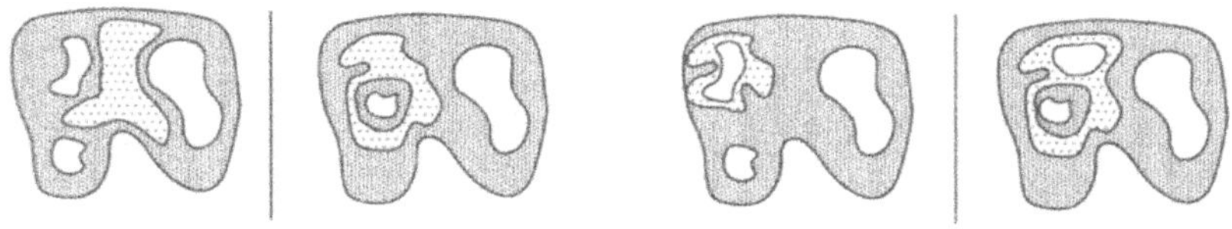

**Figure 11.3.** Examples for the predicates $inside_{srh}(F, G)$ (a) and $coveredBy_{srh}(F, G)$ (b).

$F$ is considered to be covered by $G$ if $F$ is a proper subset of $G$ and if $F$'s boundary touches $G$'s boundary (Figure 11.3b).

$$\begin{aligned} coveredBy_{srh}(F, G) := \neg((&inside_{srh}|equal_{srh})(F, G)) \wedge \\ &(inside|coveredBy|equal)(F_0, G_0) \wedge \\ &(\forall 1 \leq j \leq m : ((disjoint|meet)(F_0, G_j) \vee \\ &(\exists 1 \leq i \leq n : (inside|coveredBy|equal)(G_j, F_i)))) \end{aligned}$$

The predicate $covers_{srh}$ is symmetric to the predicate $coveredBy_{srh}$, that is, $covers_{srh}(F, G) := coveredBy_{srh}(G, F)$.

Finally, the predicate $overlap_{srh}$ (Figure 11.4) covers all remaining topological situations. This predicate can, of course, be defined directly in order to give an exact characterization of the remaining topological situations. But this makes the definition unnecessarily complicated and longish. We define instead:

$$\begin{aligned} overlap_{srh}(F, G) \ := \neg((&disjoint_{srh}|meet_{srh}|coveredBy_{srh}|covers_{srh}| \\ &inside_{srh}|contains_{srh}|equal_{srh})(F, G)) \end{aligned}$$

The set $T_{srh} = \{disjoint_{srh}, meet_{srh}, overlap_{srh}, coveredBy_{srh}, covers_{srh}, inside_{srh}, contains_{srh}, equal_{srh}\}$ provides a complete coverage of topological relationships for two simple regions with holes, and its elements are mutually exclusive. Completeness of $T_{srh}$ follows immediately from the complementary character of the definition of $overlap_{srh}$. Hence, at least one predicate must hold for any pair $F, G$ of simple regions with holes. Mutual exclusion of each pair of different topological predicates $P_1$ and $P_2$ can be proved by showing that $\neg(P_1(F, G) \wedge P_2(F, G))$ holds for any pair $F, G$. Since we have $k = 8$

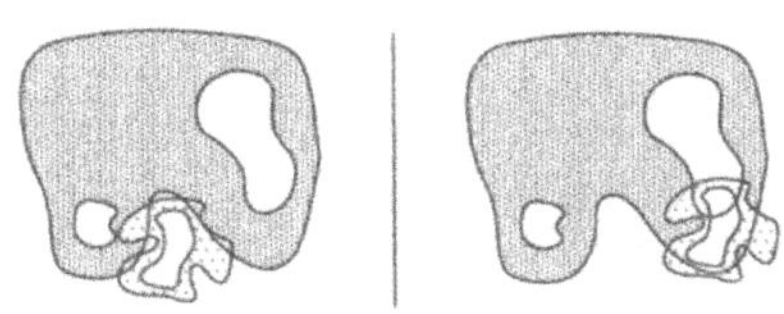

**Figure 11.4.** Examples for the predicate $overlap_{srh}(F, G)$.

predicates, we have to check the diversity of $\frac{1}{2}(k^2 - k) = 28$ predicate pairs. We will not show the validity for the predicate pairs here in detail but only tell the strategy. First, we can use the mutual exclusion of the topological relationships for simple regions employed in the definition of some predicates. For instance, $disjoint_{srh}$ and $meet_{srh}$ exclude each other since *disjoint* and *meet* as well as *inside* and *coveredBy* are mutually exclusive in the 9-intersection model. Second, several predicates use the negation of other predicates on simple regions with holes for their definition. For instance, $overlap_{srh}$ excludes all other predicates. Similarly, $coveredBy_{srh}$ excludes both $inside_{srh}$ and $equal_{srh}$. Overall, at most one predicate is valid for any pair $F, G$.

The set $T_{srh}$ of topological predicates on simple regions with holes is in two ways compatible with the set $T$ of topological predicates on simple regions obtained by the 9-intersection model. First, if both $F$ and $G$ do not have holes, then $T_{srh}$ and $T$ coincide. Second, each of the eight topological predicates on simple regions with holes has the same boolean results for the nine intersections as the corresponding predicate on simple regions (see Section 11.2).

### 11.8.2 Topological Predicates on Complex Regions

With the aid of the topological predicates on simple regions with holes we are now able to define the corresponding predicates on complex regions. Let $F = \bigcup_{i=1}^{n} F_i$ and $G = \bigcup_{j=1}^{m} G_j$ be complex regions where the $F_i$ and $G_j$ are simple regions possibly with holes. We define the following predicates:

$$\begin{aligned}
disjoint_{cr}(F,G) &:= \forall\, 1 \le i \le n\ \forall\, 1 \le j \le m : disjoint_{srh}(F_i, G_j) \\
meet_{cr}(F,G) &:= \neg disjoint_{cr}(F,G)\ \wedge\ (\forall\, 1 \le i \le n\ \forall\, 1 \le j \le m : \\
&\qquad (disjoint_{srh}|meet_{srh})(F_i, G_j)) \\
inside_{cr}(F,G) &:= \forall\, 1 \le i \le n\ \exists\, 1 \le j \le m : inside_{srh}(F_i, G_j) \\
contains_{cr}(F,G) &:= inside_{cr}(G,F) \\
equal_{cr}(F,G) &:= n = m\ \wedge\ (\exists \pi : \{1,\ldots,n\} \to \{1,\ldots,n\}, \pi \text{ bijective}, \\
&\qquad \forall\, 1 \le i \le n : equal_{srh}(F_i, G_{\pi(i)})) \\
coveredBy_{cr}(F,G) &:= \neg((inside_{cr}|equal_{cr})(F,G))\ \wedge \\
&\qquad (\forall\, 1 \le i \le n\ \exists\, 1 \le j \le m : \\
&\qquad\qquad (inside_{srh}|coveredBy_{srh}|equal_{srh})(F_i, G_j)) \\
covers_{cr}(F,G) &:= coveredBy_{cr}(G,F) \\
overlap_{cr}(F,G) &:= \neg((disjoint_{cr}|meet_{cr}|coveredBy_{cr}|covers_{cr}| \\
&\qquad inside_{cr}|contains_{cr}|equal_{cr})(F,G))
\end{aligned}$$

With similar arguments as in the last section we can recognize that two complex regions satisfy exactly one of these topological predicates. In other words, the topological predicates of the set $T_{cr} = \{disjoint_{cr}, meet_{cr}, inside_{cr},$

*contains*$_{cr}$, *equal*$_{cr}$, *coveredBy*$_{cr}$, *covers*$_{cr}$, *overlap*$_{cr}$} are mutually exclusive and complete. Note that the predicates *disjoint*$_{cr}$, *meet*$_{cr}$, *equal*$_{cr}$, and *overlap*$_{cr}$ are symmetric whereas the others are not.

## 11.9 Topological Predicates on Complex Fuzzy Regions

In this section we introduce a concept of topological predicates for fuzzy regions. In a similar way as we can generalize the characteristic function $\chi_A : X \to \{0,1\}$ to the membership function $\mu_{\tilde{A}} : X \to [0,1]$ (Section 11.3)[7], we can generalize a (binary) predicate $p_c : X \times Y \to \{0,1\}$ to a (binary) *fuzzy predicate* $p_f : \tilde{X} \times \tilde{Y} \to [0,1]$. Hence, the value of a fuzzy predicate can be interpreted as the degree to which the predicate holds for its operand objects. In our case of topological predicates, $X = Y =$ *region*, $\{0,1\} =$ *bool*, and $\tilde{X} = \tilde{Y} =$ *fregion* hold. For the set $[0,1]$ we have to introduce a new type *fbool* for *fuzzy booleansfuzzy boolean.*

For the definition of fuzzy topological predicates, we take the view of a fuzzy region as a set of $\alpha$-level regions (Section 11.7.3). We know that an $\alpha$-level region is a crisp complex region, and in the last section we have defined topological predicates on complex regions. This preparatory work now enables us to reduce topological predicates on fuzzy regions to topological predicates on collections of crisp regions[8].

The approach presented in this section is generic in the sense that any meaningful collection of topological predicates on complex crisp regions can be the basis for our definition of a collection of topological predicates on complex fuzzy regions. If the former collection additionally fulfils the properties of completeness and mutual exclusion (which is the case for $T_{cr}$), the latter collection automatically inherits these properties.

The open question now is how to compute the topological relationships of two collections of $\alpha$-level regions, each collection describing a fuzzy region. We use the concept of basic probability assignment [9] for this purpose. A *basic probability assignment* $m(F_{\alpha_i})$ can be associated with each $\alpha$-level region $F_{\alpha_i}$ and can be interpreted as the probability that $F_{\alpha_i}$ is the "true" representative of $F$. It is defined as

$$m(F_{\alpha_i}) = \alpha_i - \alpha_{i+1}$$

for $1 \leq i \leq n$ for some $n \in \mathbb{N}$ with $\alpha_1 = 1$ and $\alpha_{n+1} = 0$. That is, $m$ is built from the differences of successive $\alpha_i$'s. It is easy to see that the telescoping sum $\sum_{i=1}^{n} m(F_{\alpha_i}) = \alpha_1 - \alpha_{n+1} = 1 - 0 = 1$.

---

[7] Note that $\chi_A$ is a unary crisp predicate and that $\mu_{\tilde{A}}$ is a unary fuzzy predicate.

[8] Another great benefit of this approach is its easy implementability through well known concepts for crisp spatial objects and for crisp topological predicates.

Let $\pi_f(F, G)$ be the value that represents a (binary) property $\pi_f$ between two fuzzy regions $F$ and $G$. Based on the work in [9] property $\pi_f$ of $F$ and $G$ can be determined as the summation of weighted predicates by[9]

$$\pi_f(F, G) = \sum_{i=1}^{n} \sum_{j=1}^{n} m(F_{\alpha_i}) \cdot m(G_{\alpha_j}) \cdot \pi_{cr}(F_{\alpha_i}, G_{\alpha_j})$$

where $\pi_{cr}(F_{\alpha_i}, G_{\alpha_j})$ yields the value of the corresponding property $\pi_{cr}$ for two crisp $\alpha$-level regions $F_{\alpha_i}$ and $G_{\alpha_j}$. This formula is equivalent to

$$\pi_f(F, G) = \sum_{i=1}^{n} \sum_{j=1}^{n} (\alpha_i - \alpha_{i+1}) \cdot (\alpha_j - \alpha_{j+1}) \cdot \pi_{cr}(F_{\alpha_i}, G_{\alpha_j})$$

If $\pi_f$ is a topological predicate of $T_f = \{disjoint_f,\ meet_f,\ overlap_f,\ equal_f,\ inside_f,\ contains_f,\ covers_f,\ coveredBy_f\}$ between two fuzzy regions, we can compute the degree of the corresponding relationship with the aid of the pertaining crisp topological predicate $\pi_{cr} \in T_{cr}$. The value of $\pi_{cr}(F_{\alpha_i}, G_{\alpha_j})$ is either 1 (*true*) or 0 (*false*). Once this value has been determined for all combinations of $\alpha$-level regions from $F$ and $G$, the aggregated value of the topological predicate $\pi_f(F, G)$ can be computed as shown above. The more fine-grained the level set $\Lambda$ for the fuzzy regions $F$ and $G$ is, the more precisely the fuzziness of topological predicates can be determined.

It remains to show that $0 \le \pi_f(F, G) \le 1$ holds, that is, $\pi_f$ is really a fuzzy predicate. Since $\alpha_i - \alpha_{i+1} > 0$ for all $1 \le i \le n$ and since $\pi_{cr}(F_{\alpha_i}, G_{\alpha_j}) \ge 0$ for all $1 \le i, j \le n$, $\pi_f(F, G) \ge 0$ holds. We can show the other inequality by determining an upper bound for $\pi_f(F, G)$:

$$\begin{aligned}
\pi_f(F, G) &= \sum_{i=1}^{n} \sum_{j=1}^{n} (\alpha_i - \alpha_{i+1}) \cdot (\alpha_j - \alpha_{j+1}) \cdot \pi_{cr}(F_{\alpha_i}, G_{\alpha_j}) \\
&\le \sum_{i=1}^{n} \sum_{j=1}^{n} (\alpha_i - \alpha_{i+1}) \cdot (\alpha_j - \alpha_{j+1}) \quad (\text{since } \pi_{cr}(F_{\alpha_i}, G_{\alpha_j}) \le 1) \\
&= (\alpha_1 - \alpha_2)(\alpha_1 - \alpha_2) + \cdots + (\alpha_1 - \alpha_2)(\alpha_n - \alpha_{n+1}) + \cdots + \\
&\quad\ (\alpha_n - \alpha_{n+1})(\alpha_1 - \alpha_2) + \cdots + (\alpha_n - \alpha_{n+1})(\alpha_n - \alpha_{n+1}) \\
&= (\alpha_1 - \alpha_2)\big((\alpha_1 - \alpha_2) + \cdots + (\alpha_n - \alpha_{n+1})\big) + \cdots + \\
&\quad\ (\alpha_n - \alpha_{n+1})\big((\alpha_1 - \alpha_2) + \cdots + (\alpha_n - \alpha_{n+1})\big) \\
&= (\alpha_1 - \alpha_2) + \cdots + (\alpha_n - \alpha_{n+1}) \quad (\text{since } \sum_{i=1}^{n} (\alpha_i - \alpha_{i+1}) = 1) \\
&= 1
\end{aligned}$$

Hence, $\pi_f(F, G) \le 1$ holds.

An alternative definition of fuzzy topological predicates, which pursues a similar strategy like the one discussed so far, is based on the topological

[9] For reasons of simplicity, we assume that $\Lambda_{\tilde{F}} = \Lambda_{\tilde{G}} =: \Lambda$. Otherwise, it is not difficult to "synchronize" $\Lambda_{\tilde{F}}$ and $\Lambda_{\tilde{G}}$ by forming their union and by reordering and renumbering all levels.

predicates on simple regions possibly with holes, that is, on predicates $\pi_{srh} \in T_{srh}$. If $F_{\alpha_i}$ is an $\alpha$-level region, let us denote its faces by $F_{\alpha_{i1}}, \ldots, F_{\alpha_{if_i}}$. Similarly, we denote the faces of an $\alpha$-level region $G_{\alpha_j}$ by $G_{\alpha_{j1}}, \ldots, G_{\alpha_{jg_j}}$. We can then define a topological predicate $\pi'_f$ as

$$\pi'_f(F,G) = \sum_{i=1}^{n} \sum_{k=1}^{f_i} \sum_{j=1}^{n} \sum_{l=1}^{g_j} \frac{(\alpha_i - \alpha_{i+1}) \cdot (\alpha_j - \alpha_{j+1}) \cdot \pi_{srh}(F_{\alpha_{ik}}, G_{\alpha_{jl}})}{f_i \cdot g_j}$$

It is obvious that $\pi'_f(F,G) \geq 0$ holds since all factors have a value greater than or equal to 0. We can also show that $\pi'_f(F,G) \leq 1$ by the following transformations:

$$\begin{aligned}
\pi'_f(F,G) &\leq \sum_{i=1}^{n} \sum_{k=1}^{f_i} \sum_{j=1}^{n} \sum_{l=1}^{g_j} \frac{(\alpha_i - \alpha_{i+1}) \cdot (\alpha_j - \alpha_{j+1})}{f_i \cdot g_j} \\
&\quad (\text{since } \pi_{srh}(F_{\alpha_{ik}}, G_{\alpha_{jl}}) \leq 1) \\
&= \sum_{i=1}^{n} \sum_{j=1}^{n} \frac{(\alpha_i - \alpha_{i+1}) \cdot (\alpha_j - \alpha_{j+1})}{f_i \cdot g_j} \cdot f_i \cdot g_j \\
&= \sum_{i=1}^{n} \sum_{j=1}^{n} (\alpha_i - \alpha_{i+1}) \cdot (\alpha_j - \alpha_{j+1}) \\
&= 1
\end{aligned}$$

Hence, $\pi'_f(F,G) \leq 1$ holds. As a rule the predicates $\pi_f$ and $\pi'_f$ do not yield the same results. Assume that $F_{\alpha_i}$ and $G_{\alpha_j}$ fulfil a predicate $\pi_{cr} \in T_{cr}$. This fact contributes once to the summation process for $\pi_f$. But it does not take into account that possibly several faces $F_{\alpha_{ik}}$ (at least one) of $F_{\alpha_i}$ satisfy the corresponding predicate $\pi_{srh} \in T_{srh}$ with several faces $G_{\alpha_{jl}}$ (at least one) of $G_{\alpha_j}$. This fact contributes several times (at most $f_i \cdot g_j$) to the summation process for $\pi'_f$. Hence, the evaluation process for $\pi'_f$ is more fine-grained than for $\pi_f$.

Both generic predicate definitions reveal their quantitative character. If the predicate $\pi_{cr}(F_{\alpha_i}, G_{\alpha_j})$ and the predicate $\pi_{srh}(F_{\alpha_{ik}}, G_{\alpha_{jl}})$, respectively, is never fulfilled, the predicate $\pi_f(F,G)$ and $\pi'_f(F,G)$, respectively, yields *false*. The more $\alpha$-level regions of $F$ and $G$ (simple regions with holes of $F_{\alpha_i}$ and $G_{\alpha_j}$) fulfil the predicate $\pi_{cr}(F_{\alpha_i}, G_{\alpha_j})$ ($\pi_{srh}(F_{\alpha_{ik}}, G_{\alpha_{jl}})$), the more the validity of the predicate $\pi_f$ ($\pi'_f$) increases. The optimum is reached if all topological predicates are satisfied.

## 11.10 Querying with Fuzzy Topological Predicates

In this section we demonstrate how fuzzy topological predicates can be integrated into an SQL-like spatial query language. The fact that the membership degree yielded by a fuzzy topological predicate is a computationally determined quantification between 0 and 1, that is, a fuzzy boolean, impedes a direct integration. First, it is not very comfortable and user-friendly to use such a numeric value in a query. Second, spatial selections and spatial joins expect crisp predicates as filter conditions and are not able to cope with fuzzy predicates.

As a solution, we propose to embed adequate qualitative linguistic descriptions of nuances of topological relationships as appropriate interpretations of the membership values into a spatial query language. For instance, depending on the membership value yielded by the predicate $inside_f$, we could distinguish between *not* inside, *a little bit* inside, *somewhat* inside, *slightly* inside, *quite* inside, *mostly* inside, *nearly completely* inside, and *completely* inside. These *fuzzy linguistic terms* can then be incorporated into spatial queries together with the fuzzy predicates they modify. We call these terms *fuzzy modifiers* since their meaning is that of intensifying or relaxing the constraint expressed by the primary term to which they are applied. For instance, *somewhat inside* is a relaxation of the constraint *inside* since we can expect that it is better satisfied than *inside* even if some portions are outside. It is conceivable that a fuzzy modifier is either predefined and anchored in the query language, or user-defined.

We know that a fuzzy topological predicate $\pi_f$ is defined as $\pi_f : \mathit{fregion} \times \mathit{fregion} \to [0,1]$. The idea is now to represent each fuzzy modifier $\gamma \in \Gamma = \{$*not, a little bit, somewhat, slightly, quite, mostly, nearly completely, completely*$\}$ by an appropriate fuzzy set with a membership function $\mu_\gamma : [0,1] \to [0,1]$. Let $F, G \in \mathit{fregion}$, and let $\gamma\,\pi_f$ be a fuzzy predicate (like *somewhat inside* with $\gamma = \mathit{somewhat}$ and $\pi_f = \mathit{inside}_f$). Then we can define:

$$\gamma\,\pi_f(F,G) = \mathit{true} \quad :\Leftrightarrow \quad (\mu_\gamma \circ \pi_f)(F,G) = 1$$

That is, only for those values of $\pi_f(F,G)$ for which $\mu_\gamma$ yields 1, the predicate $\gamma\,\pi_f$ is true. A membership function that fulfils this quite strict condition is, for instance, the crisp partition of $[0,1]$ into $|\Gamma|$ disjoint or adjacent intervals completely covering $[0,1]$ and the assignment of each interval to a fuzzy modifier. If an interval $[a,b]$ is assigned to a fuzzy modifier $\gamma$, the intended meaning is that $\mu_\gamma(\pi_f(F,G)) = 1$, if $a \leq \pi_f(F,G) \leq b$, and 0 otherwise. For example, we could select the intervals $[0.0, 0.02]$ for *not*, $[0.02, 0.05]$ for *a little bit*, $[0.05, 0.2]$ for *somewhat*, $[0.2, 0.5]$ for *slightly*, $[0.5, 0.8]$ for *quite*, $[0.8, 0.95]$ for *mostly*, $[0.95, 0.98]$ for *nearly completely*, and $[0.98, 1.00]$ for *completely*.

Alternative membership functions are shown by the fuzzy sets in Figure 11.5. While we can always find a fitting fuzzy modifier for the partition due to the complete coverage of the interval $[0,1]$, this is not necessarily the case here. Each fuzzy modifier is associated with a *fuzzy number* having a trapezoidal-shaped membership function. The transition between two consecutive fuzzy modifiers is smooth and here modeled by linear functions. Within a fuzzy transition area $\mu_\gamma$ yields a value less than 1 which makes the predicate $\gamma\,\pi_f$ false. Examples in Figure 11.5 can be found at 0.2, 0.5, or 0.8. Each fuzzy number associated with a fuzzy modifier can be represented as a quadruple $(a,b,c,d)$ where the membership function starts at $(a,0)$, linearly

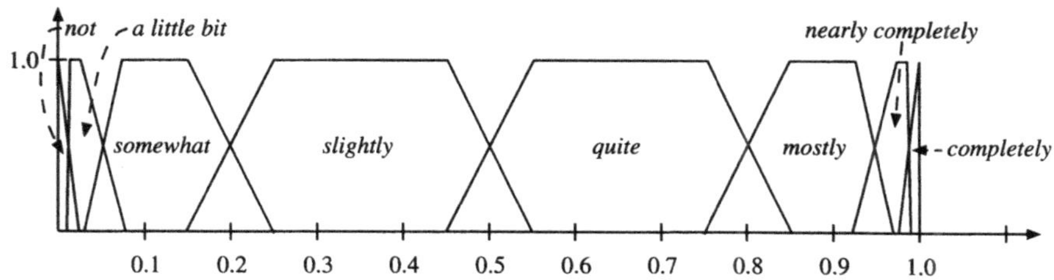


**Figure 11.5.** Membership functions for fuzzy modifiers.

increases up to $(b, 1)$, remains constant up to $(c, 1)$, and linearly decreases up to $(d, 0)$. Figure 11.5 assigns $(0.0, 0.0, 0.0, 0.02)$ to *not*, $(0.01, 0.02, 0, 03, 0.08)$ to *a little bit*, $(0.03, 0.08, 0.15, 0.25)$ to *somewhat*, $(0.15, 0.25, 0.45, 0.55)$ to *slightly*, $(0.45, 0.55, 0.75, 0.85)$ to *quite*, $(0.75, 0.85, 0.92, 0.96)$ to *mostly*, $(0.92, 0.96, 0.97, 0.99)$ to *nearly completely*, and $(0.97, 1.0, 1.0, 1.0)$ to *completely*.

So far, the predicate $\gamma\,\pi_f$ is only true if $\mu_\gamma$ yields 1. We can relax this strict condition by defining:

$$\gamma\,\pi_f(F, G) = \mathit{true} \quad :\Leftrightarrow \quad (\mu_\gamma \circ \pi_f)(F, G) > 0$$

In a crisp spatial database system this gives us the chance also to take the transition zones into account and to let them make the predicate $\gamma\,\pi_f$ true. When evaluating a fuzzy spatial selection or join in a fuzzy spatial database system, we can even set up a weighted ranking of database objects satisfying the predicate $\gamma\,\pi_f$ at all and being ordered by descending membership degree $1 \geq \mu_\gamma > 0$.

A special, optional fuzzy modifier, denoted by *at all*, checks whether a predicate $\pi_f$ can be fulfilled to any extent. An example query is: "Do regions $A$ and $B$ (*at all*) overlap?" With this modifier we can determine whether $\mu_\gamma(x) > 0$ for some value $x \in [0, 1]$.

The following few example queries demonstrate how fuzzy spatial data types and fuzzy topological predicates can be integrated into an SQL-like spatial query language. It is not our objective to give a full description of a specific language. We assume a relational data model where tables may contain fuzzy regions as attribute values.

What we need first is a mechanism to declare user-defined fuzzy modifiers and to activate predefined or user-defined fuzzy modifiers. This mechanism should allow to specify trapezoidal-shaped and triangular-shaped membership functions as well as crisp partitions. In general, this means to define a *classification*, which could be expressed in the following way:

```
create classification fq
    (not                 (0.00, 0.00, 0.00, 0.02),
     a little bit        (0.01, 0.02, 0, 03, 0.08),
     somewhat            (0.03, 0.08, 0.15, 0.25),
     slightly            (0.15, 0.25, 0.45, 0.55),
     quite               (0.45, 0.55, 0.75, 0.85),
     mostly              (0.75, 0.85, 0.92, 0.96),
     nearly completely   (0.92, 0.96, 0.97, 0.99),
     completely          (0.97, 1.0, 1.0, 1.0))
```

Such a classification could then be activated by

```
set classification fq
```

Assuming that we have a relation *pollution*, which stores among other things the blurred geometry of polluted zones as fuzzy regions, and a relation *areas*, which keeps information about the use of land areas and which stores their vague spatial extent as fuzzy regions. A query could be to find out all inhabited areas where people are rather endangered by pollution. This can be formulated in an SQL-like style as (we here use infix notation for the predicates):

```
select areas.name
from   pollution, areas
where  area.use = inhabited and
       pollution.region quite overlaps areas.region
```

This query and the following two ones represent *fuzzy spatial joins*.

Another query could ask for those inhabited areas lying almost entirely in polluted areas:

```
select areas.name
from   pollution, areas
where  areas.use = inhabited and
       areas.region nearly completely inside
       pollution.region
```

Assume that we are given living spaces of different animal species in a relation *animals* and that their vague extent is represented as a fuzzy region. Then we can search for pairs of species which share a common living space to some degree:

```
select A.name, B.name
from   animals A, animals B
where  A.region at all overlaps B.region
```

As a last example, we can ask for animals that usually live on land and seldom enter the water or for species that never leave their land area (the built-in aggregation function **sum** is applied to a set of fuzzy regions and aggregates this set by repeated application of fuzzy geometric union):

```
select name
from   animals
where  (select sum(region) from areas)
         nearly completely covers or
         completely covers region
```

## 11.11 Conclusions

In this chapter we have introduced a fuzzy spatial algebra that introduces spatial data types for fuzzy points, fuzzy lines, and fuzzy regions, and that deals with fuzzy topological predicates operating on these data types. Structure and semantics of types and predicates are formally defined. The characteristic feature of the design is the modeling of smoothness and continuity which is inherent to the objects themselves and to the transitions between different fuzzy objects.

Future work will have to deal with the design of fuzzy spatial operations and the implementation of the whole algebra.

## References


1. Altman D. (1994) Fuzzy Set Theoretic Approaches for Handling Imprecision in Spatial Analysis. Int. Journal of Geographical Information Systems 8(3):271–289.
2. Blakemore M. (1984) Generalization and Error in Spatial Databases. Cartographica, 21.
3. Burrough P.A. (1996) Natural Objects with Indeterminate Boundaries. In: Burrough P.A., Frank A.U. (Eds.) Geographic Objects with Indeterminate Boundaries, GISDATA Series, vol. 2, Taylor & Francis, pp. 3–28.
4. Burrough P.A., Frank A.U. (Eds.) (1996) Geographic Objects with Indeterminate Boundaries. GISDATA Series, vol. 2, Taylor & Francis.
5. Chang C.L. (1968) Fuzzy Topological Spaces. Journal of Mathematical Analysis and Applications 24:182–190.
6. Clementini E., Di Felice P., Califano G. (1995) Composite Regions in Topological Queries. Information Systems 20(7):579–594.
7. Clementini E., Di Felice P. (1996) An Algebraic Model for Spatial Objects with Indeterminate Boundaries. In: Burrough P.A., Frank A.U. (Eds.) Geographic Objects with Indeterminate Boundaries, GISDATA Series, vol. 2, Taylor & Francis, pp. 153–169.
8. Cohn A.G., Gotts N.M. (1996) The 'Egg-Yolk' Representation of Regions with Indeterminate Boundaries. In: Burrough P.A., Frank A.U. (Eds.) Geographic Objects with Indeterminate Boundaries, GISDATA Series, vol. 2, Taylor & Francis, pp. 171–187.
9. Dubois D., Jaulent M.-C. (1987) A General Approach to Parameter Evaluation in Fuzzy Digital Pictures. Pattern Recognition Letters, pp. 251–259.

10. Dutta S. (1989) Qualitative Spatial Reasoning: A Semi-Quantitative Approach Using Fuzzy Logic. In: Proc. of the 1st Int. Symp. on the Design and Implementation of Large Spatial Databases. LNCS 409, Springer-Verlag, pp. 345–364.
11. Dutta S. (1991) Topological Constraints: A Representational Framework for Approximate Spatial and Temporal Reasoning. In: Proc. of the 2nd Int. Symp. on Advances in Spatial Databases, LNCS 525, Springer-Verlag, pp. 161–180.
12. Egenhofer M.J. (1989) A Formal Definition of Binary Topological Relationships. In: Proc. of the Int. Conf. on Foundations of Data Organization and Algorithms, LNCS 367, Springer-Verlag, pp. 457–472.
13. Egenhofer M.J., Clementini E., Di Felice P. (1994) Topological Relations between Regions with Holes. Int. Journal of Geographical Information Systems 8(2):128–142.
14. Erwig M., Schneider M. (1997) Partition and Conquer. In: Proc. of the 3rd Int. Conf. on Spatial Information Theory, LNCS 1329, Springer-Verlag, pp. 389–408.
15. Erwig M., Schneider M. (1997) Vague Regions. In: Proc. of the 5th Int. Symp. on Advances in Spatial Databases, LNCS 1262, Springer-Verlag, pp. 298–320.
16. Finn J.T. (1993) Use of the Average Mutual Information Index in Evaluating Classification Error and Consistency. Int. Journal of Geographical Information Systems 7(4):349–366.
17. Gaal S. (1964) Point Set Topology. Academic Press.
18. Güting R.H., Schneider M. (1995) Realm-Based Spatial Data Types: The ROSE Algebra. VLDB Journal 4:100–143.
19. Kollias V.J., Voliotis A. (1991) Fuzzy Reasoning in the Development of Geographical Information Systems. Int. Journal of Geographical Information Systems 5(2):209–223.
20. Lagacherie P., Andrieux P., Bouzigues R. (1996) Fuzziness and Uncertainty of Soil Boundaries: From Reality to Coding in GIS. In: Burrough P.A., Frank A.U. (Eds.) Geographic Objects with Indeterminate Boundaries, GISDATA Series, vol. 2, Taylor & Francis, pp. 275–286.
21. Preparata F.P., Shamos M.I. (1985) Computational Geometry. Springer Verlag.
22. Schneider M. (1996) Modelling Spatial Objects with Undetermined Boundaries Using the Realm/ROSE Approach. In: Burrough P.A., Frank A.U. (Eds.) Geographic Objects with Indeterminate Boundaries, GISDATA Series, vol. 2, Taylor & Francis, pp. 141–152.
23. Schneider M. (1997) Spatial Data Types for Database Systems – Finite Resolution Geometry for Geographic Information Systems. LNCS 1288, Springer-Verlag, Berlin Heidelberg.
24. Shibasaki R. (1993) A Framework for Handling Geometric Data with Positional Uncertainty in a GIS Environment. GIS: Technology and Applications, 21–35.
25. Tilove R.B. (1980) Set Membership Classification: A Unified Approach to Geometric Intersection Problems. IEEE Trans. on Computers C-29:874–883.
26. Usery E.L. (1996) A Conceptual Framework and Fuzzy Set Implementation for Geographic Features. In: Burrough P.A., Frank A.U. (Eds.) Geographic Objects with Indeterminate Boundaries, GISDATA Series, vol. 2, Taylor & Francis, pp. 71–85.
27. Wang F. (1994) Towards a Natural Language User Interface: An Approach of Fuzzy Query. Int. Journal of Geographical Information Systems 8(2):143–162.
28. Wang F., Hall G.B., Subaryono (1990) Fuzzy Information Representation and Processing in Conventional GIS Software: Database Design and Application. Int. Journal of Geographical Information Systems 4(3):261–283.

29. Worboys M.F., Bofakos P. (1993) A Canonical Model for a Class of Areal Spatial Objects. In: Proc. of the 3rd Int. Symp. on Advances in Spatial Databases, LNCS 692, Springer-Verlag, pp. 36–52.
30. Zadeh L.A. (1965) Fuzzy Sets. Information and Control 8:338–353.
31. Zhan F.B. (1997) Topological Relations between Fuzzy Regions. In: Proc. of the ACM Symp. for Applied Computing, pp. 192–196.

# Part II

# APPLICATIONS

# 12

# A Possibility-Based Model to Index Remote Sensing Images

Paola Carrara[1], Gabriella Pasi[2], Monica Pepe[1], and Anna Rampini[1]

[1] IREA-CNR
Via Bassini 15, I-20133 Milan (Italy)
{carrara.p, pepe.m, rampini.a}@irea.cnr.it
[2] ITC-CNR
Via Bassini 15, I-20133 Milan (Italy)
gabriella.pasi@itc.cnr.it

## 12.1 Introduction

Recently, several environmental applications took advantage of the use and analysis of remote sensing images, since they are intrinsically referred to the spatial distribution of the phenomena of interest. Remote sensing is widely employed for land monitoring, land planning and risk prevention; its ultimate aim is to identify specific image contents to create thematic maps. An efficient management of large collections of remote sensing images and effective retrieval mechanisms are therefore becoming a need, so that remote sensing images have been included in the list of *Grand Challenges* application fields for visual information management systems or Content-Based Information Retrieval (CBIR) systems [24]: in fact this application area justifies the use of large computers and storage capacity necessary for visual databases.

The traditional approach to manage and retrieve remote sensing images has been for long time (and it is still in some application field) based on archiving in file systems; with each image a meta-information is associated (generally a textual description), which tries to capture some important features of the considered image; this information is stored in an auxiliary file or within the image file itself. This approach, though simple, easy to implement and use, and apparently inexpensive, limits searching to file names or to meta-information. Scalability problems soon emerge with the increase of file number and size. Also the meta-information definition, extraction, and maintenance are time consuming activities and may present drawbacks.

A complete and exhaustive textual description of images is extremely difficult to achieve, it is subjective and often inadequate if retrieval needs change or update. However many systems for the management of databases of remotely sensed images are based on this approach [1, 25, 7].

To overcome these limitations, several Content-Based Retrieval methods for image databases have been proposed in the literature [9, 11, 13, 16, 17], which are based on the extraction of significant features from images by means of interpretation processes.

This can be considered the first step of the CBIR paradigm consisting in three main steps, i.e.:

- visual feature extraction
- multi-dimensional indexing
- retrieval system design

The features can be distinguished either in general, i.e. colors, textures, shapes, or domain-specific features. Many content-based approaches and systems take into account several features of both categories: in this case, if the image collection is sufficiently large, the multi-dimensional indexing becomes a key step to increase the retrieval efficiency [20].

CBIR approaches have been proposed also to manage archives of remote sensing images; some of them are based on domain-independent features such as color or texture (see for example [26] and [23]). Remote sensing imagery is however a narrow domain of application, where the variability of images is limited and the semantic description is well-defined. It has been observed [24] that for such domains the gap between features and their semantic interpretation (the so-called *semantic gap*) is usually small, so domain-specific models may be of help. Therefore we propose a model for the indexing and retrieval of remote sensing images which considers a unique, meaningful domain-dependent feature, i.e. the spectral properties of the images. We focus on the two first steps of the above mentioned paradigm, discussing both how the feature of interest can be extracted from the images, and which image structures are identified in order to obtain an effective indexing. The contribution also addresses the problem of selecting similarity measures that mime the human perception of visual content, able to extract a similarity/dissimilarity judgement also when some information lack or many different integrated aspects have to be taken into account.

We assume that the atomic components of our information items (i.e., of the remote sensing images) are pixels, represented as vectors of values (called *spectral signatures*). In the literature some works can be found which are based on the use of spectral signatures (or, in general, spectral properties). For example, in [2] the spectral properties allow to segment images in regions to which more indexes are associated, representing both spectral and non-spectral aspects of the regions; [14] performs a clustering of the images and associates the clusters to classes by exploiting expert knowledge, while in [27] the association process is based on statistic considerations. Also the approach in [4] is based on the extraction of clusters, but they are not explicitly associated to ground cover classes.

The approach we propose is based on the classification of the images by clustering and the association of the obtained clusters to ground classes avoiding experts' supervision: it differs from previous approaches mainly because it takes into account the uncertainty intrinsic both in the extraction of significant patterns (homogeneous regions) from the images and in their assignment to ground cover classes. In the model proposed in this Chapter, we assume that each pixel in an image belongs to a spectral region, which groups a set of 'homogeneous' pixels, and which is identified by a representative spectral signature. As a consequence, an image is represented as the set of signatures of the regions identified in the image. The association of each region with a predefined, meaningful ground cover class, is performed by comparing the region signature and the spectral signatures of the reference ground cover classes. Therefore, for different applications a significant set of predefined reference classes has to be defined.

The process of associating the image classes with the reference ground classes is characterized by uncertainty, intrinsic in the result of comparing the signature of an image region with the one of the predefined ground classes. To take into account this uncertainty, we adopt the framework of possibility theory [10, 29]: for each spectral region a possibility degree of being a ground class is computed on the basis of qualitative decision rules expressed on the comparison measures.

Based on these assumptions the images are then represented by means of a set of spectral signatures, each of which constitutes the centroid of an identified spectral region in the image. Each spectral signature has also associated a possibility distribution over the reference classes; the degree associated with a given reference class represents the possibility that the considered region corresponds to that reference class.

The approach here presented takes into account uncertainty at distinct levels: in the image classification phase Fuzzy clustering is applied; in the indexing phase, which associates clusters to ground classes, possibility theory and distance measures are adopted; in the query formulation phase requests on image representations are expressed by soft constraints.

The Chapter is organized as follows: in Sect. 12.2 the spectral properties of images are analyzed and we discuss the conditions under which they can be proposed as image descriptors. In Sect. 12.3 the indexing model is presented, and in Sect. 12.4 we sketch the data structure and some ideas concerning the query language and the query evaluation mechanism. Section 12.5 illustrates a proposal for the evaluation of the method.

## 12.2 Spectral properties as image descriptors

In multi-spectral remote sensing images, each pixel is a vector of values corresponding to the electromagnetic energy reflected or emitted in different spectral bands (i.e. different wavelengths) from the corresponding portion of observed ground surface. The radiation reflected as a function of the wavelength is called the spectral signature of the surface [22]. Spectral signatures are meaningful if they are referred to ground cover classes: each surface type such as water, bare soil or vegetation reflects radiation differently in the distinct wavelengths.

In this work the spectral values have been derived by samples selected on images of the Thematic Mapper (TM) and Enhanced Thematic Mapper plus ($ETM^+$) sensors on board the Landsat 5 and 7 satellites, which record the spectral response of ground surfaces in six spectral bands from visible to shortwave infrared (TM1, TM2, TM3, TM4, TM5, TM7) with 30 m of spatial resolution, and the spectral emission in one thermal infrared band (TM6) with 120 m of spatial resolution. We consider here only the six reflective bands - from visible to shortwave infrared - have been considered.

The energy reflected from the earth surface is digitally acquired in 8 bits, hence, for each band, each element of the satellite raster image is represented by a Digital Number (DN) in the range [0..255].

Each surface type is therefore characterized by a the vector of DNs, i.e. the spectral signature. Though some variations of the surface type can occur, the overall shape of the signature is maintained. Macro-classes, such as vegetation, bare soil, water, clouds, snow, may be easily distinguished on the basis of their reflective properties. For example, the vegetation has a remarkably high reflection in the near infrared region of the electromagnetic spectrum, and a low reflection in the visible region corresponding to red; this makes it possible to distinguish vegetation areas from bare ground as the difference of reflection in visible red and near infrared is greater for vegetation areas than for bare ground (see the first and third graphs in Fig. 12.1 respectively, where the visible red and near–infrared bands correspond to the TM3 and TM4 bands). On the contrary, different types of vegetation or different conditions of growth of the same type of vegetation, present similar spectral responses with small variations, as can be observed in the graphs (a) of Fig. 12.1 which present many similar curves for the macro-class vegetation.

Since the association of each region with a predefined, meaningful ground cover class is performed by comparison of the region signatures and the reference ground cover classes, theoretically the process needs the spectral signatures to be physically homogeneous. It means that the satellite raw signal should be transformed in physical units and corrected by the attitude and atmospheric effects. This process requires information regarding the satellite sensor and platform, as well as the atmospheric conditions at the satellite overpass.

In our case it is not useful to perform the calibration of data as soon as databases of remotely sensed images are often populated by uncorrected images (this is the case of the Landasat TM and $ETM^+$ image archive which is used for the experiments). We must allow that archives are affected by the so called *sensory gap*, i.e. 'the gap between the object in the world and the information in a (computational) description derived from a recording of that scene' [24]. This gap generates uncertainty that we take into account in the proposed indexing method.

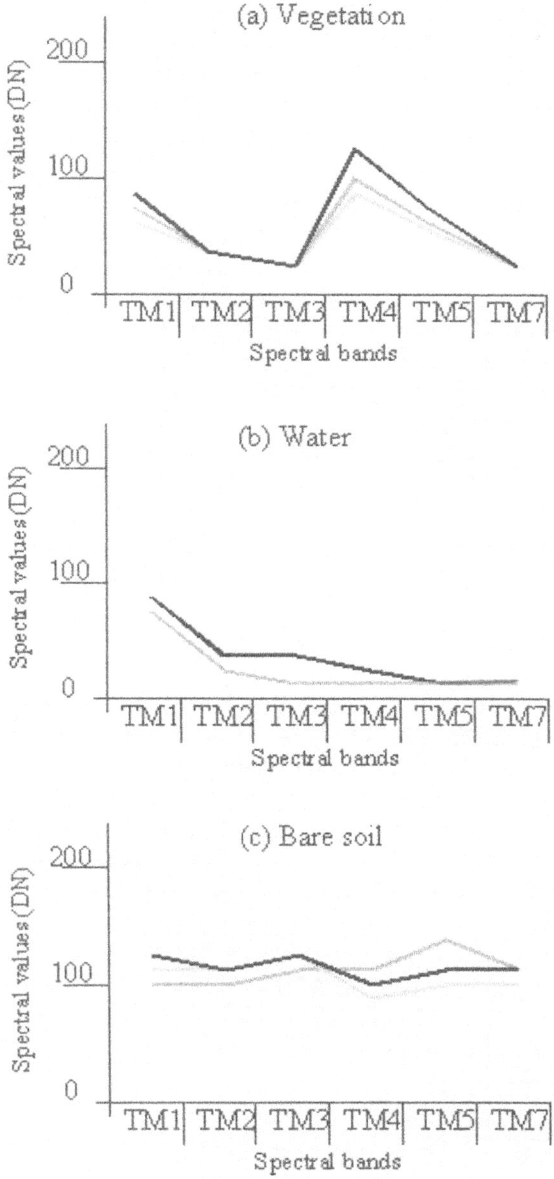


**Figure 12.1.** Graphic representations of spectral responses in digital number values of vegetation (a), water (b) and bare soil (c); on the x–axis we reported six reflective bands on the Landsat satellites.

The interpretation of remote sensing images is performed by means of a classification process that assigns each pixel to a ground cover class. The most applied approach to classify multi-spectral images is based on the use of multi-spectral information contained in each pixel as the main feature. Multi-spectral classification is an information extraction process that analyzes the spectral signatures and then groups pixels into thematic classes on the basis of similar signatures; it can be performed by unsupervised or supervised techniques depending on the available 'a priori' information [22].

The supervised classification is based on the use of 'a priori' knowledge called ground truth, gathered on a geographically representative sample of objects and used to train the classifier. When the 'a priori' knowledge lacks, an unsupervised classification can be used to group multi-spectral response patterns in separable clusters to be then labelled into ground-cover classes by an 'a posteriori' interpretation.

Although the supervised method is much more effective in terms of accuracy, it is often difficult to collect ground truth data in highly variable heterogeneous landscapes. On the other hand, the main problems of the unsupervised approach lie both in the need of defining a relation between spectral classes and ground cover, and in the implicit assumption that clusters represent actual cover patterns in the image. In both cases, the classification is a long and heavy process, requiring an examination and elaboration of heterogeneous information.

As in the context of this Chapter the classification process has to be as far as possible accurate and automatic without requiring any specific human knowledge, since it has to be used to represent and retrieve information for heterogeneous users with unknown abilities and background and it has to be independent on the increase and extension of the data volume [4]. The proposed solution consists therefore in the adoption of an unsupervised approach based on the knowledge of reference spectral responses to label clusters identified by an automatic procedure [6]. The accuracy of the association of clusters with reference classes depends both on a meaningful collection of representative spectral signatures derived from non-calibrated remote sensing images (see the Sect. 12.5 on evaluation for a discussion about this aspect) and on the definition of suitable distance measures.

## 12.3 Indexing remote sensing images by spectral properties

In Information Retrieval Systems, the indexing process is aimed at defining a formal representation of the contents of the considered information items. This formalization is necessary to make the system able to automatically compare queries and documents; the formal representations of queries and documents must be compatible, of course. In textual information retrieval systems (IRS), the information items are constituted by textual documents, which are usually represented by means of a collection of weighted terms [21]. In the case of image retrieval the information items are constituted by images. An image can be formally described by means of several and distinct features (color, texture, objects and their relative positions, etc.), which are the basic elements of the formal image representation [18, 9, 11, 13].

The main idea at the basis of the indexing method proposed in this Chapter is to represent remote sensing images as a set of spectral regions corresponding to reference land cover classes. This means to identify in an image the classes of which it is composed: this task is accomplished through a complete image classification process, which constitutes the basis of the image indexing procedure. The whole unsupervised classification is performed in several steps (see Table 12.1).

The first step consists in a clustering of the considered image: each pixel in an image belongs to a spectral region, which groups a set of homogeneous pixels, and which is also identified by a representative spectral signature. Several clustering algorithms are available in the literature; the K-Means [22] algorithm has been tested within this project. The output of the clustering algorithm is a set of clusters characterized by the spectral values of their centroids. The subsequent steps compare the spectral signature of each centroid with the spectral signatures of the reference classes, to the aim of identifying to which class each cluster corresponds. To make this comparison, a similarity measure has to be adopted. Among the several similarity measures proposed in the literature [15], we have selected two measures, i.e. the Simple Matching Coefficient (SMC), which quantifies the similarity of vectors' values, and the Correlation Coefficient (CC), which quantifies the similarity of trends represented by vectors. This choice has been done to take into account two distinct aspects of similarity between spectral signatures, i.e. a value-to-value similarity and a shape similarity. In order to identify the reference class to be associated with an image region, the selection of the class with the highest overall similarity could not be sufficient; in fact, the process of determining the reference class is intrinsically uncertain, due to various factors such as the sensory gap, the uncertainty in identifying the homogeneous regions in an image, and the variability of the signatures ought to the different conditions (in space and time) of each class. To explicitly model this uncertainty, we interpret the similarity degree between an image region and a reference class as the possibility degree that the considered region in the image corresponds to the reference class [10, 29]. Each region has then associated a possibility distribution over the considered reference classes [10].

Let us now describe and discuss the procedure that we propose to compute this possibility distribution, starting from the SMC and CC numeric values.

As previously outlined, the SMC and CC measures represent distinct aspects of the similarity between two vectors; therefore they must be aggregated to obtain an overall similarity value. To this aim we adopted a heuristic procedure justified by the following considerations, based on the behavior of the

**Table 12.1.** Steps of image classification

**Step 1 - Extraction of cluster centers:** The K-Means algorithm extracts a given number of cluster centers (the number has been heuristically set to 20 with a maximum number of iterations set to 15); this technique is applied on elements randomly selected all over the image [12]. It then assigns all the image pixels to the identified clusters.

**Step 2 - Measures of similarity:** SMC and CC measures between the spectral signatures identifying the cluster centroids and the reference spectral signatures are computed.

**Step 3 - Labelling:** Two heuristic tests based on the SMC and CC measures are evaluated to associate clusters to reference classes (with a possibility degree).

signatures of different land cover classes and on the results obtained in tests about SMC and CC.

If both the adopted measures give the maximum value in comparing a spectral region to the same reference class $i$, then the similarity between the signature of the region and the signature of the reference class is sufficient to certainly associate the region to the *i-th* class. However, there are exceptions, i.e. the classes with a low reflectivity, such as water or shadow: for different reasons these two classes represent situations in which the reflective power is low or very low, thus generating by default an uncertain situation which requires further analysis.

This analysis is also performed if the condition on the maximum of similarities is not verified, i.e. corresponding to the following situations:

- the cluster is constituted by 'mixed' pixels in which more classes coexist (borderline situations or mixtures)
- the reference class presents a strong variability and the reference signature is therefore scarcely representative and discriminant

All these cases, as well as classes with low reflectivity, require to model the decision process depending on the considered class $i$. For example more tolerant criteria have to be adopted in the comparison with a highly variable class (such as vegetation) with respect to a class with a strong discrimination power (such as snow).

From a formal point of view, the above described decision procedure consists of the two following tests:

*Test 1*: For each representative region in a given image, first of all the numeric values $\mathrm{smc}_i$ and $\mathrm{cc}_i$ are computed to assess the similarity to the *i-th* reference class (we suppose here that the reference classes are $n$ and $i \leq n$). Then the two values

$$Max_{i=1..n}(smc_i) = max.smc$$

and

$$Max_{i=1..n}(cc_i) = max.cc$$

are computed. If the two maximum values are obtained in correspondence to the same class $i$, then the region is associated to the *i-th* class with a maximum degree of possibility, while the degree with respect to the other classes is the minimum (i.e. $p_i = 1$ and $p_j = 0$ for $j \leq n$ and $j \neq i$). If either the condition on the maximum of similarities is not verified or the obtained reference class has a low reflectivity (such as water and shadow), the association is not performed and the second test must be activated.

*Test 2*: For each reference class two fuzzy sets are defined on the numeric variables SMC and CC respectively, by taking into account the specific variability of the considered reference classes. The membership functions of these fuzzy sets $\mu_{SMC}$ and $\mu_{CC}$ are heuristically defined depending on the considered class and they express the compatibility between a numeric value and an 'acceptable' similarity with the considered reference class (see Fig. 12.2). When the region of an image is analyzed, the membership degrees $\mu_{SMC}(smc_i)$ and $\mu_{CC}(cc_i)$ are computed and aggregated to obtain an overall degree of similarity, which is interpreted as the possibility that the considered image region corresponds to the reference class $i$. This aggregation is performed by applying the minimum operator:

$$p_i \quad = \quad Min(\mu_{SMC}(smc_i), \mu_{CC}(cc_i))$$

This process is applied for each reference class, so that for each spectral region a possibility distribution is computed.

## 12.4 Image representation and retrieval

In this Section a data structure is proposed, which suitably represents the chosen image indexes, i.e. the spectral signatures.

The information contained and organized in the data structure is used by the query evaluation mechanism, which is described in the following subsection. The data structure for image retrieval is therefore constituted of two main components: the dictionary of the spectral signatures of the reference ground classes and the inverted file. Each element of the dictionary is constituted by the couple (<class name>, <vector of values of the reference spectral signature>) and points to the list of all the images containing the reference class <class name>.

The inverted file contains the lists of the representations of the archived images; each list corresponds to a class of the dictionary. Each element of the list pointed by the *i-th* class is constituted by an archived image identifier $I_j$ (with $j \leq N$, and $N$ is the number of images contained in the archive), a spectral region identifier $SR_{jk}$ (with $k \leq M$, and $M$ is the number of clusters contained in an archived image $I_j$), the vector of the centroid of the spectral region in the image itself $VC_{jk}$, the two computed similarities between the

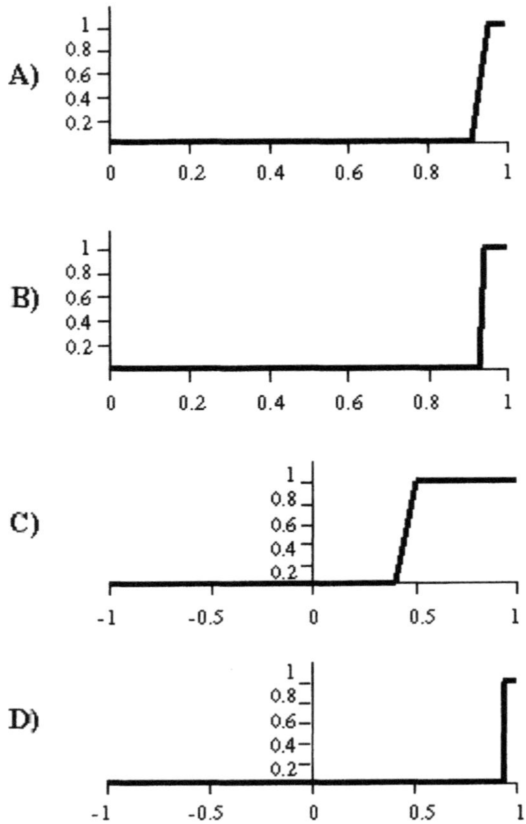


**Figure 12.2.** SMC membership functions for the class Vegetation (A) and for the class Water (B); CC membership functions for the class Vegetation (C) and for the class Water (D)

region signature and the spectral signature of the *i-th* class $smc_{ijk}$ and $cc_{ijk}$ and the degree of possibility $p_{ijk}$ that the region $k$ in image $j$ is associated with the pointing class $i$.

In order to optimize the dimensions of the data structure, we store only the regions having a possibility degree $p_i$ higher than a heuristically determined threshold. Therefore if the region has a high possibility to be associated with two classes, this information will be recorded in the indexing data structure and the region will be retrieved by queries regarding both classes.

In Figure 12.3 the structure of the dictionary and the inverted file is sketched.

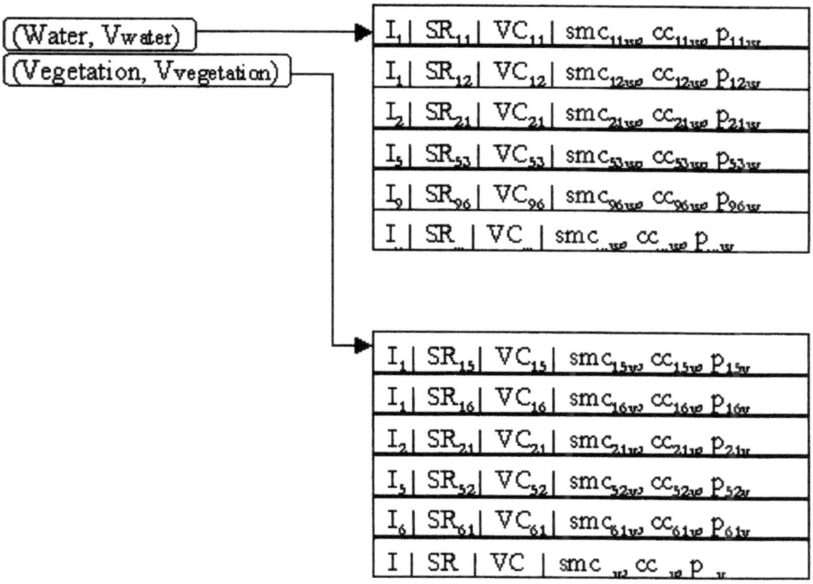


**Figure 12.3.** An illustration of the data structure; the boxes on the left are two elements of the dictionary, while on the right there are the pointed lists

## Query formulation and evaluation

The basic and more meaningful elements of the data structure are constituted by the reference classes, which are identified by their spectral signatures. As a consequence the query language allows to express constraints on these information elements. In our approach the main type of constraint in a query is the specification (either explicit or implicit) of the reference classes that have to be contained in the desired images. Furthermore, the spectral regions in the images are archived with the degree of possibility associated with the class, thus allowing the user to express also a constraint on this degree.

We propose three main types of class specifications: by textual identifier, by spectral signature and by image sub-window. For each query type an evaluation mechanism is defined. In the following we first describe the three proposed types of queries and then their evaluation procedures. A query evaluation produces a list of retrieved images with a numeric score of their relevance with respect to the considered query, called the Retrieval Status Value (RSV).

The semantics of the three types of queries are:

1. *Query by class label:* this kind of query is specified by means of one or more identifiers of the predefined reference classes. When the user specifies for example the identifier 'water', she/he is looking for images containing one or more regions possibly classified as water. In this type of query the identifier of a reference class constitutes the atomic selection criterion,

which specifies the retrieval constraint of the presence of that reference class in an image in order to consider it relevant. In other words the query evaluation mechanism selects the images which 'contain' at least a spectral region corresponding at least to some extent to the specified reference class. Like in query languages of textual IRSs the atomic selection criteria (i.e. the class identifiers) can be combined through aggregation operators such as the usual AND and OR. By these operators more complex constraints can be specified, although the query evaluation mechanism starts by evaluating the atomic selection criteria and then by combining the obtained results in accordance with the semantics associated with the aggregation operators. When connecting the class identifiers by the AND aggregation operator the user is requiring the images that contain all the specified classes. When connecting the class identifiers by the OR aggregation operator the user is requiring the images that contain at least one of the specified classes. The AND and the OR are usually interpreted as the *min* and *max* aggregation operators, respectively.

2. *Query by a vector of spectral values:* this kind of query consists in the specification of a vector of spectral values. The vector is associated with one or more reference classes on the basis of the following procedure. First the possibility degrees of the input vector with respect to the predefined reference classes are computed (by the steps 2 and 3 of the classification procedure described in Table 12.1). Second, just the classes for which these possibility degrees are greater than a given threshold are considered for retrieval. By this kind of query, the user is looking for images that possibly contain the classes with spectral values 'close' to those of the specified vector (the concept of closeness is crisply modelled by the choice of a threshold). The retrieved documents are the images contained in the lists pointed by these classes. The first images presented are those in the list associated with the class with respect to which the input vector has the highest possibility degree.
3. *Query by a homogeneous sub-region of an unknown image (query by example):* in this case the user presents an image to the system, and within this image she/he selects a sub-region which is as far as possible homogeneous. Then, the spectral signature of the sub-region is computed and used as in the query of type 2.

In all the above cases the result of a query evaluation is a list of images with an associated relevance estimate (the so called RSV). The RSV is computed on the basis of the possibility degrees $p_i$ of the image clusters pointed by the reference class $i$ identified by the query. Two cases may occur:

- just one cluster in the image is pointed by the reference class
- more than one cluster in the image is pointed by the class; this happens when, during the classification process, more than one cluster matches with the same reference class (i.e., more clusters are possibly associated to the class under consideration)

In the first, simplest case, the image RSV is given by the possibility degree $p_i$ of the unique cluster. In the second case, the $p_i$s of the different clusters have to be aggregated. This is done by applying the maximum operator, so that the RSV of the image coincides with the highest possibility degree of its clusters.

Each image is associated with some additional information, i.e.: the original, unclassified, corresponding image; the classified image; the list of all clusters in the image which 'satisfy' the query (with their associated $p_i$s).

## 12.5 An evaluation proposal

The method proposed to index and retrieve remote sensing images has been evaluated by an experiment conducted on an archive of Landsat TM images containing 263 images of 81 km$^2$ each. They record heterogeneous scenes of the Italian peninsula in different seasons and the total covered area is 18,747 km$^2$. Landscapes range from coastal zones to mountain regions, as well as agricultural and urban areas; cloud covered areas are also included. This heterogeneity improves the quality of the evaluation [4]. The collection of spectral signatures has also been created including general purpose ground cover classes. The considered reference classes are: vegetation, water, bare soil/urban, snow, cloud, shadow. The selection of the candidate reference signatures has been based on mean spectral signatures collected from images which are not included in the archive: the main selection criteria are statistical parameters and the representation power. The chosen signatures should be as far as possible independent on background conditions such as atmospheric and topographic effects, season, etc. The evaluation experiments proved that a good collection of reference signatures is a key factor in the obtained effectiveness.

The evaluation has been performed by specifying four queries by label with distinct classes which are included in the considered archive.

Once a query has been evaluated, we can establish if the retrieval method is effective, i.e. if it retrieves all the images in the archive containing the class requested, and if the collection of retrieved images does not comprise erroneous classes. The traditional measures of effectiveness in IRSs are the so-called Recall and Precision [19]. Let $S_q$ be the set of relevant images according to the query $q$ and $A_q$ the set of returned images; then the Recall and Precision measures are defined as:

$$R_q = \frac{A_q \bigcap S_q}{S_q} \tag{12.1}$$

$$P_q = \frac{A_q \bigcap S_q}{A_q} \tag{12.2}$$

These measures, heavily adopted in the text retrieval field, are sometimes reported also for image retrieval [11], but in the CBIR field common test databases (with a statistically relevant number of documents in the collection) and evaluation metrics lack, mainly due to the problem of determining whether a given image really answers a given query [24]. In fact, the determination of the relevance of retrieved images with respect to a query is in general affected by the subjectivity of human perception and by the expectations of the users. Another drawback is constituted by the *semantic gap* of CBIR systems, i.e. the gap between the high level expectations of the users and the primitive features used by the systems to index and retrieve the images in an archive [24]. The user, for example, looks for a beach scene, whereas the system interprets this request as large areas of blue on top of areas of yellow [5], thus retrieving images that can have also a different meaning.

In our specific case, the problem of the subjectivity in the determination of image relevance is reduced (or even overcome) as we rely on the strong and objective semantics provided by ground truth and photo–interpretation for deciding if an image (or one region in it) really contains a (set of) ground cover(s) or not. As far as the semantic gap, the proposed queries help in lowering its influence: the gap is absent in the case of query by label (as in textual retrieval); in both the queries by a homogeneous region and by vector the user does not express her/his expectations by a high level concept but directly through the primitive feature considered by the method.

We can therefore accept Recall and Precision as useful effectiveness measures as we can decide without ambiguity or subjectivity if an image is relevant (that is, the class specified in the query is contained in it or not).

The results of the four queries are shown in Table 12.2.

**Table 12.2.** Recall and Precision for selected queries

| Query $q$ | Vegetation | Water | Cloud | Bare soil |
|---|---|---|---|---|
| Relevant $S_q$ | 230 | 218 | 67 | 231 |
| Returned $A_q$ | 230 | 226 | 84 | 234 |
| Recall $R_q$ | 1 | 0.99 | 0.95 | 1 |
| Precision $P_q$ | 1 | 0.95 | 0.76 | 0.98 |

There are classes, such as cloud, which have a heavy discrimination power as they are contained only in some specific and well known images: the class cloud can therefore be adopted to easily verify the accuracy of the method.

On the other hand, it must be observed that some classes like vegetation are present in almost all the images of the archive: the query 'vegetation' has therefore a low discrimination power, and it can be adopted for a negative test: it is straightforward to verify that some images, such as the images of sea, are not retrieved at all. This is confirmed by the high value of Precision for this class.

## 12.6 Conclusions

The exponential growth of satellite imagery makes more and more urgent the definition of CBIRs in the field of remote sensing, due both to the huge increase of image archives and to a wide range of users with other fields of expertise and no knowledge about remote sensing; moreover, it has been observed that the deployment of efficient and effective automated repositories could be crucial in emergency and disaster management [3]. However it is necessary to develop methods related to the specific domain, such as the spectral properties of images, and evaluate their performance in archives containing frames acquired by heterogeneous sensors which can heavily influence the image representation. In this Chapter we presented an effort in this direction: in fact, we investigated the capability of spectral signatures as indexes of images from remote sensing. We also evaluated the method in a database of images representing heterogeneous landscapes.

Among the future developments we plan to define a hierarchical extension of the data structure, in order to examine not only the macro-classes, but also more specific classes such as 'forest', 'grass', etc.

Another important aspect to be investigated concerns the modelling of spatial characteristics of images. The spectral regions identified by the indexing procedure correspond to several objects spatially distributed within an image. The spatial features in terms of metric and topological relationships of the identified objects could then be included. This extension enables the definition of a richer query language, allowing the specification of spatial constraints.

## References

1. Anderson J.T., Stonebraker M. (1994) Sequoia 2000 metadata schema for satellite images. ACM SIGMOD Record 23(4):42–48.
2. Barros J., French J., Martin W., Kelly P., White J.M. (1994) Indexing Multispectral Images for Content-Based Retrieval. In: Image and Information Systems: Applications and Opportunities (23rd AIPR Workshop), Proc. of SPIE 2368, Washington DC, Oct. 1994, pp. 25–36.
3. Bergman L.D., Castelli V., Li C.-S. (1997) Progressive Content-Based Retrieval from satellite image archives. D-Lib Magazine, October 1007, [Online], <http://www.dlib.org/dlib/october97/ibm/10li.html>
4. Bretschneider T., Cavet R., Cao O. (2002) Retrieval of remotely sensed imagery using spectral information content. In: Proc. of the International Geoscience and Remote Sensing Symposium, Vol. 4, pp. 2253–2256.
5. Capodiferro L., Kiranyaz S., Gabbouj M. (2003) Evaluation criteria and evaluation report. Deliverable of the Project 'Network of excellence in Content-based semantic scene analysis and Information Retrieval', IST-2001-32795.
6. Carrara P., Galli C., Rampini A. (2000) A Database for Remote Sensing Image Retrieval by Spectral Features. ITIM-CNR Tech. Rep.

7. Chang D., Moon B., Acharya A., Shock C., Sussman A., Saltz J.H. (1997) Titan: A high-performance remote sensing database. In: Proc. of the International Conference on Data Engineering, pp. 375–384.
8. Congalton R.G. (1991) A review of Assessing the Accuracy of Classification of Remotely Sensed Data. Remote Sens. Environ. 37:35–46.
9. Del Bimbo A. (1999) Visual Information Retrieval. Morgan Kaufmann Publishers, San Francisco, CA.
10. Dubois D., Prade H. (1988) Possibility Theory: An Approach to Computerized Processing of Uncertainty. Plenum Press, New York.
11. Eakins J.P. (2001) Retrieval of Still Images by Content. In: Agosti M., Crestani F., Pasi G. (Eds.) Lectures in Information Retrieval. Springer, Berlin Heidelberg New York, pp. 110–138.
12. Ghezzi P.P., Binaghi E., Galli C., Rampini A. (1997) Integrazione di Tecniche di Clustering 'Split and Merge' e Fuzzy per la Classificazione di Immagini. In: Proc. of the First Nat. Conf. ASITA, Parma (Italy), pp. 149–150.
13. Jain A.K., Vailaya A. (1998) Shape-based retrieval: A case study with trademark image databases. Pattern Recognition 31(9):1369–1390.
14. Koperski K., Marchisio G.B. (2000) Multi-level Indexing and GIS Enhanced Learning for Satellite Imagery. In: Proc. of the Workshop on Multimedia Data Mining MDM/KDD2000, Boston (MA) USA, pp. 8–13.
15. Miyamoto S. (1990) Fuzzy Sets in Information Retrieval and Cluster Analysis. Kluwer Academic Publishers, Dordrecht.
16. Niblack W., et al. (1993) The QBIC Project: Query by image by content using color, texture, and shape. In: Niblack W., Jain R. (Eds.) Proc. of Storage and Retrieval for Image and Video Databases, Vol. 1908, SPIE Press, Bellingham, WA, pp. 173–187.
17. Petrakis E.G.M., Orphanoudakis S.C. (1993) Methodology for the Representation, Indexing and Retrieval of Images by Content. Image and Vision Computing 11(8):504–521.
18. Petrakis E.G.M., Faloutsos C. (1994) Similarity Searching in Large Image Databases. Technical Report CS-TR-3388, College Park, MD, USA.
19. Robertson S. (2001) Evaluation in Information Retrieval. In: Agosti M., Crestani F., Pasi G. (Eds.) Lectures in Information Retrieval. Springer, Berlin Heidelberg New York, pp. 81–92.
20. Rui Y., Huang T.S., Chang S.F. (1999) Image retrieval: Past, present, and future. Journal of Visual Communication and Image Representation 10:1-23.
21. Salton G., McGill M. (1983) Introduction to Modern Information Retrieval. McGraw-Hill, New York.
22. Schowengerdt R.A. (1997) Models and Methods for Image Processing, Academic Press, San Diego.
23. Sheikholeslami G., Zhang A., Bian T. (1999) A Multiresolution Content-based Retrieval Approach for Geographic Images. Geoinformatica 3(2):109–139.
24. Smeulders A.W.M., Worring M., Santini S., Gupta A., Jain R. (2000) Content-Based Image Retrieval at the end of the early years. IEEE Transactions on Pattern Analysis and Machine Intwlligence 22(12):1349–1380.
25. Stephenson T., Voorhees H. (1995) IMACT: An interactive, multiterabyte image archive. In: Proc. of the IEEE Symposium on Mass Storage Systems, pp. 146–161.

26. Val Cura L.M., Leite N.J., Bauzer Medeiros C. (2000) An Architecture for Content-Based Retrieval of Remote Sensing Images. In: Proc. of the IEEE International Conference on Multimedia and Expo, New York City, NY (USA), pp. 303–306.
27. Vellaikal A., Kuo C.C., Dao S. (1995) Content-Based Retrieval of Remote Sensed Images Using Vector Quantization. In: Proc. of SPIE Visual Information Processing, Vol. 2488, pp. 178-189.
28. Zadeh L.A. (1975) The Concept of a Linguistic Variable and its Application to Approximate Reasoning I-II. Information Sciences 8:199–249.
29. Zadeh L.A. (1978) Fuzzy Sets as a Basis for a Theory of Possibility. Fuzzy Sets and Systems 1:3–28.

# 13

# Vague Geographical Knowledge Management – A Flow-Chart Based Application to Spatial Information Analysis

Josef Benedikt[1], Sebastian Reinberg[2], and Leopold Riedl[3]

[1] GEOLOGIC
Lerchengasse 34/3
A-1080 Vienna, Austria
josef.benedikt@geologic.at
[2] im-plan-tat
Reinberg & Partner OEG
Heinrich Oeschl Gasse 56
A-3430 Tulln, Austria
sebastian.reinberg@im-plan-tat.at
[3] Vienna University of Technology
Regional Science Institute (SRF)
Karlsplatz 13 A-1040 Vienna, Austria
leopold.riedl@tuwien.ac.at

## 13.1 Introduction

Introducing fuzzy sets as an extension in modelling categories has drawn the attention on the discussion on different types of certainty as well as uncertainty. Fuzzy sets broadened the concepts on dealing with error and imprecision, which are issues well known to spatial analysis. Eliminating or tracking error and imprecision has been contributing to increase knowledge with respect to spatial phenomena. Notions in space and spatial terms became more precise in a complex decision making environment, too. The use of fuzzy sets in GIS applications triggered the question on what kind of uncertainty fuzzy logic is aiming at in the process of gaining precise knowledge. Following an approach discussed in [10] ambiguity and vagueness are addressed by fuzzy methodology. Uncertainty in knowledge (i.e. ambiguity) occurs when values are associated with multiple attributes and no reliable decision criterion is available, e.g. "Is a particular elevation a 'hill' or a 'mountain'?". Uncertain knowledge (i.e. fuzziness or vagueness) is due to our inability to make sharp and precise distinctions in our world, e.g. "What is a 'mountain'?" as opposed to "What is not a 'mountain'?". Whereas uncertainty in knowledge is being dealt with by probability and possibility measures, uncertain knowledge is addressed by measures of fuzziness [8].

This chapter is focusing on the second type of uncertainty, which comes with the linguistic representations of geographical terms. Fuzziness or vagueness is seen as different from ambiguity and the problems of making decisions under uncertainty. It rather focuses on the problem of making precise distinctions in our world. Measures of fuzziness are used to develop examples on terrain features of slope which will illustrate the difference of modelling "How steep is 30%" from "How steep is steep?" referring to ambiguity and vagueness, respectively. Vagueness should not be "tolerated" or "eliminated" but rather be treated as a main source of "knowledge" as represented in the linguistic descriptors of geographic phenomena like "steep" or "flat". In the next sections a logical background on modelling vagueness is developed: A measure of fuzziness is introduced enabling the user to model vagueness in GIS. MapModels is introduced as a sophisticated interface to do spatial analysis. The final part provides three examples on how explicit and implicit vagueness in geographical categories may be evaluated. Results of applying different parameters on measures of fuzziness and further research issues are discussed.

## 13.2 Vagueness, Fuzzy Sets & GIS

Vagueness is nothing special to geography since "vagueness is a pervasive phenomenon of human thought and language" [13]. Vagueness is, nevertheless, a ubiquous and inherent feature in spatial knowledge on geographical phenomena. This is mainly due to the usage of linguistic terms in describing geographic reality introducing semantic aspects in the categorization of percepted, experienced or measured data and information. Vagueness is associated with the problem of making sharp distinctions in the real world. "Vague knowledge is not wrong per se, but it cannot be mapped using traditional logic" Russell is quoted in [3]. Black developed a "consistency profile" because "with the provision of an adequate symbolism the need is removed for regarding vagueness as a defect of language" [4]. With the introduction of fuzzy sets in 1965 [14] as the core of a multi-valued logical model we are able to use some of the information that comes with vagueness by applying that logical extensions to spatial terms. Vagueness is becoming a major source of information rather than another description of blurriness, error or other kind of imprecision. Vagueness or fuzziness "is not a burden that makes it tedious to know about anything but is an efficient way for humans to solve problems in their daily life and use expert knowledge appropriately" [10].

The concept of truth is at the core of modelling fuzziness. The analogous meaning of a set membership to a truth value of a proposition offers new ways of dealing with vagueness that is focusing on the logical truth and its complement (Fig. 13.1). Fuzzy logic is just one example of an multi-valued logical concept that allows for modelling semantic aspects in vague categories. In classical set theory and logic we are limiting ourselves to only two truth values. A membership in a classical fuzzy set is equal to the truth value of a statement saying that "value $x$ is a member of set $A$ is true". GIS analysis using classical logic would assign "no data" to the membership value 0.8, which is further developed as an example in the application section.

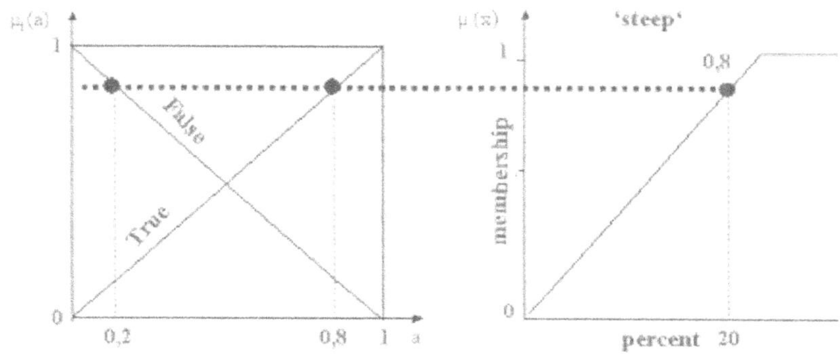


**Figure 13.1.** The logical truth (left) of membership (right) in a fuzzy set "steep" (i.e. the truth value of a proposition '20% are steep' is 0.8).

Fuzzy logic introduces a number of truth values to a proposition. The truth value itself can be modelled by any kind of function and thus allows for dealing with a new level of "preciseness". The goal is not to find an "exact" truth value but one value that represents the mapped knowledge at best (Fig. 13.2). At first sight, this is intriguing of course and seems to make assigning a truth value an arbitrary task. Classical logic, however, is argued to be arbitrary in the first place due to its limitations in representing gradual transitions of truth. Fuzzy logic is adding to the transparency of logical modelling by being much more tolerant on different "truth" values that come with a complex spatial decision making situation. Classical logic in that part is not adjustable and thus not tolerant of any error, vague concept or imprecision of any kind. We argue this kind of "inexactness" being an enhancement to spatial knowledge modelling and also adding to the modelling of imprecision.

Spatial categorization in GIS relies on classical logic. GIS use "objectified geographical entities", which are not necessarily the right means to model the "boundary poor empirical world", as Helen Couclelis puts it [5]. Geographical categories on terrain features (e.g. slope) are, however, described by spatial properties like "steep". The question in this chapter is not directed towards finding more precise means to measure "steep" ("is it 28%, 29% or 30%?") but to what extent "steep" is representing the information inherent in both the area of investigation and the knowledge of the GIS expert. In other words, we focus on how much a GIS expert is convinced that "steep" is the right descriptor of a spatial phenomenon thus expressing the quality of the linguistic

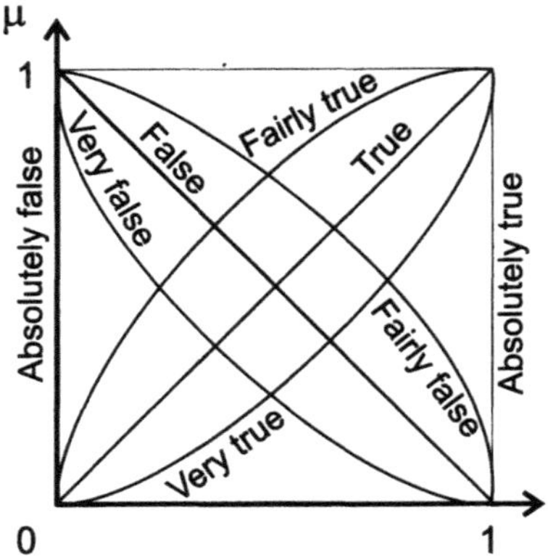


**Figure 13.2.** Examples of logical truth functions.

descriptor. This "strongness" is argued to be modeled by the most simple formal operation on a fuzzy set, the complement. Furthermore, an example will be given on how much this complement is contributing to related concepts like the seemingly opposite of a spatial notion ("flat"). Different kinds of complements have been introduced, e.g. Kaufmann's linear index of fuzziness [12]. In this chapter we employ complements of the so-called Yager-class as referred to by George Klir and Tina Folger [8].

The fuzziness we are dealing with is inherent in the linguistic category applied to geographical objects (slope) which we are using. The fuzziness is not based on the undecisiveness on whether 30% is steep or not. The question addressed is "How steep is steep?". This implies that we regard the opposite of steep not being flat (or the problem of deciding whether a pixel belongs to a city or a suburb). The opposite of steep is not 'flat' but 'not steep', which in turn adds to the quality of describing the opposite. That is, all the possibilities of an area being 'flat' lie in the possibilities of logical modelling of 'not steep'.The differences set the frame for all possible logical solutions and thus provide information on the membership of a steep slope as well as a flat slope in a decision making process. Implicit and explicit vagueness that comes with fuzzy sets is used to model the strengthness or "knowledge" of a spatial problem or issue or category.

## 13.3 Measure (Index) of Fuzziness

The extension of set theoretical notions makes it necessary to extend GIS operations on spatial terms, too. The complement is the most self-evident (and simple) operation on a single fuzzy set. In classical set theory the complement would be 0 or 1. The content on information on vagueness is equal to the set membership. A fuzzy set, however, allows for a much more diverse set of operations by introducing multi-valued truth evaluations as stated in the previous section. A measure of fuzziness in this logical context indicates the relationship between a set ('steep') and its complement ('not steep'). Following Klir and Folger ([8], p. 40ff) we employ a Yager-class of complements:

$$c_w(a) = (1 - a^w)^{\frac{1}{w}} \tag{13.1}$$

where $c_w$ assigns a membership value to $a \in A$ with $w \in ]0, \infty[$ and $A$ being a crisp or fuzzy subset (e.g. 'steep') of the universal set 'slope'. Using parameter $w$ allows to model different complements on a fuzzy set depending on its vagueness. Changing $w$ results in a deformation of the function $c_w$. With $w = 1$ the function becomes the classical complement (i.e. $c_1(0.8) = 0.2$). The following discussion is based on this classical complement. For better readability the parameter $w$ is omitted therefore throughout this section.

Given the relationship between a membership concept and truth values of a fuzzy proposition (see Fig. 13.1) it can be seen that this simple operation (see Equation (13.1)) extends our ability to model the amount of contrariness which can be of value when used e.g. in GIS overlay operations. This general class of complements is used as a basis for aggregating the "lack of distinction between a set and its complement" ([8], p. 140ff).

$$\delta_{c,A}(x) = |\mu_A(x) - c(\mu_A(x))| \qquad \text{for all } x \in X \tag{13.2}$$

where $\delta$ represents a distance measure between the set and its complement. $c(\mu_A(x))$ assigns a value to each membership grade $\mu_A(x)$ using Equation (13.1). While $\delta$ expresses local (i.e. individual) differences, the total amount of contrariness between a set and its complement is defined as

$$D_{c,r}(A, A^c) = \left[ \sum_{x \in X} \delta^r_{c,A}(x) \right]^{\frac{1}{r}} \qquad \text{for finite sets} \tag{13.3}$$

or

$$D_{c,r}(A, A^c) = \left[ \int_{x \in X} \delta^r_{c,A}(x) dx \right]^{\frac{1}{r}} \qquad \text{for infinite sets} \tag{13.4}$$

where $A^c$ denotes the complement of $A$ produced by function $c$ and $r \in [1, \infty[$. $r$ relates to the metric concept used for modelling the individual differences. If, e.g. a Eucledian distance is used, $r$ would be 2. Based on aggregating individual differences, a general measure of fuzziness $f$ is defined as an index of vagueness [8, 1]:

$$f : \wp(X) \rightarrow \mathbb{R} \tag{13.5}$$

where $\wp(X)$ denotes the fuzzy power set of $X$ (i.e. the set of all fuzzy subsets of $X$). In this application $X$ is the universal set of all slope-related linguistic terms. In our particular application $f$ assigns a value $f(A)$ to each fuzzy subset $A$ of $X$ in $\mathbb{R}$, the set of real numbers, which characterizes the degree of fuzziness of $A$. The measure (index) of fuzziness using a Minkowski class of metric distances, e.g. Euclidean distances, has the following form:

$$f_{c,r}(A) = |X|^{\frac{1}{r}} - D_{c,r}(A, A^c) \tag{13.6}$$

with $|X|$ denoting the cardinality of the base set. The normalized version of $f_{c,r}(A)$ is

$$\widehat{f}_{c,r}(A) = 1 - \frac{D_{c,r}(A, A^c)}{|X|^{\frac{1}{r}}} \tag{13.7}$$

so that

$$0 \leq \widehat{f}_{c,r}(A) \leq 1 \tag{13.8}$$

For a detailed discussion on developing a more general class of fuzzy measures we refer to [8]. Applications using other classes of complements (e.g. Sugeno class of complements) or measures of fuzziness (e.g. quadratic index of fuzziness) [12] are not discussed in this chapter.

A measure of fuzziness is addressing precisely the concept of uncertainty or fuzziness, which makes the use of crisp sets distinct from fuzzy sets. For crisp sets $\widehat{f}_{c,r}$ is 0, in the fuzzy case the value increases as the mean distance between a set and its complement decreases. $\widehat{f}_{c,r}$ equals 1 with all membership values being 0.5 throughout $X$. In the following section this measure of fuzziness is applied to geografic notions within a GIS analysis environment.

## 13.4 Application – Study Area

The study area is located in the Austrian Alps (Fig. 13.3). The region in Tyrol and Vorarlberg along the border to Switzerland is known as Silvretta Mountains. The area is digitally stored as a grid (raster) image. The used DEM (digital elevation model) has the size of $100\text{km}^2$ in reality. These $100\text{km}^2$ are divided into $1024 \times 1024$ pixels. One pixel covers about $100\text{m}^2$ ground area and the value of the elevation is saved in each cell (pixel). The study area is situated in a height of 1746 to 3306m above Adriatic sea level. The used raster image was interpolated from digitised map data of the BEV (Federal Office for Metrology and Surveying). So the basic data of the study is the most accurate available in Austria [6].

## 13.5 Application – Software

MapModels is a flexible tool for explorative spatial data analysis. It has been developed at the Regional Science Institute of the Vienna University of Technology with the intention to bridge the gap between spatial decision analysts

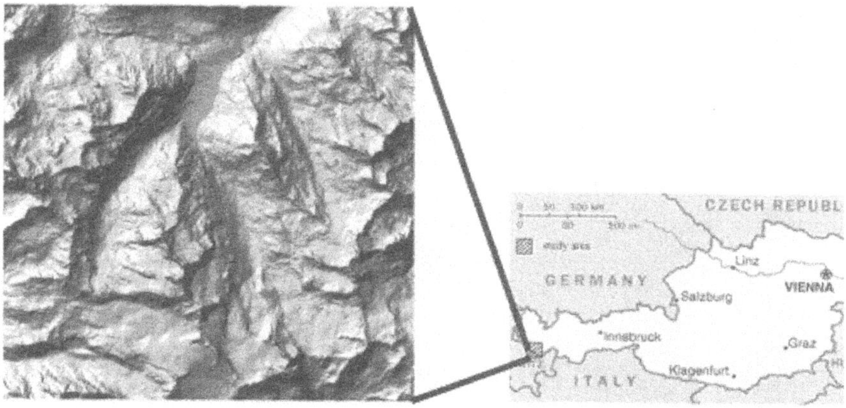


**Figure 13.3.** The study area.

and computer programmers [9]. MapModels is a visual programming language based on the widespread desktop GIS ArcView 3. It supports the development and implementation of analysis procedures based on flowchart representations in a very intuitive and user-friendly manner. It is particularly suited for extended decision making with fuzzy set modelling of geographical notions.

Flowcharts are used for the visualization of models and analysis processes in a wide range of applications. Normally this kind of graphic representation is simply focused on the illustration of the model structure and information flow but does not directly control the underlying processes. Within MapModels the nodes of a flowchart are in fact active elements of the model. They provide a visual encapsulation of real analysis procedures and data objects where input data and analysis operations are represented by labelled icons connected by edges which characterize the dataflow (Fig. 13.4).

Since MapModels flowcharts contain executable code, the specification and the implementation of the analysis model is just one single step. It is a kind of a drawing-process where flowchart elements are inserted into the model environment and connected by means of drag-and-drop operations with the mouse.

Currently a basic function library is available which contains a steadily increasing number of flowchart elements for a wide range of analysis operations including e.g. the application of fuzzy logic [2]. Standard MapModels-users usually will apply functions out of this set. Users with basic Avenue$^{TM}$ programming skills can extend and/or customize the existing set according to their specific needs with low effort.

The recently implemented option of "nesting" models enhances the flexibility of MapModels [9]. This concept is very similar to the routine-subroutine design of conventional programming languages. A submodel is defined as a col-

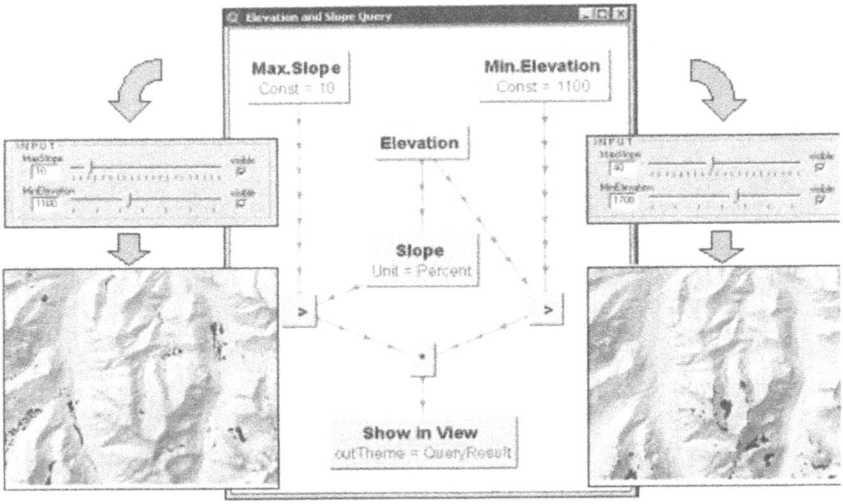


**Figure 13.4.** A simple spatial query: "find all relatively flat areas with an elevation higher than a given threshold" (center: MapModel; left: slope¡10% and elevation¿1100m; right: slope¡40% and elevation¿1700m with corresponding results).

lection of MapModels-functions which act as a single flowchart element when used within other models. An input/output interface has to be defined for each submodel in order to address the submodel on higher levels of the model (Fig. 13.5).

This modular design enables users to build customized basic analytical units for their applications and reuse them wherever they want and need. This is especially suitable for tasks that are performed repeatedly using different parameters. Compared to "All-In-One" concepts (i.e. the whole analysis model is represented within a large single model) the modular approach in the recent version 2 of MapModels[4] helps to keep models small and self-explanatory and avoids redundancies within the whole analysis-structure as well (i.e. duplication of programming code).

## 13.6 Application – Results

The first MapModel (Fig. 13.6) is designed to show the impact of implicit vagueness on visualizing uncertain knowledge in GIS layers that eventually become part of spatial decision making systems. The fuzziness is reduced automatically as a linguistic hedge (very) is applied. This is due to maximizing

[4] A trial-version of the of MapModels (as an Extension for ArcView 3 GIS) including examples, literature and documentation can be downloaded on the authors webpage "http://www.srf.tuwien.ac.at/MapModels/MapModels.htm".

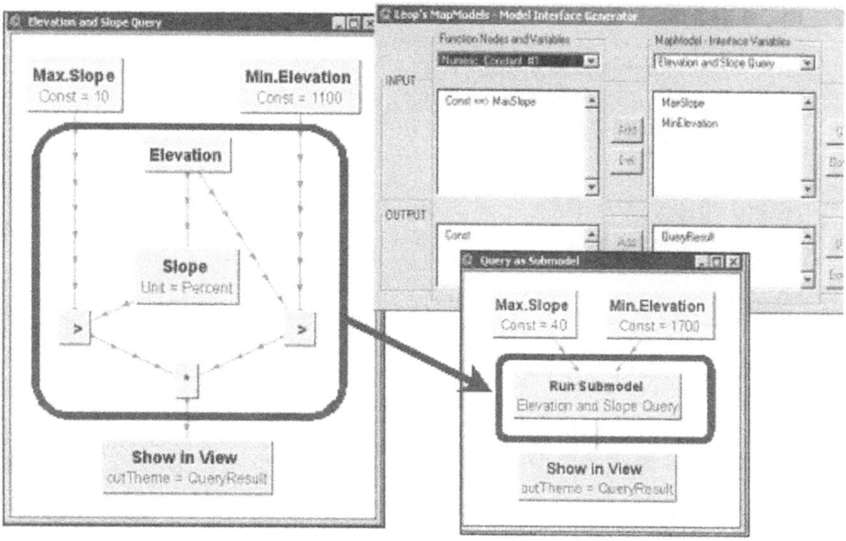


**Figure 13.5.** Concept of submodels: the flowchart inside the dark box (left model) is merged by means of a specialized dialog box (model interface generator, upper right picture) in a single node (lower right model).

the distance between elements in 'steep' and 'not steep' just by using a different membership function on steep. Note, that there is no vagueness in a crisp set 'steep' (all values are 0).

Due to the complete GIS functionality of MapModels the results of applying MapModels are shown as GIS layers in Fig. 13.7. We can see the vagueness of steep being very low at the ground and on top of the mountains, whereas the vagueness in between is pretty high related to the uncertainty that comes with modelling slopes. Note: The results do not refer to the transition zones between steep and more or less steep (or city and suburb and similar investigations of transition zones) but focus uniquely on the logical implication of memberships as represented by the linguistic variable steep (or city, respectively). The top is considered definitely steep. The grounds are definitely not steep. But in between the implicit amount of vagueness gives the opportunity for people (or in logical models) to discuss whether its steep or not by making explicit use of vagueness and measures of fuzziness, which is seen as a major advancement of using fuzzy sets in representing geographical notions [2]. The calculation steps, based on equations (13.1) to (13.7), are used to measure fuzziness or vagueness. The results are summarized in Table 13.1.

Based on the first example we extend our modelling efforts towards making explicit use of the complement (Fig. 13.8). In the first example we just took the complement "as is". Now we draw our attention to explicitly changing

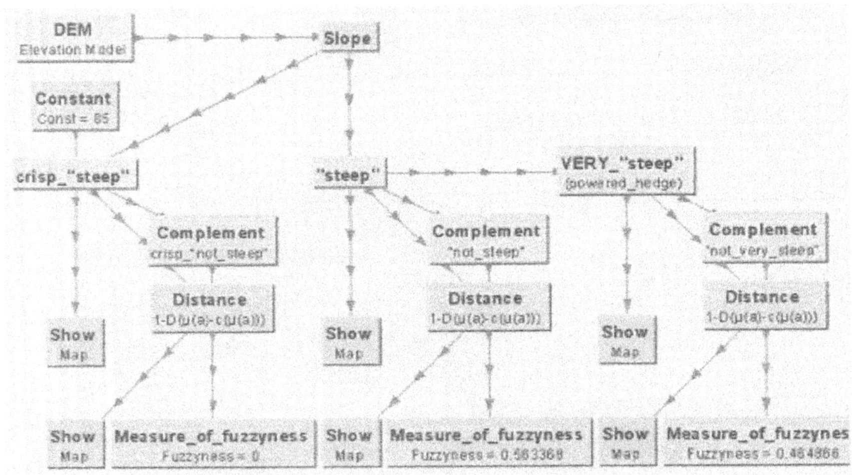


**Figure 13.6.** MapModel of implicit vagueness of steep (crisp), steep (fuzzy) and very steep (fuzzy).

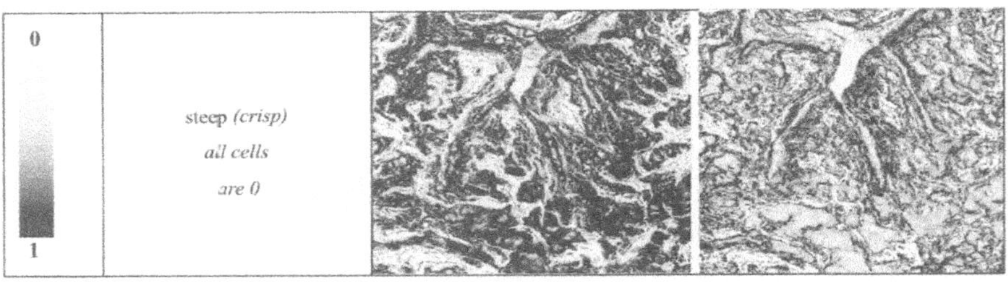


**Figure 13.7.** Mapping implicit vagueness of steep (crisp), steep (fuzzy) and very steep (fuzzy).

**Table 13.1.** Degree of fuzziness in linguistic categories by using classic complements.

| Measures of fuzziness: $c_w(a) = (1 - a^w)^{\frac{1}{w}}$, $w = 1$ | | | |
|---|---|---|---|
| | Crisp steep | Steep | Very steep |
| $r = 1$ | 0.00 | 0.56 | 0.46 |

the truth functions (see also Fig. 13.2). In Fig. 13.8 this is indicated by the model nodes "How vague is...". Using explicit fuzziness extends the possibility to explore the vagueness using the concept of a complement as a measure of fuzziness with respect to a particular situation. Within MapModels we can interactively change our opinion on the (un)certainty of 'steep (crisp)', 'steep (fuzzy)' or 'very steep (fuzzy)'. In addition our change of attitude can be seen immediately on maps. Note: In case of a crisp categorization of steep the results of using explicit vagueness by modelling the complement are "No data" (Fig. 13.9). Therefore crisp sets are suboptimal models to handle uncertainty in geographical analysis when vagueness comes into the picture.

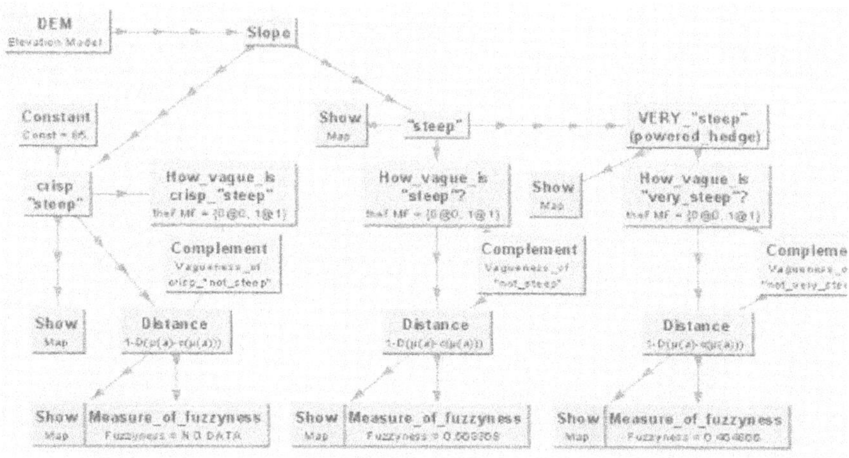


**Figure 13.8.** MapModel of explicit vagueness of steep (crisp), steep (fuzzy) and very steep (fuzzy).

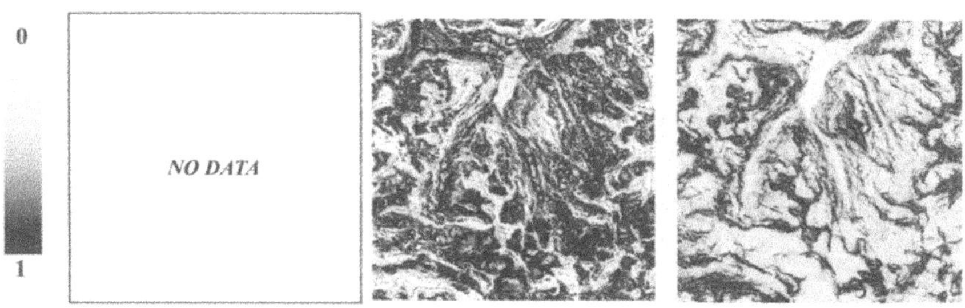


**Figure 13.9.** Mapping explicit vagueness of steep (crisp), steep (fuzzy) and very steep (fuzzy).

Based on these applications the analysis of vagueness in a linguistic variable can be generalized to address the overlay of different fuzzy sets representing different variables. The idea behind this generalization is using explicit vagueness on a fuzzy set like *steep* to broaden the solution space of *flat* by exploring the relations between *non-steep* and *flat* (Fig. 13.10). The mapped results (Fig. 13.11) show the alternative approach of how another linguistic variable (fuzzy set) *flat* is contributing to the reduction of vagueness of another set (*steep*) by using the complement and employing different $r$ as in (13.3). Table 13.2 is summarizing the measures of fuzziness achieved with different $r$.

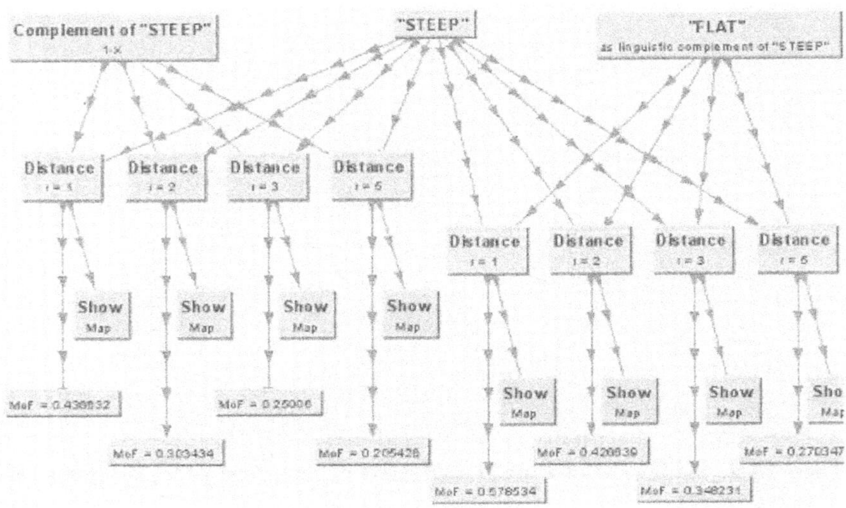


**Figure 13.10.** MapModel of calculating (explicit) measures of vagueness for different measures of distance on arithmetic versus linguistic complements.

In general, using a higher $r$ results in a set with more precise boundaries, which is not surprising, since the difference between a set and its complement is getting larger, that is, the set elements share a greater distance. What is intriguing, though, of this kind of application is the possibility of using a logical model that allows for greater variations in modelling linguistic descriptors of geographical data rather than focusing on a single arithmetic calculation. Fuzziness of a variable contributes to a better understanding of another linguistic variable or at least allows for making differences in fuzzy set overlays more visible.

## 13.7 Conclusion

The overall goal of this chapter was to explore the power of multi-valued logic in modelling semantic aspects of linguistic expressions used to describe spatial categories. Our applications focus on information which is argued to be inherent in linguistic terms and variables respresenting uncertain spatial

**Table 13.2.** Degree of fuzziness in linguistic categories by using *steep/non-steep* and *flat*.

| Measures of fuzziness: $c_w(a) = (1 - a^w)^{\frac{1}{w}}$, $w = 1$ | | |
|---|---|---|
| | Steep (non-steep) | Flat (non-steep) |
| $r = 1$ | 0.437 | 0.579 |
| $r = 2$ | 0.303 | 0.427 |
| $r = 3$ | 0.250 | 0.348 |
| $r = 5$ | 0.205 | 0.270 |

knowledge. The truth values of slope measurements being "steep" and "non-steep" at the same time are examined by employing measures of fuzziness. To put the focus on logical truth of building linguistic categories offers possibilities in modelling semantic interpretations of data. The results do not refer to the transition zones between steep and more or less steep, or city and suburb or forest and meadow but focus uniquely on the logical modelling of memberships as represented by linguistic terms and variables such as steep, city, forest etc. This chapter also shows the immediate impact on resulting maps by using MapModels, an easy-to-use modelling language.

The main issue of using fuzzy sets and MapModels is to make uncertain knowledge "visible" or "explicit", not to eliminate it. So far, uncertain knowledge has been eliminated (i.e mostly ignored) or treated as statistical error in general. Using MapModels we incorporated fuzziness to the spatial modelling process. Our models do not provide unique solutions which are valid in every instance but help to add more transparency in a complex decision making environment with many actors expressing different opinions.

This work and the application's results are meant to stimulate research on a fuzzy spatial logic regarding issues like tracking truth values along the overlay process in GIS analysis. Different logical models of the complement may also lead towards a greater flexibility. The results are promising in making advanced use of measures of fuzziness and in contributing to a better understanding of complex spatial phenomena.

## References


1. Benedikt J., Nishiwaki Y. (1998) Fuzzy Clustering & Remote Sensing. In: Ruan, D. (Ed.) Proc. of the FLINS 1998 conference, World Scientific.
2. Benedikt J., Reinberg S., Riedl L. (2002) A GIS application to enhance cell-based information modelling. Information Sciences 142:151–160.
3. Biewer B. (1997) Fuzzy Methoden. Springer-Verlag.
4. Black M. (1937) Vagueness: an exercise in Logical Analysis. Philosophy of Science 4(4):427–455.
5. Couclelis H. (1993) The last frontier. Editorial, Environment and Planning B: Planning and Design 20(1):1–4.

6. Hemetsberger M., Klinger G., Niederer S., Benedikt J. (2002) Risk Assessment of Avalnaches – A fuzzy GIS application. In: Ruan D. (Ed.) Proc. of the FLINS 2002 conference, World Scientific.
7. Netherer S., Pennerstorfer J., Kalasek R., Riedl L. (2002) Spatial Analysis in Forest Protection using the visual modelling tool MapModels. In: Pillmann W., Tochtermann K. (Eds.) Environmental Communication in the Information Society – Proc. of EnviroInfo Vienna 2002, IGU/ISEP, Vienna, Vol. 1, pp. 574–581.
8. Klir G.J., Folger T. (1988) Fuzzy Sets, Uncertainty and Information. Prentice Hall, Englewood Cliffs, NJ.
9. Riedl L., Vacik H., Kalasek R. (2000) MapModels: a new approach for spatial decision support in silvicultural decision making. Computers and Electronics in Agriculture 27:407–412.
10. Spies M. (1993) Unsicheres Wissen. Spektrum Akademischer Verlag GesmbH, Heidelberg.
11. SRF (2002) http://www.srf.tuwien.ac.at/MapModels/MapModels.htm — MapModels-homepage with examples, literature, documentation and download of a trial-version.
12. Tizhoos H. (1997) Fuzzy Image Processing. Springer.
13. Varzi A.C. (2001) Vagueness in Geography. Philosophy & Geography 4(1):49–65.
14. Zadeh L.A. (1965) Fuzzy Sets. Information and Control 8:338–353.

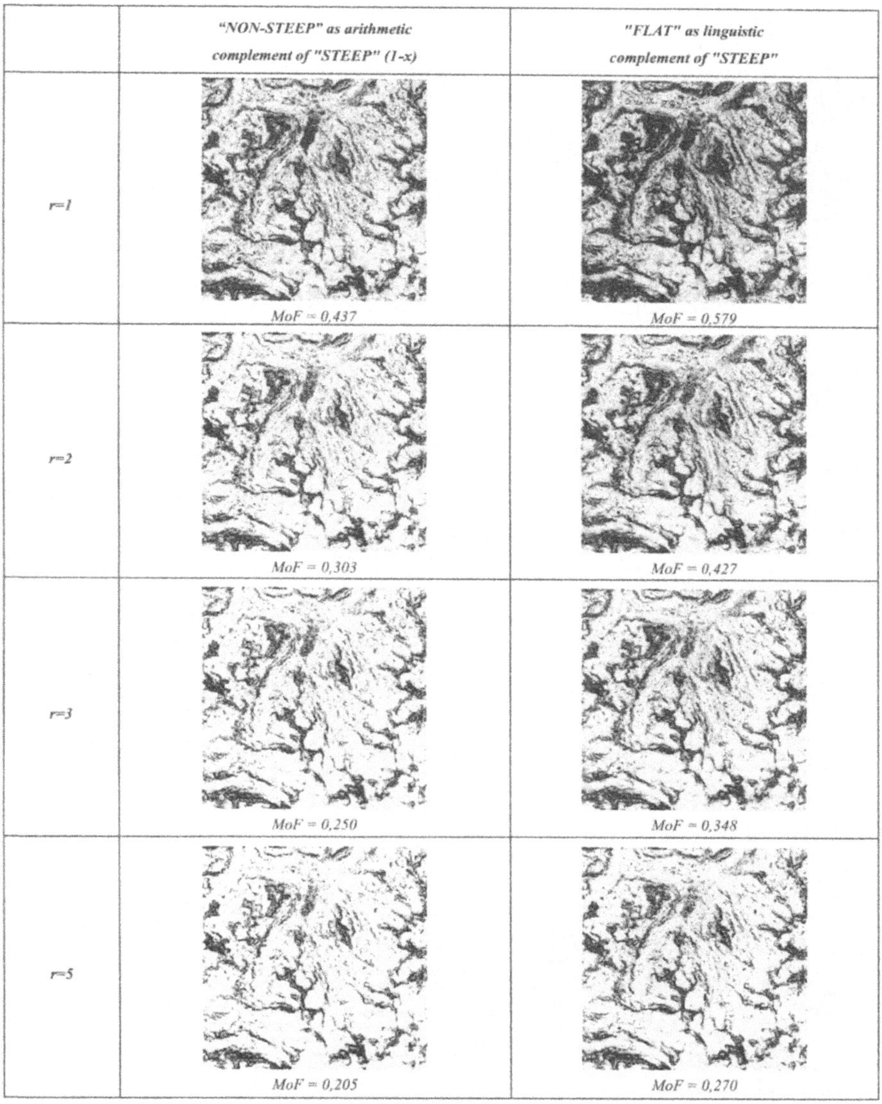


**Figure 13.11.** Mapping vagueness of steep (fuzzy) by using explicit vagueness and complements; effect of complement and distance measure on the measure (index) of fuzziness expressed by equation (13.7).

# 14

# Fuzzy Classifications in Large Geographical Databases: Assessing Vagueness in Less-Connected Nodes of the World City Network

Ben Derudder and Frank Witlox

Department of Geography
Ghent University
Krijgslaan 281/S8, B-9000 Gent (Belgium)
{ben.derudder,frank.witlox}@ugent.be

## 14.1 Introduction

Although the lineage of world city studies can be traced back to well before 1980 [26], it is only in the last two decades that a concerted research effort has emerged. The most seminal contributions to the recent upsurge in world city research are Friedmann and Wolff's [21] and Friedmann's [20] identification of 'command centers' that control and articulate the 'new international division of labour' being created by multinational corporations. This reflected the later recognition by Amin and Thrift [1] of a shift from an international to a global economy, characterized by increasingly integrated global networks of production and services. World cities are then the basing points in these multifarious networks, and their specification therefore relates to the identification of a 'global network of cities' [31].

Apart from the on-going discussions on more formal definitions and conceptualisations (see for instance [50]), one the most pressing problems in world city research is the lack of assessments of less connected nodes in this global urban network [4, 56]. While it is obvious that cities like London and New York are world cities, there has hardly been a consensus as to the status of less significant cities in this context. As a consequence, somewhat vague discourses on patterns of global competence in the outer reaches of the world city network have dominated the debates. Some authors have for instance reified the lower ranks of the hierarchy as 'sub-global cities' (e.g. [13, 42]), while this is particularly problematic given the pervasive nature of globalization. As Knox [33] has pointed out, these 'medium cities' have just as much need to respond to globalization trends as their larger neighbours. It is evident that every city operates as an integral part of the global system as a producer and marketplace for global goods and services, and as a hub in the flow of people, remittances, finance and ideas. A lower amount of global capital served does not imply a lack of global connections, and it is thus indispensable to include a large number of cities in an analysis of the world city network.

Previous assessments of lower-ranked world cities have however been limited to ad hoc taxonomies that all in all remain limited to the higher ranks of the hierarchy (e.g. [19, 20, 32, 4]). Apart from the lack of an undisputed definition of world cities and the associated dearth of data, the main reason for these somewhat eclectic approaches is the fact that classical data analysis techniques cannot assess the sparse information on the less connected nodes in the world city network. Taylor et al. ([56], p. 2378), for instance, restricted the exploratory analysis of 316 cities to the top 123 cities "because the data matrix becomes increasingly sparse [...] and thus becomes increasingly unsuitable for analysis". In the present chapter, we aim to show how a fuzzy data-mining approach copes better with the sparse objects in the data matrix, thus enabling us to analyse less connected nodes in the world city network, most of which never figured in discussions in this context. The resulting classification provides a comprehensive and detailed taxonomy of the world city network, where both general tendencies and more specific profiles can be explored.

This chapter is divided in three sections. First, we examine the theoretical and empirical rationale behind explorations of the world city network. It is demonstrated that assessments of less connected nodes have been largely absent from previous explorations, which is due to the sparseness of information on their world city-ness. The second section discusses the outset of a classical clustering algorithm and its fuzzy counterpart. This discussion of classification algorithms commences with the outset of the more classical crisp algorithm for two reasons. First, applying the crisp version of the algorithm enables us to show why more multivariate techniques fail to grasp the paucity of information on less connected nodes in the world city network, while the outset of the methodology will also serve as a guide to the fuzzy clustering algorithm. The third section, then, ascertains a detailed taxonomy of the world city network. Both general tendencies and more specific outcomes of the fuzzy set analysis will be discussed.

## 14.2 Assessment of a network of world cities

### 14.2.1 From central place system to world city network

Urban geographers have long sought to unravel and describe the systematic nature of the spatial arrangement of urban centers, as Christaller's central place-theory [12] and Lösch's [37] extensions of this central place-theory show. Most of the studies oriented towards the description of the spatial arrangement of such an urban system inherit their physical boundaries from their definition as an integrated economy. Since the beginning of the twentieth century, the world-economy is truly global [59], and hence all cities can be thought of as participating in a single central place system [10]. However, although the predicaments of these classical central place systems may still do a reasonably good job in describing the urbanization patterns on a regional scale, they are not suited to explain patterns of global urbanization. At the most basic level, there are at least three (heavily intertwined) alterations that should be taken into account with respect to the assessment of a global urban system:

1. national and sub-national urban hierarchies need to be supplemented by some additional levels;

2. there are new and hence previously unassessed central place functions, which imply that cities are increasingly defined by mutual relations rather than by relations to their immediate hinterland;

3. the presumed functional equivalence induced by a comparable hierarchical position in the urban system is downplayed in favour of functional and regional specialisations among cities.

The first and most obvious difference between more classical conceptions of urban systems and urbanization processes in a global economy lies in the fact that the original hierarchy needs to be supplemented by some additional levels, i.e. the world cities this chapter is concerned with [27]. An exact definition of a world city is difficult to pin down, since it is itself subject to a great deal of controversy [32, 28, 50]. Friedmann [20] has however pointed out that a formal definition of a 'world city' is not necessary, since it is a paradigm rather than a well-defined theoretical object. As such, a general approach of world cities as the 'command and control centers' and 'basing points' of the world-economy is a good starting point for its operationalization, for this notion captures the importance of assessing interdependencies and flows between world cities.

Second, the recent evolution of technology and economy implies that there are new services that should fit in the model of central place functions described by Christaller [12]. One of the most important features of these new central place functions lies in the steady decline or slowdown in manufacturing, which contrasts with both a sharp rise in computing and telecommunication and a sharp growth in producer services such as accounting, legal, advertising and consulting services. The fundamental reason for this explosion of producer services can be traced back to the increased service intensity in the organization of all industries. Whether in manufacturing or in warehousing, all firms are increasingly outsourcing legal, financial, advertising, consulting and accounting services, which stimulate the growth of a whole gamut of new businesses with their proper location strategies [47]. The combined and mutual reinforcing effects of (i) an explosive growth in computing, telecommunication and knowledge production, and (ii) a sharp rise in producer services, have led to different spatial patterns than those implied in classical conceptions of central place systems. That is, the new global economy is more than merely another layer of economic activity on top of the existing production process: it has induced a profound change in location strategies, for it restructures all economic activities based on goals and values introduced by the aggressive exploitation of new productivity potentials of advanced information technology. Producer service firms do not follow the demand side of the market; they have their own global strategies for selling their products in a world service market [51]. These changing geographies have been explored in more detail by other researchers (e.g. [14, 40]), but it was Sassen [45, 47] who asserted the explicit link to world city formation. This association between fragmentation and agglomeration is the articulation of the reorganisation of existing processes into new patterns and the repositioning vis-à-vis the new highly productive sectors. Moreover, these evolutions in technology and economy induce a growth dynamic that seems to be somewhat independent of the broader regional economy, a sharp change from the past where a city was deeply articulated with its hinterland. It seems likely that a strengthening of transnational ties between leading financial and business centers is accompanied by a weakening of the ties between each of these cities and its hinterland and national urban system.

Third, apart from hierarchical tendencies at the global level, world cities have the tendency to expose functional specialisations. This is very different from Christallerian patterns, where cities occupying the same rank in the hierarchy by definition exhibit functional equivalence. Camagni [8] has reminded us that the relations between cities provide more insight when assessments of these relations are based on functional specializations rather than mere hierarchies, while Knox [34] has argued that different world cities fulfil different functions within the world-system, and will hence differ both in nature and degree of their world-city-ness. For instance, in the United States, New York leads in banking, accounting and advertising, while Washington D.C. leads in legal services, R&D and membership organizations [47]. Moreover, it has been shown that cities that are less thoroughly connected in the world city network will assume more regional roles, while the upper rungs of the network will have global specificities that prevent them from being very similar to nearby cities [53, 16]. For instance, British cities appeared very similar in the way they respond to globalization with the exception of London, which have a totally different profile as one of the most connected cities in the world city network.

This brief overview of the most salient features of an emerging network of world cities has a profound impact on its assessment, i.e. empirical explorations of its spatial outline should focus on the following elements:

- world cities constitute a network rather than a clear-cut hierarchy, and analyses of world city patterns should therefore concentrate on relations between cities;
- these relations are primarily generated through the location strategies of advanced producer services;

- the global patterns that arise through the aggregation of these networked strategies are diverse, and possibly include interwining hierarchical, functional and regional and tendencies;
- less connected nodes in the world city network are not typified by a lack of global connections, and should therefore not be omitted from an analysis of the world city network.

### 14.2.2 Specification and measurement of the world city network

Drawing on the guidelines set out in the previous section, the Globalization and World Cities Group and Network (GaWC) at Loughborough University (UK) has developed a methodology for studying world city network formation by analysing the location strategies of advanced producer firms [4, 52, 55]. The world city network is hereby conceptualised as an inter-locking network with three levels: a network level (the world city network), a nodal level (world cities), and a sub-nodal level (the firms providing the advanced producer services). It is at the latter level that world city network formation takes place, and GaWC thus follows Sassen [45, 46, 47] in her treatise of world cities as global service centres. Based on this formal specification, Taylor [52] has outlined the construction of connectivity matrices that systematically measure relations between each pair of world cities through the aggregation of the cross products of all of the firms for any pair of cities. Starting from a service value matrix $V_{ij}$, which contains 'service values' $v_{ij}$ ranging from 0 to 5 for all cities $i$ and firms $j$, the basic relational element for each pair $(a, b)$ of cities is $r_{ab,j}$, which can be aggregated to produce an inter-city link $r_{ab}$:

$$r_{ab} = \sum_j r_{ab,j} = \sum_j v_{aj}.v_{bj} \tag{14.1}$$

These inter-city links are a good surrogate for measuring the actual flows of information and knowledge between the cities when two assumptions are made. First, offices generate more flows within a firm's network than to other firms in the sector. Although not empirically tested, this assumption is plausible, for flows of information and knowledge are indispensable for a seamless service. Second, the larger the office, the more flows will be generated, which will have a multiplicative effect on inter-city relations.

The advantage of this precise specification of the world city network is that elementary network analysis can be used. The most basic measure is city's connectivity in relation to all other cities in the matrix. This measure departs from the observation that each city has a link with every other city, and aggregating all the links of a city produces the global network connectivity ($GNC$) of a city:

$$GNC_a = \sum_i r_{ai} \qquad (a \neq i) \tag{14.2}$$

The limiting case is a city that shares no firms with any other city, so that all of its elemental links are 0 and it has zero connectivity. These overall $GNC$ measures will be used below to order the results hierarchically, where they will be expressed as proportions of the largest computed connectivity in the data, thus creating a scale from 0 to 1.

This theoretical rationale has been used to analyse the location strategies of a total of 46 advanced producer services across 55 world cities [4]. A new data collection along these lines covered 100 firms and 316 cities [55], and was used to investigate patterns in the 123 most connected cities [56]. The latter study was confined to the most connected cities because as the size of the data matrix increases (i.e. inclusion of more cities) it becomes relatively 'sparse' (lots of zero entries) which makes "[the matrix] increasingly unsuitable for analysis" ([56], p. 2378).

Although this analysis of 123 cities is a huge step forward compared to earlier ad hoc classifications [19, 32], it had a number of serious flaws. First, the analysis still leaves several regions relatively unrepresented, e.g. there are only two cities in inter-tropical Africa (Lagos and Nairobi) that qualify in the top 123 connected cities. Second, the analysis was based on the presence scores for firms rather than the inter-city matrices constructed in [52]. As such, the various groups derived in [56] infer relations rather than using the relational measures deemed crucial in world city network analysis. Third, the analysis was based on a principal component analysis, and this factor analytic approach implied that a significant number of the 123 cities were *not* allocated to any of the components.

The present chapter aims to move beyond this and other analyses by (i) explicitly drawing on the matrices that assert the relations between cities, and by (ii) including many more cities (234) than has been the case thus far. Moreover, by using a cluster analysis, every city will be classified. Combining Taylor's [52] specification and Taylor et al.'s [56] measurement in a single framework, we have computed measures of inter-locking connectivity between 234 cities across the world. This matrix then gives way "to various forms of analysis available to simpler types of network. This means the wide repertoire of network techniques from elementary derivation of indices to scaling, ordinating, factoring, clustering and blocking" ([52], p. 192).

## 14.3 Searching for clusters: crisp and fuzzy classifications

### 14.3.1 Crisp classification

The main purpose of unsupervised classification (clustering) of a set of objects is to detect subgroups (clusters) based on the (dis)similarity between objects. All members belonging to the same group or cluster have certain properties in common, and the classification has the effect of reducing the dimensionality of a data table by reducing the number of rows. The aim of a cluster analysis is thus to partition a given set of $n$ data or objects into $C$ clusters, with the following properties [18]:

- homogeneity within the clusters: data belonging to the same cluster should be as similar as possible;
- heterogeneity between clusters: data belonging to different clusters should be as different as possible.

In this classical crisp classification approach, the state of clustering is expressed by an $n \times C$ matrix $U = (u_{ki})$, where $u_{ki} = 1$ if object $k$ belongs to the cluster $i$, otherwise $u_{ki} = 0$. To ensure that the clusters are disjoint and non-empty, $u_{ki}$ must then satisfy the following condition [48]:

$$\forall\, k = 1, 2, \ldots, n,\ \forall\, i = 1, 2, \ldots, C : u_{ki} \in \{0, 1\} \ \wedge \ \sum_{i=1}^{C} u_{ki} = 1 \tag{14.3}$$

where $n$ is the total number of cities in the database and $C$ is the total number of clusters.

This classification scheme has certain distinct advantages. For one thing, results are clear-cut, and possible cumbersome interpretations of in-between values are expelled from any analysis since there is no overlap in cluster membership. When applied to the classification of cities, regions or countries based on certain criteria, this implies that the only admissible spatial boundaries are unambiguous ones [3, 39, 58, 17].

Defined more formally, the outset of the clustering problem can be stated as follows [11]. Let:

$$X = \{x_1, x_2, \ldots, x_n\} \tag{14.4}$$

be a set of samples to be clustered into $C$ clusters $C_1, C_2, \ldots, C_C$. The clustering process can be considered as an iterative optimization procedure. Suppose that the samples have already been partitioned into $c$ classes, be it by random assigning the data points to clusters or through theoretical considerations on potential clusters. The task at hand, then, is to adjust the partition so that the similarity measure, based on an Euclidean distance function, is optimized. The criterion function for this optimisation procedure is equal to:

$$J(V) = \sum_{k=1}^{n} \sum_{x_k \in C_i} (v_i - x_k)^2 \tag{14.5}$$

where $v_i$ is the center of the samples in cluster $i$, and

$$V = \{v_1, v_2, \ldots, v_C\} \tag{14.6}$$

The classification of the data is thus based upon a measure of dissimilarity between the different data points in the matrix, where the Euclidean distance is the most simple and common measure of dissimilarity.

In order to improve the similarity of the samples in each cluster, we can minimize this criterion function so that all samples are more compactly distributed around their cluster centers. Setting the derivative of $J(V)$ with respect to $v_i$ to zero, we obtain

$$\frac{d\,J(V)}{d\,v_i} = \sum_{k=1}^{n} \sum_{x_k \in C_i} (v_i - x_k) = 0 \tag{14.7}$$

Thus, the optimal cluster center of cluster center $v_i$ is

$$v_i = \frac{1}{n_i} \sum_{x_k \in C_i} x_k \tag{14.8}$$

where $n_i$ is the number of samples in class $i$ and $C_i$ contains all samples in class $i$.

Starting with the initial clusters and their center positions, the samples can now iteratively be regrouped so that the criterion function $J(V)$ is minimized. Once the samples have been regrouped, the cluster centers need to be recomputed to minimize $J(V)$. This process then continues for the new cluster centers: the samples are regrouped in order to reduce $J(V)$, which yields a new classification with associated cluster centers, and so forth. This iterative process can be repeated until $J(V)$ cannot be further reduced or drops below a pre-defined small number. Obviously, the criterion function is minimized if each sample is associated with its closest cluster center. This means that $x_k$ will be reassigned to cluster $i$ so that $(v_j - x_k)^2$ is minimum when $j = i$. Up to this point, each sample $x_k$ appears only once, that is, it is associated with only one cluster center.

Applying this crisp c-means algorithm to the $234 \times 234$ matrix for $C = 20$, the classification result reveals primarily hierarchical patterns (Figure 14.1). By ordering the clusters based on the average $GNC$ of the cluster members, it can be observed that because of the smaller $GNC$ gaps between lower-ranked world cities the latter tend to be classified in larger groups. In fact, while the 49 most connected cities are scattered over 16 clusters, the 185 cities least connected cities in the world city network are assigned to merely four clusters. Apart from the remarkable distinctiveness of US cities (clusters 5, 9, 12, 15, 18) and, albeit to a lesser degree, German cities (cluster 16), the outer reaches of the world city network are thus classified based on their overall connectivity. Taylor et al.'s [56] remark that sparse evidence on world city formation prevents a defendable classification thus holds sway, since this classification of lower-ranked world cities does not add information to the most elementary network measure of overall connectivity. Moreover, Figure 14.1 unveils that this crisp classification also brings about problematic results in the upper rungs of the world city network. Since top ranked world cities have very distinctive patterns, they tend to be classified in a large number of very small clusters, which makes them in turn very hard to interpret soundly. The meaning of dyads such as Amsterdam-São Paulo (cluster 6), Zurich-Buenos Aires

| CLUSTER 1 | CLUSTER 2 | CLUSTER 3 | CLUSTER 4 | CLUSTER 5 | CLUSTER 6 | CLUSTER 7 | CLUSTER 8 |
|---|---|---|---|---|---|---|---|
| NEW YORK | HONG KONG | MADRID | SYDNEY | LOS ANGELES | AMSTERDAM | FRANKFURT | BUENOS AIRES |
| LONDON | TOKYO | MILAN | TORONTO | SAN FRANCISCO | SAO PAULO | BRUSSELS | ZURICH |
| | | PARIS | | CHICAGO | | | |
| | | SINGAPORE | | | | | |

| CLUSTER 9 | CLUSTER 10 | CLUSTER 11 | CLUSTER 12 | CLUSTER 13 | CLUSTER 14 | CLUSTER 15 | CLUSTER 16 |
|---|---|---|---|---|---|---|---|
| MIAMI | MELBOURNE | BANGKOK | WASHINGTON | COPENHAGEN | BARCELONA | ATLANTA | BERLIN |
| | MEXICO CITY | BEIJING | | STOCKHOLM | HOUSTON | BOSTON | DUSSELDORF |
| | MONTREAL | JAKARTA | | ROME | | DALLAS | HAMBURG |
| | AUCKLAND | KUALA LUMP | | | | | MUNICH |
| | CALGARY | SEOUL | | | | | STUTTGART |
| | VANCOUVER | SHANGHAI | | | | | |
| | | TAIPEI | | | | | |

| CLUSTER 17 | | CLUSTER 18 | | CLUSTER 19 | | CLUSTER 20 | |
|---|---|---|---|---|---|---|---|
| BUDAPEST | MONTEVIDEO | BALTIMORE | CHARLOTTE | ABU DUBAI | HOBART | OSAKA | DOULA |
| CARACAS | MUMBAI | CLEVELAND | ST LOUIS | ADELAIDE | LA PAZ | ALMATY | ESSEN |
| GENEVA | NEW DELHI | COLUMBUS | CINCINATTI | BIRMINGHAM | LEIPZIG | ANTWERP | GABERONE |
| ISTANBUL | OSLO | PHILADELPHIA | DENVER | BRISBANE | LIMASSOL | BOLOGNA | ISLAMABAD |
| JOHANNESB | RIO DE JAN | KANSAS CIT | HARTFORD | CAPE TOWN | LJUBJANA | BRAZILIA | KAMPALA |
| MANILA | TEL AVIV | SAN JOSE CA | HONOLULU | COLOGNE | NICOSIA | DRESDEN | KINGSTON |
| MOSCOW | VIENNA | MINNEAPOLIS | SAN DIEGO | LEEDS | PERTH | EDINBURGH | LABUAN |
| PRAGUE | WELLINGTON | PITTSBURG | TAMPA | LYON | PORTO ALEG | SACRAMENTO | LAGOS |
| SANTIAGO | ABIJAN | PORTLAND | PHOENIX | MANCHESTER | RUWI | GOTHENBURG | LAHORE |
| WARSAW | ACCRA | SEATTLE | | MARSEILLE | TUNIS | HANOI | LIVERPOOL |
| ATHENS | AMMAN | | | ROTTERDAM | BELFAST | LILLE | MANAGUA |
| BANGALORE | ASUNCION | | | CHRISTCHURCH | HAMILTON | ROCHESTER | MAPUTU |
| BOGOTO | BEIRUT | | | DHAKA | INDIANAPOLIS | RIYADH | MONTERREY |
| BRATISLAVI | CALCUTTA | | | DURBAN | LAUSANNE | ST PETERSBURG | NAGOYA |
| BUCHAREST | CASABLANCA | | | EDMONTON | PORT LOUIS | TASHKENT | NEWCASTLE |
| CAIRO | CHENNAI | | | MALMO | TEGUCIGALPA | TEHRAN | NEW ORLEANS |
| COLOMBO | GUATEMALA | | | SOUTHAMPTON | | THE HAGUE | NOTTINGHAM |
| DETROIT | GUAYAQUIL | | | | | TURIN | OTTAWA |
| DUBAI | HARARE | | | | | UTRECHT | PORT OF SPAIN |
| DUBLIN | JEDDAH | | | | | ANKARA | SEVILLE |
| GUANGZHOU | KARACHI | | | | | BAKU | SHENZEN |
| HELSINKI | KUWAIT | | | | | BASEL | STRASBOURG |
| HO CHI MINH | LUSAKA | | | | | BELO HORIZONTE | TIANJIN |
| KIEV | MANAMA | | | | | BILBAO | VALENCIA |
| LIMA | NAIROBI | | | | | BONN | WINNIPEG |
| LISBON | PANAMA CITY | | | | | BRISTOL | YANGON |
| LUXEMBOURG | QUITO | | | | | CANBERRA | YOKOHAMA |
| SOFIA | SANJOSECR | | | | | CURITABA | ABERDEEN |
| WINDHOEK | SAN SALVAD | | | | | DAKAR | BERN |
| ZAGREB | SANTO DOMIN | | | | | DALIAN | BORDEAUX |
| NASSAU | | | | | | DAR ES SALAAM | GUADALAJARA |
| | | | | | | DOHA | HANNOVER |
| | | | | | | TALLIN | LAS VEGAS |
| | | | | | | VILNIUS | MEDELLIN |
| | | | | | | GLASGOW | NURMBERG |
| | | | | | | RICHMOND | RIGA |
| | | | | | | BUFFALO | PALO ALTO |

**Figure 14.1.** Classification result.

(cluster 8) and Barcelona-Houston (cluster 14) is for instance hard to infer, while it is unclear to what degree Singapore is different from Hong Kong and Tokyo (clusters 2-3). And finally, in the first section we have stressed that classifications should focus on intertwining hierarchical, regional and functional tendencies, while the crisp separation purported by a classical clustering algorithm can hardly taken these possible hybrid tendencies into account.

### 14.3.2 A fuzzy-set approach towards classification

The theory of fuzzy sets has been formally introduced by Zadeh [64], and addressed problems in which the absence of sharply defined criteria is involved. In particular, fuzzy sets aim at mathematically representing the vagueness and lack of preciseness, which are intrinsic in linguistic terms and approximate reasoning. As such, through the use of the fuzzy set theory, ill-defined and imprecise knowledge and concepts can be treated in an exact mathematical way [57]. However, this does not imply that fuzziness is mere ambiguity or stems from total or partial ignorance. Rather, fuzzy sets aim to provide the mathematical underpinnings for the specification of the inherent vagueness, and are therefore formally defined as "classes of objects with a continuum of grades of membership" ([64], p. 338). Fuzzy sets are thus characterized by a membership function which assigns to each object of the set a grade of membership ranging from zero (non-membership of the set) to one (full membership of the set). Apart from the apparent fuzziness in standard linguistic terminology and everyday events, vagueness is also a problem in classification schemes framed upon the unravelling of patterns in large data sets [5, 43, 6, 44]).

As MacMillan [39] has pointed out, fuzzy thinking has been around in geography for as far back as the 1970s. MacMillan himself [38] and Gale [22, 23], for instance, applied fuzzy set theory with respect to locational decision-making and behavioural geography. However, at that stage, it did not become fashionable in geographical circles [39]. More recent examples of applications of the use of fuzzy sets in geography can be found in the domains of spatial analysis (e.g. [35, 36]), site selection (e.g. [60, 61]), and land-use planning (e.g. [49, 62]). Although there are, then, quite a few examples of the use of fuzzy set theory, research topics and methodology issues relying on the use fuzzy set theory are as yet not a part of mainstream geography. Furthermore, the outset of the basic premises of fuzzy set theory itself was merely the start for a myriad of studies leading to an explosive growth of both the original core ideas and possible extensions, such as research of expert knowledge systems and neural networks.

One of the major advantages of the use of a fuzzy set-theoretical approach lies in the fact that it is possible to model various aspects of vagueness and uncertainty in the same way [18]. Fuzzy sets can for instance model both the vagueness due to the sparsity of data as well as the lack of theoretically defined pre-existent categories. In general, four of the main useful features of fuzzy set methodologies are [11, 29]:

(i) Fuzzy set theory provides a systematic basis for quantifying vagueness due to incompleteness of information.
(ii) Classes with unsharp boundaries can be easily modelled using fuzzy sets.

(iii) Fuzzy reasoning is a formalism that allows the use of expert knowledge, and is able to process this expertise in a structured and consistent way.
(iv) There is no broad assumption of complete independence of the evidence to be combined as required for probabilistic approaches.

A fuzzy set approach to clustering thus dispenses with unambiguous mapping of the data to clusters, and computes degrees of membership specifying to what extent objects belong to clusters. With the understanding that $\mu_{ki}$ denotes the degree of membership of object $k$ in cluster $i$, this implies that the condition $u_{ki} \in \{0,1\}$ in Equation [14.3] is replaced by

$$\mu_{ki} \in [0,1] \tag{14.9}$$

When this general premise is introduced in the criterion function for the crisp clustering algorithm, Equation [14.5] can be replaced by its fuzzy notion counterpart ([11, 29], all drawing on the seminal work by [5]), based on the iterative minimization of

$$J(U,V) = \sum_{k=1}^{n} \sum_{i=1}^{C} \mu_{ki}^{m} (v_i - x_k)^2 \tag{14.10}$$

where

- $x_1, x_2, \ldots, x_n$ are $n$ data samples
- $V = \{v_1, v_2, \ldots, v_n\}$ are cluster centers
- $U = [\mu_{ki}]$ is a $n \times C$ matrix, where $\mu_{ki}$ is the membership grade of the $k^{th}$ input sample $x_k$ in the $i^{th}$ cluster $C_i$, and the membership grades satisfy the following condition:

$$\forall\, k = 1,2,\ldots,n,\ \forall\, i = 1,2,\ldots,C : \mu_{ki} \in [0,1] \ \wedge\ \sum_{i=1}^{C} \mu_{ki} = 1$$

- $m \in\, ]1,\infty[$ is an exponent weight factor. This weight factor $m$ reduces the influence of small membership values. The larger the value of $m$, the smaller the influence of samples with small membership values in the optimization procedure outlined below.

The altered objective function is the sum of the squared Euclidean distances between each input sample and its corresponding cluster center, with the distances weighted by the fuzzy memberships. The algorithm is iterative and makes use of the following equations [29]:

$$v_i = \frac{1}{\sum_{k=1}^{n} \mu_{ki}^{m}} \sum_{k=1}^{n} \mu_{ki}^{m} x_k \tag{14.11}$$

$$\mu_{ki} = \frac{\left(\frac{1}{(v_i - x_k)^2}\right)^{\frac{1}{m-1}}}{\sum_{j=1}^{C}\left(\frac{1}{(v_i - x_k)^2}\right)^{\frac{1}{m-1}}} \tag{14.12}$$

For the calculation of a cluster center, all input samples are considered in accordance with their membership value. For each sample, its membership value in each cluster depends on its distance to the corresponding cluster center. Following [11], the clustering procedure consists of the following steps:

1. Initialize $U^{(0)}$ randomly or based on an approximation (for instance, the results of the crisp c-means clustering) by initializing $V^{(0)}$ and calculating $U^{(0)}$. The iteration counter $\alpha$ is set to 1, and the number of clusters $C$ and the exponent weight $m$ are chosen.
2. Using the criterion function, the cluster centers ($V^{(\alpha)}$) can be computed based on the values of the membership values ($U^{(\alpha-1)}$).
3. The membership values ($U^{(\alpha)}$) are then calculated based on the new cluster centers ($V^{(\alpha)}$).
4. This iteration is stopped if $\max_{k,i}(|\mu_{ki}^{(\alpha)} - \mu_{ki}^{(\alpha-1)}|) \leq \varepsilon$, where $\varepsilon$ is a pre-specified small number representing the smallest acceptable change in $U^{(\alpha)}$, else let $\alpha = \alpha + 1$ and go to step 2.

Note that the crisp c-means clustering algorithm can be considered as a special case of the fuzzy c-means clustering algorithms. If $\mu_{ki}$ is 1 for only one class and zero for all other classes in Equation [14.11], then the criterion function $J(U, V)$ used in the fuzzy c-means clustering algorithm is the same as the criterion function $J(V)$ used in the crisp c-means cluster algorithm.

## 14.4 A detailed taxonomy of the world city network

Analysing the 54756 pieces of relational information with the fuzzy c-means algorithm (with $C = 20$) requires a comprehensive look at the content of the clusters, and we therefore summarize our results by designating cities based on their membership degrees in various clusters:

- *distinctive cities* have a membership degree above 0.6 in a cluster, and are therefore unambiguously assigned to this cluster;
- *hybrid cities* have an in-between profile, and are defined as those cities that have two membership degrees between 0.3 and 0.6;
- *singular cities* are members that have no affiliation as high as 0.6 in any cluster, and have maximum one membership degree above 0.3.

The diverse tendencies in the world city network are summarized in Figure 14.2 and Figure 14.3. Figure 14.2 asserts the affiliations of each of the 234

| | Cluster 1 | Cluster 2 | Cluster 3 | Cluster 4 | Cluster 5 | Cluster 6 | Cluster 7 | Cluster 8 | Cluster 9 | Cluster 10 |
|---|---|---|---|---|---|---|---|---|---|---|
| cluster nucleus | London<br>New York | Frankfurt<br>Hong Kong<br>Milan<br>Paris<br>Singapore<br>Tokyo | Chicago<br>Los Angeles<br>San Francisco | Amsterdam<br>Buenos Aires<br>Madrid<br>Mexico City<br>Sao Paulo<br>Sydney<br>Toronto | Bangkok<br>Beijing<br>Jakarta<br>Kuala Lumpur<br>Manila<br>Seoul<br>Shanghai<br>Taipei | Berlin<br>Dusseldorf<br>Hamburg<br>Munich | Atlanta<br>Boston<br>Dallas | Copenhagen<br>Dublin<br>Helsinki<br>Lisbon<br>Oslo<br>Rome | Cairo<br>Istanbul | Budapest<br>Moscow<br>Prague<br>St. Petersburg<br>Warsaw |
| hybrid members | | Barcelona<br>Brussel<br>Zurich | | Barcelona<br>Brussels<br>Zurich | | | Miami<br>Seattle<br>St. Louis | Bratislava<br>Sofia<br>Stockholm | Bratislava<br>Sofia | Stockholm |
| singular members | | | | | | Geneva | Washington<br>Houston | Athens<br>Helsinki<br>Oslo | Dubai<br>Mumbai | Kiev<br>Curitaba<br>Vienna |

| | Cluster 11 | Cluster 12 | Cluster 13 | Cluster 14 | Cluster 15 | Cluster 16 | Cluster 17 | Cluster 18 | Cluster 19 | Cluster 20 |
|---|---|---|---|---|---|---|---|---|---|---|
| cluster nucleus | Bogota<br>Caracas<br>La Paz<br>Santiago | Adelaide<br>Auckland<br>Brisbane<br>Christchurch<br>Edmonton<br>Hobart<br>Johannesburg<br>Melbourne<br>Montreal<br>Ottawa<br>Perth<br>Vancouver | Bangalore<br>Dhaka<br>Islamabad<br>Karachi<br>Nairobi<br>New Delhi | Baltimore<br>Charlotte<br>Cincinatti<br>Columbus<br>Honolulu<br>Indianapolis<br>Kansas City<br>Philadelphia<br>Phoenix<br>Pittsburgh<br>Portland<br>Richmond<br>Sacramento<br>San Diego<br>Tampa | Asuncion<br>Belo Horizonte<br>Montevideo<br>Porto Alegre | Casablanca<br>Dalian<br>Kuwait<br>Tunis<br>Zagreb | Bonn<br>Dresden<br>Nurnburg<br>Utrecht | Aberdeen<br>Leeds<br>Nottingham | Basel<br>Lausanne<br>Lille<br>Strasbourg | Abijan<br>Accra<br>D E Salaam<br>Dakar<br>Doha<br>Gaberone<br>Harare<br>Kampala<br>Labuan<br>Lagos<br>Lusaka<br>Windhoek |
| hybrid members | Lima<br>Miami | Birmingham<br>Manchester<br>Southampton | | St Louis<br>Seattle | Guatemala City<br>Lima<br>Managua | Tehran<br>Almaty<br>Ankara<br>Baku<br>Tashkent | Antwerp<br>Bilbao<br>Gothenburg<br>Malmo<br>Riga<br>Rotterdam<br>Tallinn<br>The Hague<br>Turin<br>Valencia | Bilbao<br>Birmingham<br>Manchester<br>Southampton<br>Valencia | Antwerp<br>Gothenburg<br>Malmo<br>Rotterdam<br>The Hague<br>Turin<br>Valencia<br>Vilnius | Almaty<br>Ankara<br>Baku<br>Riga<br>Tallinn<br>Tashkent<br>Teheran<br>Vilnius |
| singular members | Kingston<br>Quito<br>Rio de Janeiro | Calgary<br>Canberra<br>Cape Town<br>Durban<br>Hamilton (BD)<br>Manama<br>Monterrey<br>Port of Spain<br>Ruwi<br>Wellington<br>Winnipeg | Bucharest<br>Calcutta<br>Chennai<br>Hanoi<br>Ho Chi Minh<br>Jeddah<br>Lahore<br>Riyadh | Buffalo<br>Buffalo<br>Cleveland<br>Denver<br>Detroit<br>Hartford<br>Las Vegas<br>Minneapolis<br>New Orleans<br>Palo Alto<br>Rochester<br>Sacramento<br>San Jose, CA | Guadalajara<br>Guayaquil<br>Port Louis<br>San Jose<br>Panama<br>Medellin | Abu Dhabi<br>Amman<br>Beirut<br>Nicosia<br>Osaka<br>Tel Aviv<br>Yangon | Luxembourg<br>Cologne<br>Essen<br>Hannover<br>Leipzig<br>Stuttgart | Belfast<br>Bristol<br>Edinburgh<br>Glasgow<br>Liverpool<br>Newcastle | Bern<br>Bologna<br>Bordeaux<br>Ljubljana<br>Lyons<br>Marseilles<br>Seville | Brazilia<br>Colombo<br>Doula<br>Guangzhou<br>Guatemala City<br>Managua<br>Maputo<br>Nagoya<br>San Salvador<br>Santo Domingo<br>Shenzen<br>Tegucigalpa<br>Tianjin<br>Yokohama |

**Figure 14.2.** Diverse tendencies in the world city network.

cities, while Figure 14.3 visualizes these results through ordering the clusters both hierarchically and regionally, i.e. based on their average connectivity and their approximate geographical latitude. In order to aid in the initial interpretation of the results, Figure 14.3 also shows two distinctive cities for each cluster, in addition to a number of hybrid cities.

Cluster 1 unveils London and New York as a wholly distinctive dyad that ascertains the most important link in the world city network in terms of connectivities. Clusters 2 and 4 show important cross-regional links between major European, Pacific Asian and Latin American cities, which replaced the somewhat odd dyads in the crisp classification. These trans-regional links

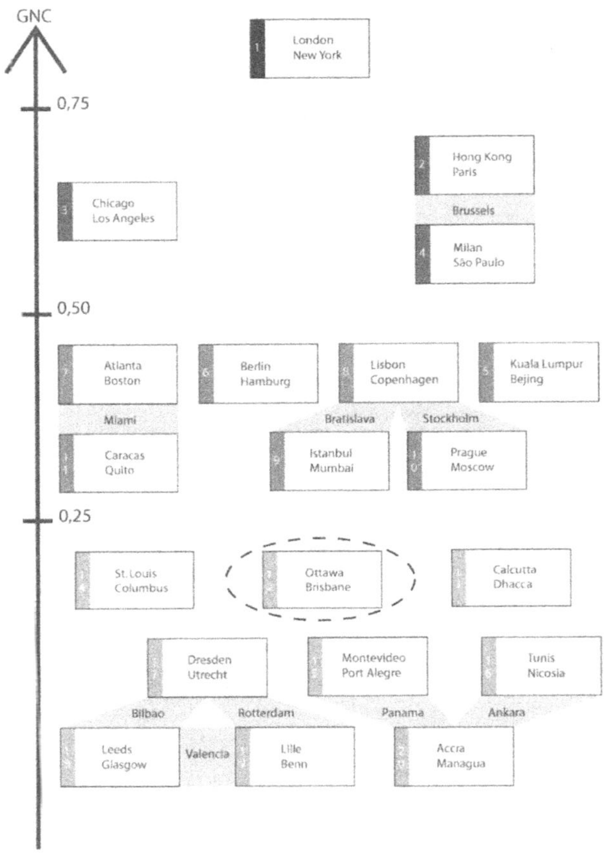


**Figure 14.3.** Diverse tendencies in the world city network (visualization).

show that European cities are pivotal in linking to other world regions, which contrasts enormously with the relational patterns of US cities. Cluster 3, for instance, includes just the three US cities that rank below New York. This cluster thus only contains US cities, and articulates an idiosyncratic pattern that re-appears for lower-ranked world US cities (clusters 7 and 14). Apart from New York, as half of the NY-LON dyad, and Miami, as an important hybrid city linking to Latin America (clusters 7 and 11), there are no continental US cities that have important relations beyond their own country. We interpret this as relating to the sheer scale of the US economy and its long-developed, massive market in financial and business services that provides less of an incentive for firms 'to go global' to the same degree as global service forms from other world regions. Although it will become clear that most clusters that bring together lower-ranked world cities retain an important regional dimension, this tendency is nowhere as distinct as in the case of US cities. Lower-ranked European cities, for instance, appear to be thoroughly interlinked. With the exception of a cluster that only contains secondary German cities (cluster 6), lower-ranked European clusters share a lot of hybrid members (clusters 17, 18, 19). Moreover, a number of European cities seem to perform gateway functions to other world regions, e.g. Stockholm as gateway city between Western Europe and Eastern Europe (clusters 8 and 10), and Bratislava and Sophia between western Europe and the Middle East (clusters 8 and 9). Note that in the classical clustering algorithm all cities are lumped together in a single cluster, which does not allow assessing these gateway functions.

Cluster 5 brings together the most important Pacific Asian cities beyond Hong Kong, Tokyo and Singapore. The average connectivity of this cluster unveils that Pacific Asia is a third main 'globalization arena' next to Europe and the US, while the lack of hybrid cities and trans-regional links furthermore shows the systematic interconnectedness between second-ranked Pacific Asian cities. It is clear that cities in southern Asia and the Middle East (clusters 9, 13, and 16) play a far less important role in this transnational urban network.

Beyond the three dominant world-regions, the relational patterning of Latin American cities reveal the ambiguous role it has come to play in the world-economy. There are a limited number of Latin American cities that can be found among the most important world cities (Mexico City, São Paulo, and Buenos Aires), but the relational prowess of these cities contrasts with the marginal position of the other Latin American cities, especially Central American cities, which are found in lower-ranked clusters (clusters 11, 15, and 20). This implies that although these major Latin American cities are found along the likes of Zurich and Brussels, their position is embedded in a totally different regional context. That is, the significant connectivity gap between major and minor Latin American cities is very distinct from Europe, the USA and Pacific Asia, where there is a markedly even distribution of cities across most connectivity levels. Following the initial suggestions in [19] on world city categories, this suggests a semi-peripheral pattern as nodes that articulate surplus transfers to the core of the world-economy.

The remarkable regional dimension beyond the highest rungs of the world city network is broken by a cross-regional cluster that covers the old British Commonwealth (cluster 12). This 'cultural' historical throwback arena replicates previous findings with fewer cities [56], and is in fact the only urban arena that adds a politico-cultural configuration by providing a slightly different historical twist to the story of the otherwise dominating hierarchical and regional tendencies in the world city network. Membership covers Australian, Canadian, New Zealand and South African cities not found in the clusters with the most connected cities, in addition to a number of British cities that link up with lower-ranked European cities.

Cluster 20 brings together the less connected nodes in the world city network. With the exception of Nairobi, all inter-tropical African cities are distinctive members of this cluster. Singular cities include a number of relatively unimportant Asian and Latin American cities, hybrid cities include Baltic cities, which link up with minor European cities (clusters 17 and 19), and a number of South Asian cities that have a hybrid pattern with cluster 16. Nairobi is the only sub-Saharan city that is not a contained in cluster 20. Its affiliation in cluster with South Asian cities may seem quite odd from the perspective of the strong regional tendency of the arenas, but it is the only sub-Saharan African city that moves a marginal connections to the wider world city network. Likewise, Mumbai is the only South Asian city that escapes the otherwise marginal position of this region through its affiliation with a cluster with most important cities in the Middle. This also indicates that Mumbai rather than New Delhi is the leading Indian city in the world-economy.

Our fuzzy-set approach has allowed unveiling a number of hierarchical and regional tendencies beyond the top-ranked cities, but it also sets aside to 'individual' profiles. Miami is the exemplary case here with its hybrid profile in clusters 7 and 11. The former contains secondary US cities, the latter is the arena of major Latin American cities. Although Miami is not a major world city according to [4], our analysis is picking up the suggestion that it does perform a very important regional role through its articulation role between the USA and Latin America. As a city it has indeed been designated as unusual before, the 'most foreign city' in the USA [41], a contingent political (CIA) creation [25], a sort of 'extra-mural capital' of Latin America [7], with totally distinctive connections [54], and our analysis lends further empirical support for these theses. Furthermore, it can be read from Figure 14.2 that the odd linkage Houston/Barcelona and Washington's distinctive pattern are now replaced with a pattern which makes intuitively more sense, i.e. both North American cities now linking up with their 'hinterland'.

## 14.5 Conclusion

The present chapter has moved beyond previous attempts of world city categorization by addressing the methodological problems related to assessments of less-connected nodes in a global urban network. The end-result complements previous specifications and measurements of the world city network with a detailed taxonomy of relational profiles of cities across the globe.

The use of a fuzzy c-means algorithm, which computes membership degrees rather than mere membership and allows influencing the results through choosing the exponent weight factor $m$, provides more flexibility for classifying objects. A comparison between Figures 14.1 and 14.2 clearly unveils that the latter is more fitted to serve as a detailed taxonomy of the world city network. In addition to being less sensitive to bias in the dataset, the fuzzy classification reveals more than just hierarchical patterns for less connected cities, while the idiosyncratic patterns in more thoroughly connected nodes are replaced with more knowledgeable patterns (i.e. larger clusters). And finally, the fuzzy clustering result is more useful since the identification of multiple patterns has been the specific goal of our analysis.

Controlling the degree of fuzziness through manipulation of m has thus allowed transforming the small clusters/large clusters dichotomy of the crisp results to a distinct clusters/hybrid clusters dichotomy. Although this is of course the whole gist of the alternative approach we have taken in the present chapter, it induces the problem of the 'right' choice of $m$. There are no guidelines for an optimal choice of this parameter; a good ad hoc guideline may be exploring different results for different values of $m$ (and $C$). This commonsensical guideline is of course by no means an objective operation.

The detailed taxonomy presented in Figures 14.2 and 14.3 are not 'explanations', neither statistical nor so-called causal. Rather, this classification is merely the result of a data-mining approach that allows exploring the patterns that emerge in this global urban network. By discerning the patterns in the complex relational patterns between cities, it can be seen that world cities are tied together in three different ways. First, there is a strong hierarchical dimension to the clusters. Cities with similar levels of overall connectivity tend to be classified together, e.g. New York and London as the dominant nodes are distinct members of a single cluster. Second, there is a strong regional dimension to the clusters. Cities from the same part of the world tend to be classified together. US cities are the exemplary case here, but beyond this clear-cut example there remains an important regional dimension to intercity relations. Third, there is a tendency for interaction between these two dimensions. Clusters with low average connectivity tend to be more regionally restricted in membership. The theoretical implications of the latter lay in the observation that instantaneous communications and a global network do not imply the 'end of geography'. Although distance no longer features as a factor in financial transactions, our analysis is picking up the suggestion that cities and their networks do not exist in some sort of abstract service space. Rather, there is a multifaceted geography of world regions through which cities operate as basing points for global capital. Hence, as well as the commonplace notion that individual world cities represent critical local-global nexuses, there are also world regions that represent regional-global nexuses in 'real' geographical space.

## References


1. Amin A., Thrift N. (1992) Neo-Marshallian nodes in global networks. International Journal of Urban and Regional Research 16(3):571–587.
2. Arrighi G. (1994) The Long Twentieth Century: Money, Power, and the Origins of Our Times. Verso, London.
3. Arrighi G., Drangel J. (1986) The Stratification of the World-Economy: an Exploration of the Semiperipheral Zone. Review X(1):9–74.
4. Beaverstock J.V., Smith R.G., Taylor P.J. (1999) A Roster of World Cities. Cities 16(6):445–458.

5. Bezdek J.C. (1981) Pattern Recognition with Fuzzy Objective Function Algorithms. New York: Plenum Press.
6. Bezdek J.C., Sankar P.K. (1992) Fuzzy Models For Pattern Recognition. In: Methods That Search for Structures in Data, IEEE Press, New York.
7. Brown E., Catalano G., Taylor P.J. (2002) Beyond World Cities. In: Central America in a space of flows, Area 34, pp. 139–148.
8. Camagni R. (1998) Beyond complexity in urban development studies. In: Bertuglia C., Bianchi G., Mela A. (Eds.) The City and its sciences, Physica-Verlag, Heidelberg.
9. Castells M. (1996) The Information Age: Economy, Society, and Culture (Vol. I): The Rise of the Network Society. Blackwell, Oxford.
10. Chase-Dunn C. (1985) The System of World Cities, A.D. 800- 1975. In: Timberlake M. (Ed.) Urbanization and the World-Economy, Academic Press, New York, pp. 269–292.
11. Chi Z., Yan H., Pham T. (1996) Fuzzy Algorithms with Applications to Image Processing and Pattern Recognition. World Scientific, Singapore.
12. Christaller W. (1933) Die Zentralen Orte in Suddeutschland. Gustav Fischer, Jena. Translated by Baskin C.W. (1966) Central Places in Southern Germany. Prentice Hall, Englewood Cliffs, New Jersey.
13. Daly M.T., Stimson R.J. (1992) Sydney: Australia's gateway and financial capital. In: Blakely E.J., Stimson R.J. (Eds.) New Cities of the Pacific Rim, Institute for Urban and Regional Development (University of California), Berkeley, CA.
14. Daniels P., Moulaert F. (1993) The Changing Geography of Advanced Producer Services. Economic Geography 69:101–102.
15. Derudder B., Witlox F. (2002) Classification Techniques in Complex spatial Databases: Assessment of Vagueness and Sparsity in a Network of World Cities. Solstice: An Electronic Journal of Geography and Mathematics XIII:1–32.
16. Derudder B., Taylor P.J., Witlox F., Catalano G. (2003) Hierarchical Tendencies and Regional Patterns in the World City Network: a global urban analysis of 234 cities. Regional Studies 37(9):875–886.
17. Dezzani R. (2001) Classification Analysis of World Economic Regions. Geographical Analysis 33(4):330–352.
18. Everitt B.S., Landau S., Leese M. (2001) Cluster Analysis. 4th edition, Arnold, London.
19. Friedmann J. (1986) The World City Hypothesis. Development and Change 17(1):69–84.
20. Friedmann J. (1995) Where We Stand: a Decade of World City Research. In: Knox P.L., Taylor P.J. (Eds.) World Cities in a World-System, Cambridge University Press, Cambridge, UK, pp. 21–47.
21. Friedmann J., Wolff G. (1982) World City Formation. An Agenda for Research and Action. International Journal of Urban and Regional Research 6(3):309–344.
22. Gale S. (1972) Inexactness, Fuzzy Sets and the Foundations of Behavioural Geography. Geographical Analysis 4:337–349.
23. Gale S. (1972) Remarks on the Foundations of Locational Decision-Making. Antipode 4:41–79.
24. Geddes P. (1915) Cities in Evolution. Williams and Norgate, London. Reprinted in: LeGates R., Stout F. (Eds.) (1998) Early Urban Planning 1870–1940, Vol. 4, Routledge, London.

25. Grosfoguel R. (1995) Global logics in the Caribbean city system: the case of Miami. In: Knox P.L., Taylor P.J. (Eds.) World Cities in a World-System, Cambridge University Press, Cambridge, UK, pp. 156–170.
26. Hall P. (1966) The World Cities. Weidenfeld and Nicolson, London.
27. Hall P. (2002) Christaller for a Global Age: Redrawing the Urban Hierarchy. In: Mayr A., Meurer M., Vogt J. (Eds.) Stadt und Region: Dynamik von Lebenswelten, Deutsche Gesellschaft fr Geographie, Leipzig, pp. 110–128.
28. Hill R.C., Kim J.W. (2000) Global Cities and Developmental States: New York, Tokyo and Seoul. Urban Studies 37(12):2167–2196.
29. Höppner F., Klawonn F., Kruse R., Runkler T. (1999) Fuzzy Cluster Analysis. John Wiley & Sons, Chichester.
30. Kaufman L., Rousseeuw P.J. (1990) Finding Groups in Data: an Introduction to Cluster Analysis. John Wiley & Sons, New York.
31. King A.D. (1990) Global Cities: Post-Imperialism and the Internationalism of London, Routledge, London.
32. Knox P.L. (1995) World Cities in a World-System. In: Knox P.L., Taylor P.J. (Eds.) World Cities in a World-System, Cambridge University Press, Cambridge, UK, pp. 3–20.
33. Knox P.L. (1996) Globalization and Urban Change. Urban Geography 17:115–117.
34. Knox P.L. (2002) World Cities and the Organization of Global Space. In: Johnston R.J., Taylor P.J., Watts M.J. (Eds.) Geographies of Global Change, 2nd edition, Blackwell, Oxford, pp. 328–338.
35. Leung Y. (1987) On the Imprecision of Boundaries. Geographical Analysis 19:125–151.
36. Leung Y. (1988) Spatial Analysis and Planning under Imprecision. North-Holland, Amsterdam.
37. Lösch A. (1954) The Economics of Location. Yale University Press, New Haven, CT.
38. MacMillan W. (1978) An Introduction to the Theory of Fuzzy Sets in the Context of the Construction of Representational Spatial Economic Theory. In: Martin R.L., Thrift N.J., Bennett R.J. (Eds.) Towards the Dynamic Analysis of Spatial Systems, Pion, London, pp. 37–52.
39. Macmillan W. (1995) Modelling: Fuzziness Revisited. Progress in Human Geography 19(3):404–413.
40. Moulaert F., Todtling F. (1995) The geography of advanced producer services in Europe. Progress in Planning 43(2-3):90–274.
41. Nijman J. (1997) Globalization to a Latin beat: the Miami growth machine. Annals of the American Academy of Political and Social Sciences 551:163–176.
42. O'Connor K., Stimson R.J. (1995) The Economic Role of Cities: Economic Change and City Development, Australia 1971–1991. AGPS, Canberra.
43. Pal S.K., Majumder D. (1986) Fuzzy Mathematical Approach to Pattern Recognition. John Wiley & Sons, New York.
44. Pal S.K., Mitra S. (1999) Pattern Recognition: Methods in Soft Computing. John Wiley & Sons, New York.
45. Sassen S. (1991) The Global City: New York, London, Tokyo. Princeton University Press, Princeton.
46. Sassen S. (1995) On concentration and centrality in the global city. In: Knox P.L., Taylor P.J. (Eds.) World Cities in a World-System, Cambridge University Press, Cambridge, UK, pp. 67–78.

47. Sassen S. (2000) Cities in a World Economy. Pine Forge, Thousand Oaks, CA.
48. Sato M., Sato Y., Jain L.C. (1997) Fuzzy Clustering Models and Applications. Physica-Verlag, Heidelberg.
49. Smith P.N. (1992) Fuzzy Evaluation of Land-Use and Transportation Options. Environment and Planning B: Planning and Design 19:525–544.
50. Smith R.G. (2003) World city actor-networks. Progress in Human Geography 27(1):25–44.
51. Taylor P.J. (2000) World Cities and Territorial States under Conditions of Contemporary Globalization. Political Geography 19(1):5–32.
52. Taylor P.J. (2001) Specification of the World City network. Geographical analysis 33(2):181–194.
53. Taylor P.J., Hoyler M. (2000) The spatial order of European Cities under conditions of contemporary globalization. Tijdschrift voor Economische en Sociale Geografie 91:176–189.
54. Taylor P.J., Walker D.R.F. (2001) World cities: a first multivariate analysis of their service complexes. Urban Studies 38:23–47.
55. Taylor P.J., Catalano G., Walker D.R.F. (2002) Measurement of the World City Network. Urban Studies 39(13):2367–2376.
56. Taylor P.J., Catalano G., Walker D.R.F. (2002) Exploratory Analysis of the World City Network. Urban Studies 39(13):2377–2394.
57. Tzafestas S.G. (1994) Fuzzy Systems and Fuzzy Expert Control: an Overview. The Knowledge Engineering Review 9(3):229–268.
58. Van Rossem R. (1996) The World System Paradigm as General Theory of Development: A Cross-National Test. American Sociological Review 61:508–527.
59. Wallerstein I. (1983) Historical capitalism. Verso, London.
60. Witlox F. (1998) Modelling Site Selection. A Relational Matching Approach Based on Fuzzy Decision Tables. TU Eindhoven, Eindhoven.
61. Witlox F. (2000) Modelling site suitability of harbour-related companies using multidimensional fuzzy decision tables: the case of Antwerp. Belgian journal of operations research, statistics and computer science 40(3–4):148–168.
62. Xiang W.N., Gross M., Fabos J., MacDougall E.B. (1992) A Fuzzy-Group, Multicriteria Decision-Making Model and its Application to Land-Use Planning. Environment and Planning B: Planning and Design 19:61–84.
63. Yates A. (1987) Multivariate Exploratory Data Analysis. State University of New York Press, Albany.
64. Zadeh L.A. (1965) Fuzzy Sets. Information and Control 8:338–353.

# 15

# Image-Based Change Detection of Geographic Information Using Spatial Constraints

Sidharta Gautama, Johan D'Haeyer, and Wilfried Philips

Department of Telecommunications and Information Processing
Ghent University
Sint-Pietersnieuwstraat 41, B-9000 Gent (BELGIUM)
sidharta.gautama@UGent.be

## 15.1 Introduction

A major challenge in the production and use of geographic information is assessment and control of the quality of the spatial data. The rapid growing number of sources of geospatial data, ranging from high-resolution satellite and airborne sensors, GPS, and derivative geospatial products, pose severe problems for integrating data. Content providers face the problem of continuously ensuring that the information they produce is reliable, accurate and up-to-date. Integrity constraints are able to resolve certain issues in the data, like valid attribute values or relationships between data objects. The main issue is however the consistency of the data with respect to the current "real-world" situation. Today the industry still relies on human operators, who collect and interpret aerial photos and field data to check and correct the current state of the data. Especially for detailed data over large regions, like digital road maps or topographic maps, this is a very labor-intensive and costly process. In addition, human processing is a source of error and inconsistency. Automated detection of change and anomalies in the existing databases using image information can form an essential tool to support quality control and maintenance of spatial information.

The main problem however are the differences in data representation. To be able to compare geospatial vector data with images, the information in the images needs to be described in terms of object features. Automatic detection of man-made objects is a difficult problem. Shadow, occlusion and variety in appearance all give rise to a fragmented and imprecise description of the image content, especially if consistent detection is required over large datasets. Making a reliable statement about the quality of the GIS data requires knowledge about the performance of the different object detection techniques. Our focus has been the development of reliable and predictable techniques. Rather than producing results on a selected number of images, we wish to be able to characterize the performance of the detection. This performance characterization is an essential step towards reliable tools within an operational environment.

The proposed system for change detection is based on feature based spatial registration, where detected features in the image are registered to corresponding features in the vector data. The system consists out of two stages: 1) a low-level feature detection process, which extracts roads and junctions using an improved ridge detector, and 2) a high-level matching process, which uses graph matching to find correspondences between the detected image information and the road vector data (cf. figure 15.1). The graph matching process, based on continuous relaxation labeling, is driven by the spatial relations between the features and takes into account different errors that can occur (e.g. spatial inaccuracy, data inconsistencies between image and vector data). The matched features can be used to calculate a local transformation between image and vector data. This transformation is able to compensate for the local distortions that can occur between the datasets. Additionally the object-to-object mapping is useful to define measures of change between datasets.

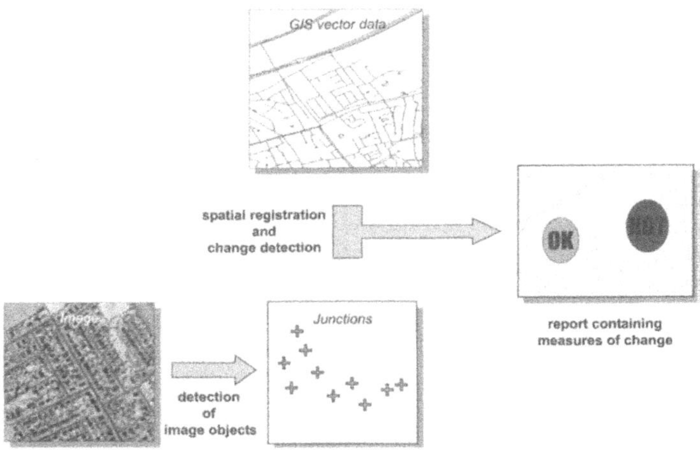


**Figure 15.1.** Overview of the system for change detection

The presented work focuses on the high-level matching process of a system for automated detection of change and anomalies in GIS road networks using very high resolution satellite images. It will be shown how error-tolerant graph matching can be used to find correspondences between the detected image information and the road vector data. The novel aspect is that spatial relations between objects are used to find a reliable object-to-object mapping, where relaxation labeling is applied as flexible query mechanism to solve the spatial query. Spatial relations between objects prove to be more reliable for detecting change when compared to the distance measure used in traditional global transformation schemes.

**Figure 15.2.** Example satellite images with the road data in overlay. Local inconsistencies between the two datasets are apparent.

The contribution is organized as follows. Section 15.2 gives an overview of related work in spatial querying of image information. Section 15.3 introduces how an image is described using road junctions. Section 15.4 explains error-tolerant graph matching as flexible spatial query mechanism. Experimental results are shown on artificial and real data. Section 15.5 describes how the matching result can be interpreted as measures of change. In section 15.6 conclusions are drawn.

## 15.2 Related work on spatial querying of image information

Although the terminology is slightly confusing, the aim of this work is not to query the GIS database. The spatial vector data as such pose a query to a set of images about the consistency with the image information. In this respect, the work in this chapter is related to the work on content-based image retrieval (CBIR) in the sense that a similarity needs to be determined between part of the vector data and the corresponding image. CBIR techniques can be divided into two classes depending on the way the content of the image is described [8].

The first class of techniques uses measurements that capture the global characteristics of the image based on color, texture or local geometry. Often the spatial information in the image (e.g. size, location and/or relationships between objects or regions) is ignored as these techniques generate a single signature of an image which is used to calculate similarities between images or to select images based on query predicates. Examples of this work are the use of color and texture histograms to find similar images based on a query image [11].

The second class of techniques uses a higher-level description of the image by forming an abstract representation based on image features. Image features like regions, lines and other primitives are obtained through segmentation. Between the features, spatial relations can be measured. The features combined with their relations form a high-level description of the image, which allows for spatial queries to be made. A spatial query can consist out of a spatial query predicate or a query-by-sketch. A spatial query predicate formed using spatial query languages poses certain difficulties due to the fact that geographic concepts are often vague and imprecise [3]. In addition, some queries are difficult to capture in lexical terms. An alternative method is query-by-sketch. In this case instead of using a verbal query language, a graphical representation of the query containing a spatial configuration is presented. Examples of this line of work are indexing schemes like spatial quad-trees, R-trees and graphs which are built on regions detected in the image (e.g. [7, 9, 1]). Tree-based indexing schemes have the problem that they encode the absolute spatial position of a spatial configuration. In our application more importance is put on the relative position of features as a measure of quality. For example, an industrial site may be slightly translated with respect to an existing road database but with the correct roads collected. This translation can cause significant absolute spatial errors while in relative terms, since the road structure of the site is good, this error is modest (i.e. a single translation). Therefore the position of the features is of less importance than the spatial relations between features, which are able to characterize relative spatial quality. This relational information can be more efficiently used through graph structures.

The focus of our development is reliability in detecting change. Reliability means that although the positive detection rate should be as high as possible, false detections should be kept at a minimum. This is an added challenge compared to the CBIR context where a larger tolerance for false detection is often permitted (i.e. a retrieval can be considered good by the user if images with the correct content are retrieved, even if a significant number of wrong images are retrieved beside it). In addition, due to the nature of comparing dissimilar representation, there is an urgent need for flexibility in the sense of being capable to be tolerant for imprecision, partial truth, uncertainty and approximation present in the image and vector data. Combining reliability and flexibility in this context implies the need for precise control over the bounds of tolerance.

## 15.3 Road junctions as image features

The road network that can be detected using a road detection scheme is often not of sufficient quality to be useful for comparison with a road database. Road detection is often performed using a low-level detector of curvilinear structures or lines (for an overview see [10]). An example is the class of differential ridge detectors which is used in this work [4]. The detection leads to a binary image of pixels belonging to lines of a certain width. Additionally, in each pixel the approximate orientation of the line is measured. This information is used to chain together pixels which show a similar orientation. The main difficulty when comparing the image information with the road database is the difference in representation between the pixel chains that can be detected from the image and the polyline vectors that represent roads in the database. This difference makes it difficult to reliably compare the information in the image with the database. More robust image objects are necessary. We found road junctions to be good candidates since in their abstract form, they can be represented as point objects both in the image as well as in the database.

We model a road junction as a point where lines meet. This means that, built upon the road network that is detected using ridge detection, we look for pixels in the pixel chains which have three or more neighbours. We choose this strategy above corner detection, because corner detection gives many spurious responses not belonging to junctions and are not easy to filter out. Our method stays closer to the road network logic. A major problem however is that the detection of the road network fails in the vicinity of road junctions since the ridge model does not hold anymore. At the junction, the intensity surface will not be modeled as a valley or a ridge but as a flat spot. As a consequence, the road network will often be broken around road junctions.

We implement a simple region growing scheme [4] which extends the initial road fragments with regions showing a similar intensity value. For roads which are adequately detected, this proves to be sufficient in many cases to bridge the bad spots at road junctions. The simple scheme is of course not foolproof. A cheap and efficient verification to filter out false alarms is to check if in the vicinity of a hypothetical road junction a flat spot exists. Junctions are seen by the ridge detector as a flat region. By simply applying different thresholds on the gradient and the curvature, which have already been calculated for the detection of lines, we can detect the flat spots and verify the road junctions. First results have shown a low false alarm rate (typical less than 10% [5]). Figure 15.3 shows an example of road and junction detection, performed on an IKONOS very-high-resolution satellite image with a spatial resolution of one meter. The top figure shows the road network that is detected in the image. The bottom figure shows the road junctions that are detected based on the initial road network. The boxes are detected junctions using the neighbour model. Crosses are junctions which are verified and retained using the flat spot check.

## 15.4 Flexible spatial queries using error-tolerant graph matching

In the previous section, road junctions have been found to be stable image features. The problem can be represented as finding the correspondence between

**Figure 15.3.** Detected road network (above) and detected junctions (below). Boxes show initial junction detection; crosses show the result after verification.

two sets of features: one set originating from the database and one set originating from the image. This can of course be generalized to situations other than image-to-GIS registration, like image-to-image or GIS-to-GIS. Given these features an abstract representation can be built as an attributed graph [2]. The vertices of the graph represent image features and the vertex attributes can contain measurements on these features. The edges of the graph represent relations between features and the edge attributes can contain measurements on spatial relations. A similar graph can be built on the vector data, using data objects as vertices and relations between objects as edges. The problem of registration is represented as a graph matching problem, which seeks the correspondence of similar vertices between two attributed graphs.

In solving the correspondence problem, one should allow tolerance to imprecision and inconsistencies. Errors can occur on the location of the junction due to inaccurate detection, differences in spatial resolution of the data and data inconsistency. In addition, false positives can be present in both datasets. Specific to the problem is that the matching technique should be able to distinguish between errors due to severe data inconsistencies and errors due to spatial inaccuracy. In the latter category the position of the roads can differ but the general structure of the road network is preserved. In the first category, the structure of the road network differs significantly.

In computer vision error tolerant graph matching techniques form an important class of techniques. These techniques seek a graph or subgraph morphism, which allows for distortions. A general distortion model defines the deletion and addition of graph vertices and edges, and replacement of attribute values. A similarity measure or distance function between two graphs is used that models the occurring distortions using heuristics. Early techniques were a generalization of string matching. More recent models are based on information theoretic principles or Bayesian modeling. Solving the correspondence problem is difficult and several optimal and approximate techniques have been proposed. These include, among others, search trees, dynamic programming, annealing and genetic algorithms. In this work, we examine relaxation labeling, a popular approximate technique which has low, polynomial time complexity.

### 15.4.1 Continuous relaxation labeling

The matching problem can be defined as a graph labeling problem, which consists out of the following elements:

1. a set of objects $i \in \Omega_i$, corresponding to image features;
2. a set of labels $\lambda \in \Omega_\lambda$, corresponding to GIS features;
3. a neighbour relationship over the objects;
4. restrictions on possible labels between pairs of neighbouring objects.

Relaxation labeling techniques use an iterative process to determine the probabilities of each object. Different update rules have been proposed. In [6] the relation between different update rules is analytically shown. The problem of finding consistent solutions is shown to be equivalent to solving a variational inequality which is based on the mathematical concept of "consistency". This concept, which is defined below, is interesting because by using it, the labeling process can be redefined as a quadratic optimization process. This offers guidance in determining good compatibility coefficients.

A solution to the problem is a labeling which is consistent with all the restrictions that are imposed.

To each object $i$ a probability distribution $\{p_i(\lambda)\}_{\lambda \in \Omega_\lambda}$ is associated that expresses that object $i$ has label $\lambda$:

$$0 \leq p_i(\lambda) \leq 1, \quad \sum_{\lambda \in \Omega_\lambda} p_i(\lambda) = 1 \tag{15.1}$$

A labeling for the problem is specified by $\mathbf{p} = \{p_i(\lambda)\}_{i \in \Omega_i, \lambda \in \Omega_\lambda}$. For each pair of neighbouring objects $i$ and $j$ and for each pair of labels $\lambda$ and $\lambda'$, a compatibility coefficient $\mathrm{r_{ij}}(\lambda,\lambda')$ is defined. These coefficients express the compatibility of assigning label $\lambda$ to object $i$ in combination with assigning label $\lambda'$ to object $j$. Negative values express incompatibility, positive values compatibility. Given these quantities, the support of a label $\lambda$ for the object $i$ given by the correspondence $\mathbf{p}$ is defined by

$$s_i(\lambda) = s_i(\lambda, \mathbf{p}) = \sum_{j \in \Omega_i} \sum_{\lambda' \in \Omega_\lambda} r_{ij}(\lambda, \lambda') p_j(\lambda') \tag{15.2}$$

To find a consistent labeling, we optimize the average local consistency, given by

$$A(\mathbf{p}) = \sum_{i \in \Omega_i} \sum_{\lambda \in \Omega_\lambda} p_i(\lambda) s_i(\lambda, \mathbf{p}) \tag{15.3}$$

This is a quadratic function in the variables $p_i(\lambda)$, which is optimized using a constrained gradient descent method taking into account the restrictions of equation (15.1).

To take into account the possibility that an object $i$ does not have a correspondent, the set of possible labels and the associated distribution $\mathbf{p}$ is extended with a null label $\lambda_\emptyset$ and probability $p_i(\lambda_\emptyset)$. As each object $i$ can be mapped on the null label, a small negative weight $w_\emptyset$ should be assigned to the compatibility coefficient $\mathrm{r_{i\emptyset}}(\lambda,\lambda_\emptyset)$ to avoid that a full null mapping comes out as the most consistent solution.

In [5], it is shown that the value for the weight $w_\emptyset$ and the weight $w_-$, attributed to inconsistent assignments, is controlled by a simple condition:

$$(1 - f^\emptyset) w_\emptyset < f^- w_- \tag{15.4}$$

whereby $f^-$ and $f^\emptyset$ are the fraction of expected inconsistent and null assignments in the neighbourhood of object $i$. This condition can be used to determine the weights for the compatibility matrix given the expected graph error. It allows to make a distinction between distortions which should find a correspondent in the other dataset and large changes which should be assigned the null label. As previous research usually relied on rules-of-thumb to determine these weights (e.g. [12]), the importance of this equation is that it allows precise definition of the weights of the graph matching problem with respect to the expected spatial error of the system.

### 15.4.2 Spatial constraints

Relaxation labeling is a generic technique to solve constraint satisfaction problems. To apply this technique to matching sets of road junctions, we need to introduce the constraints which apply to the system. There are several possibilities, among others:

1. the number of road arms should be the same for corresponding junctions;
2. interconnecting roads between pairs of corresponding junctions should be the same;
3. geometric invariants under a given spatial transformation between subsets of corresponding junctions should be preserved.

The initial detection of roads that can be achieved is not of sufficient quality to use the first two options as information for the correspondence process [4]. Fragmentation and false detections can frequently occur in the detected road network and are difficult to control. We therefore opt to use geometric invariants between subsets of corresponding junctions. The most simple constraints are binary relations like geometric relations (e.g. angle, distance) between a junction and its neighbours to find correspondences. These are much more stable features, given the detection quality which can realistically be expected from road detection. In these experiments we rely only on the relative angle between pairs of points. Figure 15.4 shows an illustration of this constraint. The black points show object points $i$ and the grey points show the label points $\lambda$. In mapping a pair of points $i$ and $j$ on $\lambda$ and $\lambda'$ the relative angle between the lines $ij$ and $\lambda\lambda'$ does not exceed a given $\Delta\alpha$. (e.g. $\pi/4$). If this constraint is violated, the compatibility coefficient $r_{ij}(\lambda, \lambda')$ is assigned a negative weight $w_-$.

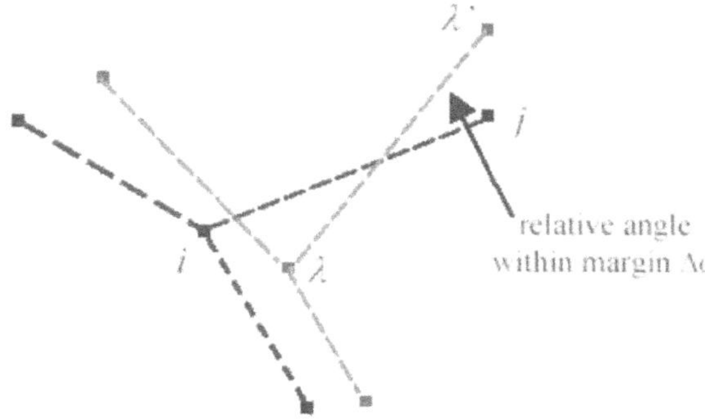


**Figure 15.4.** Illustration of the spatial constraint, based on consistent relative angle between pairs of correspondences $(i, \lambda)$ and $(j, \lambda')$

### 15.4.3 Experimental results on artificial data

A first set of experiments has been performed on random sets of points on which different perturbations are introduced. Relaxation labeling was found to give a good performance for large datasets (order of 100-200 points) within a reasonable processing time (less than 1 minute for sets of 100 points on a Pentium IV 1 GHz, 256 Mb). Performance was initially measured for datasets having the same number of points. The points are randomly scattered within an image of 512x512 pixels and gaussian noise is introduced on the coordinates with a standard deviation between one and eight pixels. A standard deviation of eight pixels already introduces significant errors due to the dense scattering within the image. In figure 15.5 an example is shown for a set of 10 pixels. Under these conditions, for a 100/100 correspondence (i.e. finding a correspondence for 100 points in a second dataset of 100 points) we can expect a correct correspondence rate of 70%. The parameters of relaxation have been set at $\triangle\alpha = \pi/4$, $w_- = -0.5$, $w_\emptyset = -0.1$. This allows for a graph relational error $f^- = 20\%$, as in this example the datasets contain no spurious points, i.e. $f^\emptyset = 0\%$. Figure 15.6 illustrates the performance. In this plot, the average correct and false correspondences of a 100/100 labeling are set out for different amounts of spatial noise. Every measurement is carried out for 100 runs of random points. The radius $R$ determines the neighbourhood relation of each point, i.e. each point which lies within this radius from point $i$ is considered an immediate neighbour of point $i$ and is joined by a graph edge with point $i$. Figure 15.6 shows that by increasing $R$, the performance improves. For $R$ =200 pixels, a correct labeling of 73% and a false labeling of 10% is achieved; the remaining 17% are mapped on the null label.

The performance however degrades considerably when we move on to datasets which differ in size. For a 10/50 correspondence an average performance of only 30% correct correspondences is found, whereas a 10/10 correspondence is almost always solved perfectly. On further investigation, the problem was found to be that when the number of points in the second set increases, the number of possible solutions which are consistent with the imposed constraints increases too. This means that there are several possible optimal solutions for the correspondence technique to choose from and that in an unbiased system the choice that is made will not necessarily be the solution which is aimed at. The solution is either to increase the number of constraints in the system or to make them more strict, thereby making the desired solution unique.

In our case, the constraint which defines the solution is given by the relative angle $\triangle\alpha$ between pairs of points. If this angle becomes too large, the constraint is violated. Initially the tolerated margin $\triangle\alpha$ was taken to be $\pi/4$. This is of course a very wide margin but in our initial experiments of 100/100 labelings this worked quite well. The difference between this situation and a 10/50 labeling is that, with hundred points, the context of a point defined by the relations with its immediate neighbours is very rich, simply because the number of immediate neighbours is high. In a 10/50 labeling, the problem lies with the smallest dataset of 10 points where the context of a point is much smaller and replication of this context consequently becomes easier. This increases the possible ambiguity in the solution space.

Subsequently reducing the allowed margin on the relative angle increases the performance considerably. Figure 15.7 shows this performance plotted for increased angle quantization $q$. At the lowest quantization $q = 4$, corresponding to a margin of $\pi/4$, the number of false correspondences largely exceeds the number of good correspondences. At the other end, for a quantization $q = 32$ (a margin of $\pi/32$) the correct correspondences are 85%, false correspondences 10%. Increasing angle quantization reduces ambiguity which results in a better performance.

In this context, we should note that the underlying transformation model at the moment is a local (rubbersheeting) transformation, i.e. each point has its counterpart in the second dataset and after triangulation, a local affine transformation of triangular patches can be performed. This is much more flexible than imposing a parametric transformation model (e.g. a global affine transformation) as it allows local deformations to exist. The downside is that it is much harder to detect outliers since the model is so flexible. To get a good performance, one should analyse the deformations that the application should allow and find a combination of constraints which optimally define the solution. This can be a combination of the graph matching scheme as is explained and of a parametric transformation model to filter out significant errors.

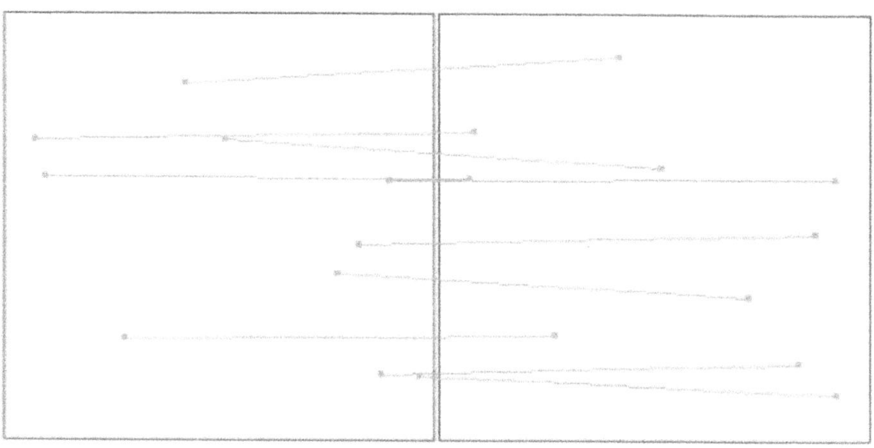

**Figure 15.5.** Example of a 10/10 correspondence: (left) original set of points, (right) points with added noise $\sigma = 8$ pixels. All points have been correctly matched as shown by connecting lines.

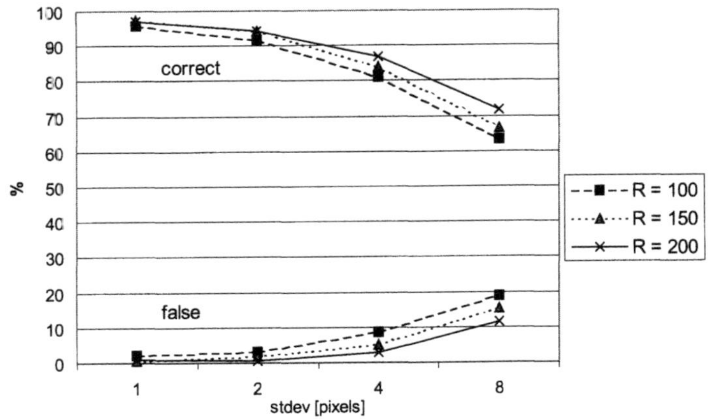


**Figure 15.6.** Performance of a 100/100 correspondence with different neighbourhood radius $R$

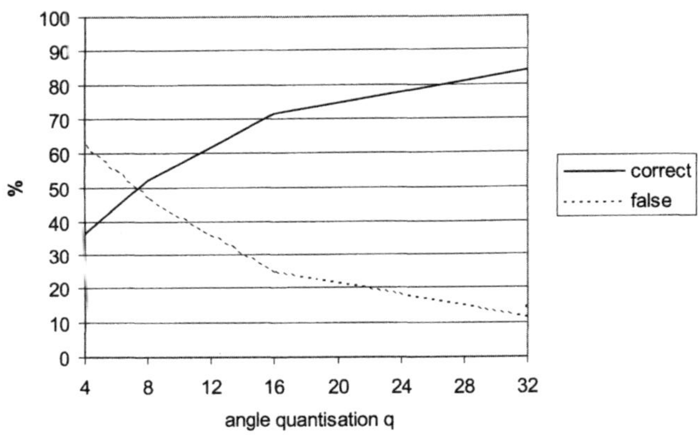


**Figure 15.7.** Performance of a 10/50 correspondence

### 15.4.4 Experimental results on real data

In a second set of experiments we investigate the automatic registration of the IKONOS image of Ghent with a road vector database. The image is a standard panchromatic GEO product. Extracts of the region under investigation is shown in figure 15.2. This figure shows examples of the IKONOS image with the road data in overlay and illustrates some of the problems that can occur. Although the observed terrain is flat, relatively large inconsistencies exist between the two data sources. In addition the database that has been used, does not contain an explicit junction class. We have taken the feature points describing the break points of the vector polylines as junctions. For the region under examination this approximates the junctions in the scene since the region contains mainly straight streets. A number of spurious points is falsely introduced in the vector set of points, due to points which do not belong to junctions (e.g. points defining a roundabout). These points however have been left in the dataset as an additional source of noisy structures.

Using the detected junctions in the image and the set of feature points from the vector data, relaxation labeling has been applied to find the corresponding junctions in the two datasets. The image dataset contains 82 points, the vector dataset contains 205 points. The correspondent of every image point is searched within a radius of 150 meter around the position of the image point in the vector dataset. To compare, the spatial accuracy of the road database is reported to be in the range of 100 meter. In this case, the radius amounts to an ambiguity of around 10 candidate vector points per image point. By performing a manual registration of the image points with the corresponding vector points, we calculated an RMS of 7 meter with a maximum error of 50 meter. Again, the parameters of relaxation were set at $\triangle\alpha = \pi/16$, $w_{-} = -0.5$, $w_{\emptyset} = -0.1$. After convergence of the relaxation labelling process, we obtain 71 correct correspondences, 11 null correspondences and no false correspondences at all.

Based on the matched points, a local rubbersheeting transformation is performed. In this way, local distortions can be compensated, which is not possible if a global transformation is used. Figure 15.8 shows the result of the registration process.

**Figure 15.8.** Registration result after matching and rubbersheeting transformation

## 15.5 Detecting change using matched objects

Whereas global transformation schemes (GTS) look for the best global transformation which optimizes the mean error between two sets of points, graph matching searches directly for the best object-to-object mapping that satisfies all local constraints. This makes graph matching more suited for change detection since local constraints are able to more finely characterize the correspondence of points than using the total mean error does. Figure 15.9 illustrates the basic problem. A fragment is shown containing the image road junctions and the vector road junctions, represented by resp. crosses and circles in the image. The two datasets, of which a small extract is shown, have been registered using the best possible affine transformation after manual registration. The mean error has been minimalized but, due to inconsistencies, some errors cannot be avoided and acceptable errors remain present even after registration.

**Figure 15.9.** Illustration of the problem in using distance as correspondence criterium.

A standard practice is to select for each feature point the nearest corresponding point in the other dataset. This correspondence is shown by the solid white arrows. This example shows that the distance between points, which is implicitely used by global transformation schemes in the error function to be optimized, is not always a good criterium to select corresponding points. It is adequate to calculate the affine parameters since these distortions have a minor impact on the calculation if sufficient points are used. Graph matching in this case is able to find the correct correspondence, since the error that is made in the previous case now violates the local constraints. The correct correspondence is shown by the dashed arrow.

The main difference between GTS and graph matching however lies not in the mistakes that are made but in the mistakes that reasonably can be tolerated. To illustrate the difference, a test area has been chosen, containing 77 road junctions in both datasets. Selected parts of the data show severe inconsistencies, while the overall consistency is adequate. The mean RMS calculated after performing a manual registration using all 77 points is $m_d =$ 21 meter.

Relaxation labeling has been performed on these data, using the same parameter set $\triangle\alpha = \pi/16$, $w_- = -0.5$, $w_\emptyset = -0.1$. The result after convergence has been summarized in table 15.1. It shows a reasonable detection rate of 59.8%. More important however is the small number of false detections, namely 2.6% or two points out of 77, which is important with respect to reliability. The condition in equation (15.4) makes it possible to predict when a null label will be assigned. Namely, a vector junction is mapped on the null label if it shows a relational neighbourhood error larger than 20% when compared to the neighbourhood of the best corresponding point. With respect to change detection this means that in the absence of spurious points, the number of null assignments can be used as a (negative) measure of the quality of the vector data. This measure is useful in addition to mean distance measured in the object-to-object mapping. As more null assignments occur, more junctions can be expected to have a relational error larger than 20%. Figure 15.10 shows a detail of the matching result. The crosses on the image show the location of the image junctions. The white squares show the vector junctions which are correctly mapped with a line drawn to the corresponding image junction. The white triangles show the false positive correspondences. The black dots show the null labels. The example shows a concentration of null labels in the area of the roundabout where severe inconsistencies exist between image and vector data.

The same data have been processed using GTS, where correspondences based on distance have been accepted using a threshold $T_d = m_d$ and $Td = 2 \times m_d$. The performance in Table 1 shows a high false detection rate of 23.4% (18 points out of 77) which remains unchanged even after reducing the distance threshold $T_d$. This again shows the inadequacy of distance as a correspondence criterium. Due to the large number of false positives, the number of null assignments is not usable as a measure of change in the case of GTS. In addition, due to the false positives, the mean distance will be an unreliable measure of change.

**Table 15.1.** Matching performance

| | positive | false positive | null |
|---|---|---|---|
| RL | 59.8% | 2.6% | 37.6% |
| GTS $T_d$=42m | 71.4% | 23.4% | 5.2% |
| GTS $T_d$=21m | 49.4% | 23.4% | 27.3% |

**Figure 15.10.** Matching result using relaxation labeling (square=correct, dots=null, triangle=wrong). In this case, the null label corresponds to relative data inconsistencies.

## 15.6 Conclusion

In this contribution, a system for automated detection of change and anomalies in GIS road networks using very high resolution satellite images has been discussed. The system uses error-tolerant graph matching to find correspondences between the detected image information and the road vector data. It has been shown how spatial relations between objects can be used to find a reliable object-to-object mapping. Relaxation labeling is applied as a flexible query mechanism to answer the spatial query. A condition has been presented which allows to determine the bounds of error tolerance and in this way characterizes acceptable over inacceptable data inconsistencies. The mapping is useful to define a measure of change based on the number of positive and null assignments. This measure in combination with relaxation labeling has been found to be more reliable for detecting change than the use of global transformation schemes.

## References

1. Carson M., Thomas S., Belongie J., Hellerstein M., Malik J. (1999) Blobworld: A system for region-based image indexing and retrieval. In: Proc. of the Third Intl. Conf. on Visual Information Systems. Amsterdam, Netherland, pp. 509–516.
2. Diestel R. (1997) Graph Theory. Springer-Verlag, New York.

3. Egenhofer M. (1997) Query Processing in Spatial-Query-by-Sketch. Journal of Visual Languages and Computing 8(4):403–424.
4. Gautama S., Borghgraef A., Bruyland I. (2002) Automatic registration of satellite images with GIS databases. In: Proc. of Advanced Concepts for Intelligent Vision Systems ACIVS 2002, Gent, Belgium, 7 p.
5. Gautama S., Borghgraef A. (2003) Using graph matching to compare VHR satellite images with GIS data. In: Proc. of the IEEE Intl. Geoscience and Remote Sensing Symposium IGARSS 2003, Toulouse, France, 3 p.
6. Hummel R., Zucker S. (1983) On the foundations of relaxation labeling processes. IEEE Transactions on Pattern Analysis Machine Intelligence 5(3):742–776.
7. Samet H. (1995) Spatial data structures. In: Kim W. (Ed.) Modern Database Systems, The Object Model, Interoperability and Beyond. ACM Press, pp. 361–385.
8. Smeulders A., Worring L., Santini S., Gupta A., Jain R. (2000) Content based image retrieval at the end of the early years. IEEE Transactions on Pattern Analysis Machine Intelligence 22(12):1349–1380.
9. Smith J., Chang S. (1996) Visualseek: a fully automated content-based image query system. In: Proc. of ACM Multimedia 96, Boston MA USA, pp. 87–98.
10. Steger C. (1998) An unbiased detector of curvilinear structures. IEEE Transactions on Pattern Analysis Machine Intelligence 20(2):113–125.
11. Stricker L., Orengo M. (1995) Similarity of color images. In: SPIE Proc. Storage and Retrieval for Image and Video Databases, vol. 2420, pp. 381–392.
12. Wilson R. (1996) Inexact Graph Matching Using Symbolic Constraints. Ph. D. thesis, Department of Computer Science, University of York.

# 16

# Noise Annoyance Mapping

Tom De Muer, Andy Verkeyn, and Dick Botteldooren

Department of Information Technology
Ghent University
Sint-Pietersnieuwstraat 41, B-9000 Gent (Belgium)
tom.demuer|andy.verkeyn|dick.botteldooren@intec.ugent.be

## 16.1 Introduction

Geographical Information Systems (GIS) allow attaching different types of information to geographical primitives. Compared to general-purpose databases, these spatial databases are augmented with a number of functions directly related to geography. Queries based on distance, surface, and spatial aggregation are but a few examples. Even more important are the extended possibilities for visualization that a typical GIS application offers. Because of these, GIS are now becoming very popular in applications related to different fields and in particular play an important role in policy decision support. The policy maker has all available information at his or her fingertips and can identify "black point" or areas of potential interest at a glimpse. This chapter focuses on applications where human perception plays an important role and on EIA(**E**nvironmental **I**mpact **A**ssessment). It is build around a particular case: the assessment of noise annoyance, but concepts and methods introduced are obviously much broader. There are two great points of concern when addressing the aforementioned application areas using classical GIS. Firstly, working with human perception of for example the quality of life, the quality of the living environment, safety, or disturbance by noise or odor, quickly introduces a kind of vagueness, that classical GIS cannot easily store nor process. Secondly, models for environmental impact assessment in general and for perception related or health impacts in particular, are often quite complex and require huge amounts of input data. Both the models and the data are imperfect. The impact of many activities of man on the environment, health, and quality of life is still unclear or at least there is considerable doubt concerning the exact magnitude of the effect. There are plenty of examples: the effect of CO2 emission on global warming, the role of PM10 in the increase in asthmatic treatment of children living in urban area, the contribution of frequent nightly noise events to prevalence of ischemic heart disease, etc. On top of this model imperfection, even the best models need accurate and reliable input data to yield acceptable results. Some data may be lacking, especially if they are related to personal factors (e.g. alcohol or tobacco consumption), or they may be of poor quality. In this chapter we will propose an extension of a typical GIS database to include imperfect data. As different flavors of imperfection can be identified, different frameworks (section 16.2) may be used to represent this imperfection. Fuzzy noise exposure maps and fuzzy noise annoyance maps are typical examples in the case study at hand (sections 16.5 and 16.7). The chain of models linking activities (section 16.3) to impact, which governs the noise annoyance problem, will be considered as an example of an environmental impact model that is extended with imperfection handling. The more interesting links in this chain are the propagation model (section 16.4) and the annoyance model (section 16.6). The propagation model relates emission at a finite number of locations to exposure over a wide area. The imperfection on exposure obtained using possibility and probability theory will be compared in section 16.4. This comparison forms the foundation for proposing the possibility approach implemented at the level of fundamental calculation (addition, multiplication, etc.) for future noise mapping enriched with a quality indication. Knowledge about the construct and prevalence of noise annoyance is so vague and uncertain that a specialized model is preferred over an extension of classical calculus. In section 16.6 a fuzzy rule based (FRB) model is proposed. Such models might be the only way that impact can be assessed in EIA without making unacceptable assumptions and approximations. The FRB has the additional advantage that it can handle the eminent imperfection in the exposure data. Vagueness in modeling results can be formulated in almost natural language, making them easy to compare to the results obtained through social surveys, which is about the only possibility for validating noise annoyance prediction. Section 16.8 discusses the results and draws the conclusions of this chapter.

## 16.2 General Methodology

### 16.2.1 Layered Environmental Impact Assessment Model

The complex task of modeling the impact of human action and activities on the quality of the (urban) living is best divided into small steps that are easier to model. A common model for environmental impact assessment is the DPSIR chain model (**D**rivingForces**P**ressure**S**tate**I**mpact**R**esponse) [16]. It is a layered model where the output of one layer is the input of the next. We will use this framework to model the noise annoyance (Figure 16.1). The response layer is not modeled in this contribution.

The first layer located at the bottom is the layer of the Driving Forces and represents all activities of interest for the noise annoyance modeling process. The focus here will be on road and rail traffic as the major contributors to urban environmental noise and annoyance. This layer is fed with raw data from spatial databases.

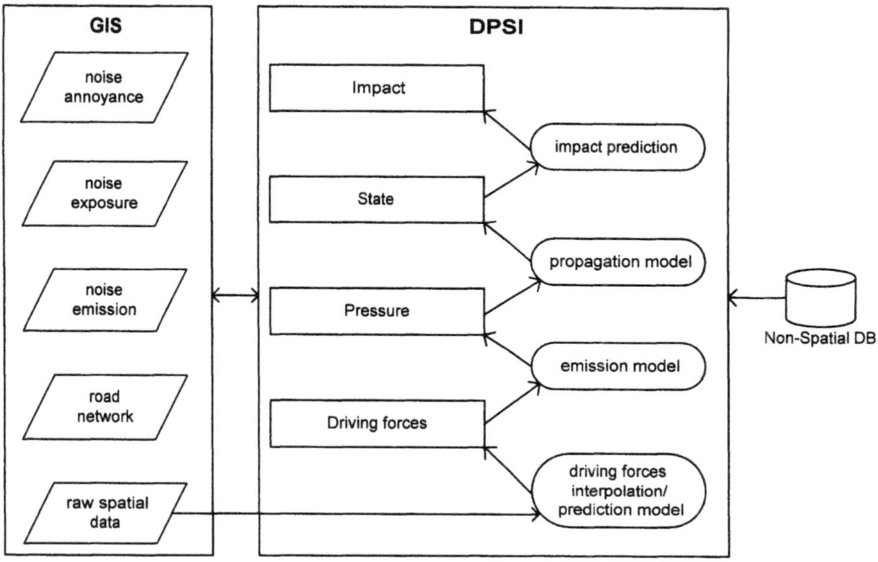


**Figure 16.1.** DPSI framework

The Pressure layer contains the emissions caused by the Driving Forces and represents the noise sources in the mapping process. Based on these emissions, noise exposure maps are computed and stored in the State layer.

With the use of the exposure in the State layer and additional spatial and non-spatial data the impact on inhabitants can be calculated and saved in the Impact layer.

The layered model lends itself naturally to be stored in a GIS. By using this general framework it is also straightforward to treat other environmental impacts using the approaches presented in this chapter. State layers of other exposure like air pollution can then be readily included in the noise annoyance impact modeling on the assessment of general quality of the living environment.

Here we have only presented the layers and not the models that compute the consecutive layers. These will also draw information from other domain specific databases. They will be discussed in more detail in sections 16.3 to 16.7.

### 16.2.2 Modeling and Reasoning

#### Imperfection

Imperfection of data and models pops up at several locations in the DPSI framework. Imperfection is a general term denoting that data or models are not ideal and show some deficiencies. We can distinguish between uncertainty, imprecision and inconsistency [17]. When data is not recorded accurately, then it has the property of imprecision. Uncertainty is related to the lack of knowledge. Inconsistency is a property of the result of the consultation of different data sources or the application of different models.

Imprecision in information is commonly modeled using probability theory and uncertainty by possibility theory. Information can however be tainted by both types of imperfection and a continuous spectrum exists between both extreme cases. Using transformations on the data, it is however possible to rely on a single framework to model imperfection.

A multiplicity of methods other then classical probability and possibility theory is also available. We will limit the discussion to the two theories mentioned above because they are the most common in environmental science. Other viable methods are (subjective) Bayesian probability, Logical theory, Sugeno's Fuzzy Measures, Upper and Lower probability models, Belief Models, ... [17].

To model imperfect numerical data the concept of a number is extended. In case of the probability approach a probability density function replaces the classical single value. For a possibilistic approach a possibility density function is used. The latter is closely related to a fuzzy set membership function, $\mu$, which explains the name *fuzzy numbers* [21]. Distributions can be described by their density function or by a set of $\alpha$-cuts. An $\alpha$-cut is defined as $A_\alpha = \{x \in X : \mu_A(x) \geq \alpha\}$.

Fuzzification of semantics and representation of numbers are illustrated in Figure 16.2.

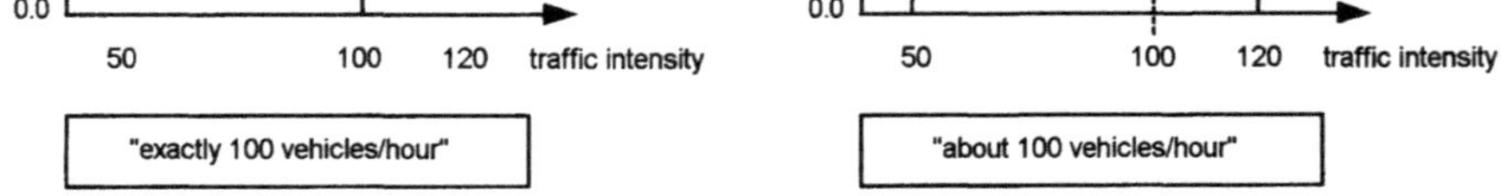


**Figure 16.2.** Fuzzification

## Arithmetic on Imperfect Numbers

To compute with imperfect data, arithmetic must be defined that can handle imperfect numbers. Each classical operator $(=, +, -, *, \div, \dots)$ used on the crisp numbers must be overloaded to operate on the extended numbers.

In the probability approach, the extended arithmetic can be defined in terms of the cumulative distribution:

$$\begin{aligned} F_{A \otimes B}(x) &= \int_{-\infty}^{+\infty} Prob[A \leq x \oslash y \wedge b < B \leq y + dy]dy \\ &= \int_{-\infty}^{+\infty} F_A(x \oslash y) f_B(y) dy \end{aligned} \tag{16.1}$$

where $\otimes$ represents a mathematical binary operator and $\oslash$ its inverse operator.

For fuzzy numbers the extension principle by Zadeh can be used to extend mathematical relation between crisp numbers to the corresponding possibility distributions [22]:

$$\mu_{A \otimes B}(x) \equiv \mu_C(x) = \sup_y \min \{\mu_A(x \oslash y), \mu_B(y)\} \tag{16.2}$$

For both frameworks unary operators are defined in an analogue way.

When spatial data is used for environmental impact assessment, the amount of data and the number of operations to be performed on them is often huge. It is therefore imperative to reduce storage requirements and optimize implementation of arithmetic on imperfect numbers. The choice of number representation is crucial:

- **Parameterized Analytical Functions**. Each distribution is modeled by a small set of parameters. All operations on the imperfect numbers must be redefined in terms of these parameters. An example are Gaussian distributions defined by their mean and variance. This is a storage efficient approach but the results of some operations can be hard or impossible to approximate by a member of the chosen family of functions. The pointwise maximum of two Gaussian distributions with different mean is such an operation.
- **Piecewise Linear Functions**. Distributions are approximated by a piecewise linear curve and all operations are performed in terms of the control points defining the curve. Depending on the number of control points a targeted precision can be achieved. This precision also relates directly to the computational cost and the amount of storage required. Simplifications can be made when convex curves are used, especially in the case of fuzzy numbers where the arithmetic collapses to interval arithmetic on the $\alpha$-cuts [14]. The cost of computation is then very low.
- **Sampled Distributions**. The distributions are stored in a sampled form. Arithmetic is performed on samples taken from the distributions of the arguments and a new sampled distribution for the result is built. Although this method is conceptually very simple it is computationally very expensive and of little practical use except for verificational purposes.

### Natural Language and Rule Bases

When a model is backed by some physical, mechanical or electromagnetical laws such as propagation models, calculation procedures are usually well known. If we are dealing with a "humanistic system", a system whose behavior is determined by human judgement, perception or emotions [19], modeling becomes much harder. In these cases, knowledge about the relations between the variables is often vague. Even experts in the field can only give vague descriptions of some assumed relations. A framework that allows to exploit this type of knowledge is a *fuzzy rule base*.

A fuzzy rule base contains a collection of IF-THEN rules. Each rule relates some antecedent to a consequent rule and can be written in the general form "IF $X$ is $A$ THEN $Y$ is $B$". $X$ and $Y$ are variables over the universes $U$ and $V$ respectively, and $A$ and $B$ are fuzzy sets over $U$ and $V$. By assigning a linguistic term to all the involved fuzzy sets, each rule can be expressed verbally. This makes the rules easy to interpret by humans. It also allows experts to formulate their knowledge in almost natural language, which is of course a far more easier approach to deal with imperfect knowledge.

Fuzzy logic provides the necessary mathematical tools to infer a conclusion based on a fuzzy rule and vague input. As a primary inference engine, the generalized modus ponens is used, where $B'$, the inferred possibility distribution of $Y$, is calculated with the compositional rule of inference [20],

$$(\forall v \in V) \left( B'(v) = \sup_{u \in U} \min \left( A'(u), R(u,v) \right) \right) \tag{16.3}$$

where $R$ is the representation of the rule and $A'$ is the fuzzy input given to the rule.

Depending on the intended semantics of the rule, $R$ should be modeled with appropriate operators [6]. For rules formulated with a possibility interpretation in mind, the antecedent is required to make the consequent possible, $R$ should be modeled using a t-norm (as an extension to the classical conjunction [22]). In particular the minimum norm is most frequently used.

Usually, a fuzzy rule based system models a single output variable for which a fuzzy value $H$ is required. Yet, the collection of rules generally involves many different input variables and also contains multiple rules describing the relation between one input variable and the desired output variable. The results inferred by rules articulated between the same variables are aggregated with the maximum conorm (as an extension to the classical disjunction [22]). This is required when a t-norm is used to model the rules. The aggregated result, $H_n$, can be seen as an estimate for the outcome of the fuzzy rule base model $H$, based solely on that input variable. Since each $H_n$ can be interpreted as a possible range for $H$, the intersection of all $H_n$ is used as the best estimate for $H$. As a model for the intersection, the product t-norm is used. Compared to the minimum t-norm, a product conserves more of the information in each of the individual sets. Alternatively, every $H_n$ can also be regarded as an expert opinion based on the input data of a particular variable. The final result is a consensus between experts. Consensus can be implemented using an aggregation operator. As a triangular norm is also considered to be an aggregator, again the product of membership degrees qualifies as a valid operation. The fuzzy set outcome of the rule base is thus obtained as $H(u) = \prod_n H_n(u)$.

Of course, not all rules will have equal impact on the final result. Therefore, every rule is assigned a sufficiency degree $\lambda \in [0, 1]$. This degree expresses to what extent it is sufficient that the antecedent is true for having the consequent true. Actually, this sufficiency degree will be applied to the rule consequent instead of to the rule itself. Although both interpretations only coincide when the antecedent is not fuzzy, this is common practice. In [5], a sufficiency qualified proposition "$Y = B$ is $\lambda$-sufficient" is linked to a necessity measure, given "$Y = B^*$"

$$N(B) = \inf_{v \in V} \mathcal{I}\,(B^*(v), B(v)) \geq \lambda \tag{16.4}$$

where $\mathcal{I}$ is an implicator [22]. Solving this inequality leads to the underlying fuzzy set $B^*$. Choosing for $\mathcal{I}$ the reciprocal of a residual implication results in

$$(\forall v \in V)(B^*(v) \leq \mathcal{I}_S(\lambda, B(v))) \tag{16.5}$$

with $\mathcal{I}_S$ an S-implicator. More specific, when $\mathcal{I}$ is the reciprocal of the Gödel implicator, the Kleene-Dienes $\mathcal{I}_S$ implicator has to be used

$$B^*(v) = \max(1 - \lambda, B(v)) \tag{16.6}$$

As can be seen from Equation 16.6, the consequent of a rule with $\lambda = 1$ will remain unchanged, having full impact as intended. On the other hand, a rule with sufficiency $\lambda = 0$ will not have any effect because the consequent will have possibility 1 over the whole universe (concluding from the antecedent that everything is possible and thus providing no additional information).

## 16.3 Fuzzifying GIS Data: the Activities and Pressure Layer

Spatial data is imperfect for several reasons. Firstly, the data stored in GIS is obtained with finite precision. In general, the quality of the data will be assessed at the level of the complete GIS-layer. For further processing this imprecision has to be represented at the level of the individual GIS objects. Secondly, data may be missing in some areas or at some level of detail (e.g. traffic intensity on smaller streets). In this case rough estimates may be used to fill the gaps, obviously introducing high levels of uncertainty. Finally, preprocessing of the raw data may involve imprecise non-spatial information.

This fuzzification of spatial data is illustrated below for the special case of noise mapping.

### 16.3.1 Noise Producing Activities

Road traffic is the most dominant noise producing activity in many urban areas. Calculating noise emission caused by traffic, requires detailed knowledge on traffic volume, vehicle speed, and vehicle fleet composition. Traffic intensity is counted on a permanent basis at a limited number of sites. Automatic intensity counts generate relatively large errors leading to imprecision in the traffic volume data. Only very rough statistical data on vehicle speed is stored by some of the counting stations and some indication may be extracted on the percentage of heavy vehicles in the traffic flow. Detailed fleet composition could be estimated in more detail from country-wide statistics.

For roads where no traffic intensity counts and vehicle speed distributions are available, one has to rely on traffic models to predict the traffic flow. These models work at different spatial resolutions and provide different levels of accuracy depending on the size of the roads. For highway traffic the region of interest is transformed into a graph. Each vertex of the topological graph represents a large city and is assigned a given traffic demand and generation. The model used is calibrated using the available intensity counts. Imprecision emerging from traffic counts and from traffic models is aggregated and stored in imprecise traffic intensity, vehicle fleet and vehicle speed.

Traffic on local roads is very rarely counted, so it is estimated using statistics for the whole city, suburb or rural village and distinguishes between a few categories of roads only. Vehicle speed distribution is roughly estimated on the basis of road category and speed limit. This procedure introduces a high level of uncertainty in the traffic attributed to local roads. The final result of the traffic interpolation and estimation is a GIS layer which contains road segments, each with their own traffic intensity, vehicle fleet and speed distribution, stored as fuzzy numbers. Fortunately, in general, imprecision and uncertainty are smaller on large roads with high intensity than on smaller roads.

### 16.3.2 Spatial Emission Data

Based on the activities stored in the Driving Forces layer, the pressure variable is computed. Again focussing on road traffic, noise emission is calculated for every road segment. A standardized emission model [13] which gives the noise emission for one vehicle of a certain type passing by at a given speed over a given road surface, is used. This emission model is assumed to be perfect as it is a fixed standard, but fuzzy arithmetic is used to propagate the imperfection from the activity data to the emission.

The result is a GIS layer containing fuzzy emission values. The noise emission of road vehicles under given driving conditions is sound frequency dependent. Usually 8 to 10 frequency bands are used to describe this frequency dependency. Hence the fuzzy noise emission value is a set of 8 to 10 fuzzy numbers. Figure 16.3 shows an example of the emission associated with a road segment. The X-axis shows the frequency band, the Y-axis the emission level and the shade of grey represents the likeliness of the value.

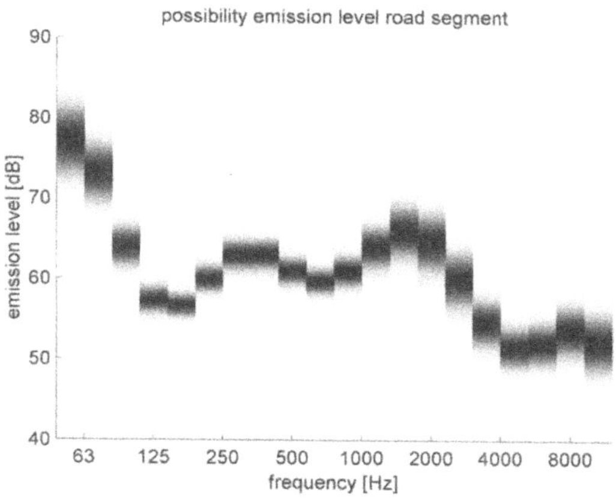


**Figure 16.3.** Fuzzy Emission

## 16.4 Fuzzy Models Operating on Spatial Data

Propagation models are generally speaking models distributing the emission of a spatially located source through the region of interest. In the application studied in this chapter, the sound (noise) of a point source is distributed over three dimensional space. During this propagation the sound will interact with the environment. The environment for noise mapping consists of buildings, noise barriers, terrain and air. Full three dimensional data for buildings is not available for large regions so a projected 2.5 dimensional version is used: the two dimensions originate from the planar view and the half dimension from the fact that the height and floor level of buildings is directly stored in the planar geometric types. Hence the third dimension is not fully developed.

Acoustical properties such as reflection coefficients are needed but difficult to sample at the large scale required for regional noise mapping. Fuzzy numbers are used to store an estimate of the reflection coefficient made by expert in the field. Position and height of buildings, and in general geometric information, is not fuzzified because the propagation model, which searches for paths between the sources and the receivers, can only cope with exact geometric data. This allows us to use a regular GIS after tables have been augmented with fields capable of storing imperfect attributes.

### 16.4.1 Meta Modeling

The model that results from introducing the possibility to handle imperfection in a classical crisp model, will be referred to as the meta model. The introduction and analysis of imperfection can be done at several levels, resulting in different types of meta models.

### Macro Meta Model

A macro meta model is a meta model that uses the original crisp model as a sub model. A shell is built around the original model while the original model is left untouched. The imperfect inputs of the meta model are sampled according to the distribution representing imperfection and the sample is fed to the original crisp model. This process is repeated many times and the output corresponding to each sample is gathered. The distribution for every output variable is constructed and translated to the required representation.

### Micro Meta Model

A micro model results from replacing all basic crisp computation blocks by their imperfect counterpart. The original model is altered at source code level. The imperfect data are directly used and propagated through the micro meta model and result in an imperfect output.

## 16.4.2 Experiment

In the process of obtaining an impact layer (noise map) we have identified two flavors of imperfection in data and models: uncertainty and imprecision. From a theoretical point of view possibility theory is best suited to handle uncertainty while probability theory is a more natural framework for working with imprecision. From a practical point of view one of these frameworks will have to be selected and implemented. Since one of the flavors of imperfection will have to be handled in a context that is less suited for it anyhow, this gives us some freedom of choice. The model for imperfection and the meta level are independent choices that can be represented on orthogonal axes in a two-dimensional design space. Figure 16.4 shows this design space. A probabilistic approach implemented at macro level and a possibilistic approach implemented at micro level are compared [4].

For implementing the macro probabilistic approach Monte Carlo sampling of the input variables is used. A large number of simulations is required to obtain an accurately sampled output distribution. This number depends exponentially on the number of input variables.

The possibilistic approach at micro level is much faster, requiring only one run of the model with the modified basic computational blocks. Fuzzy numbers with 8 control points have been used resulting in an overall slowdown of the software by a factor 3. The expected factor 8 is not found because not all processing time is spent in the basic computation blocks. Moreover, current micro processors deal very well with computations on data that are close together in memory, such as series of numbers.

Input imperfection is converted between frameworks using uncertainty preserving techniques [15, 7]. Experiments using basic operations such as multiplication and addition have indicated that care must be taken when interpreting the results because imperfection grows at a different rate in the two frameworks [4]. Output uncertainty obtained by two implementations can only be compared qualitatively. Uncertainty obtained by both models is high in the same areas of the map indicating zones where noise levels are not obtained in an accurate way because of model or input data imperfection.

Another important advantage of the fuzzy approach is that it can handle linguistic information such as "the intensity will be lower than about 100". This translates directly into a possibility distribution while no probability distribution with the intensity as variable can be found representing the same knowledge. A probability based interpretation tends to attribute more information to the density function than is actually available. Using a rectangular density function from 0 to 100 means that all values are equally probable which may not be the case. Data obtained from estimates made by experts in the field are more fuzzy in nature which make this approach an especially suitable framework.

Two combination types have not been included in the experiment: a macro possibilistic approach and a micro probabilistic approach. The first technique can only simulate monotonic transformation from input to output in an efficient way while the noise mapping application is highly non-monotonic.

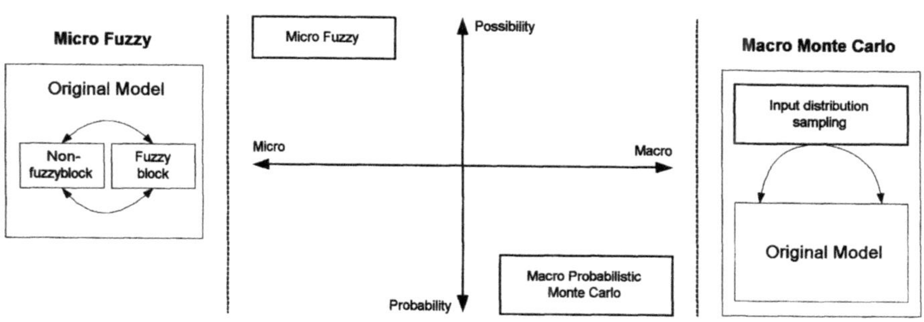


**Figure 16.4.** Design space imperfection modeling

### 16.4.3 Side Products

An interesting feature of including model imperfection and data imperfection in the calculation is that it can be used to speed up computation. Consider a sound source that is located far from the observer and this will contribute little to the local noise level. Computing the accurate attenuation from source to receiver is very costly. Instead, one would like to approximate the attenuation to a given accuracy.

A different, less detailed, model is then used generating the attenuation with larger imperfection but a few times faster. The framework can however directly take care of this imperfection and propagate it up to the final exposure map. When the final result of the noise map does not meet the desired accuracy, the more detailed models can be used.

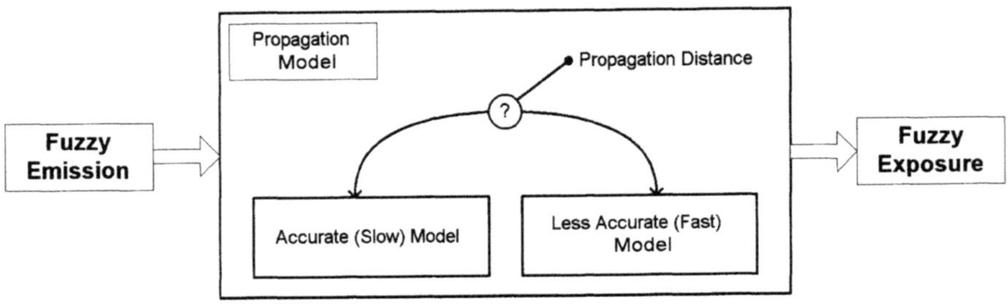


**Figure 16.5.** Accelerated computation with transparent sub modeling

## 16.5 Representing Imperfection in Spatial Exposure Data

Spatial exposure data is a first product of impact mapping. For many policy support applications it is even the final product because it is this map that is used for verifying limits, guidelines, and regulations. Specifically for noise mapping, the environmental noise directive issued by the European Committee [11] requires such spatial noise exposure maps to be built for big cities and large infrastructure. Action plans are often triggered and evaluated by predicted noise levels in a given region.

Traditionally, noise maps give no indication of the accuracy of the noise levels presented they represent. There is however a growing concern about the quality of noise maps. It is generally recognized that this quality changes over the simulation region. Using a fuzzy framework a lot of information is obtained about the imperfection of the result that is shown in the map. This information is stored in a spatial grid of fuzzy numbers. Classical color scales do not allow to visualize these two dimensional data. Choices have to be made.

The value that is displayed in the map depends on the application. A map showing the "most likely noise level" corresponds most closely to classical noise maps. The most likely value can be identified to the center of gravity of the fuzzy set, the center of its core, the center of its support, etc. Figure 16.6 shows an example. In addition, map quality can be included in an accompanying map giving the imperfection in the calculated fuzzy noise levels. Several measures of imprecision, nonspecificity, and uncertainty carried by a fuzzy number are available. Here the ratio of the supremum over the infinum of the 0.5 $\alpha$-cut is used because it behaves well under the logarithm operator used to obtain results in dB. After conversion into the dB scale it represents the difference between the supremum and infinum of the 0.5 $\alpha$-cut. Figure 16.7 shows the uncertainty on the most likely values in Figure 16.6. Highlighted regions indicate that either the input data is insufficient to accurately compute the noise level or the models are not fully applicable. In the region shown, there are important access roads with a high traffic intensity in the center. The noise exposure is high but the uncertainty low. In areas with more houses (shown as black blocks) and more local traffic, lower levels but higher uncertainty is observed.

**Figure 16.6.** Most likely noise exposure level

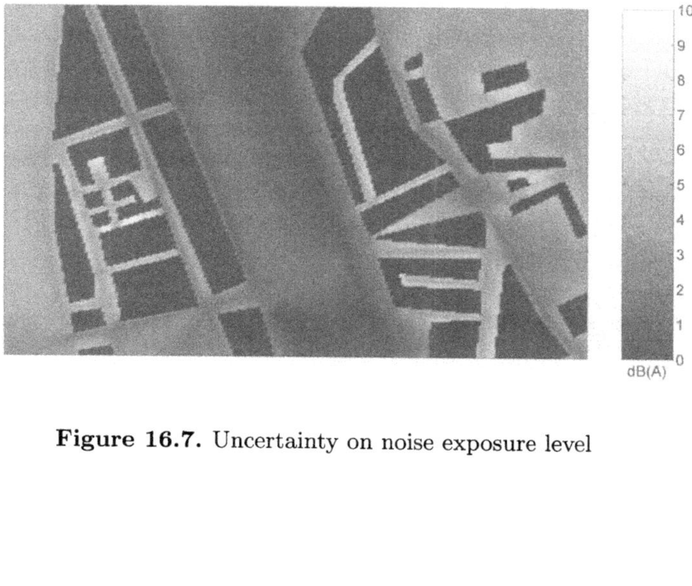


**Figure 16.7.** Uncertainty on noise exposure level

In other applications other crisp quantities may be extracted from the fuzzy numbers. If the purpose of the map is to verify upper noise limits, one may wish to have some certainty about the fulfillment of the requirement. The map could then show a value that is chosen in such a way that the possibility for the crisp but unknown exact noise level to lie above this limit is less than say 0.1. In the same way, when the goal is to find regions that are possibly quiet, so called "quiet areas", then one may wish to visualize a value for which the possibility that the exact noise level is lower, is less than say 0.1.

If we are only interested in strict regulations for granting building permissions for instance, the noise map showing the highest noise level in the support of the noise level fuzzy set is an interesting product. This map can be contrasted to regulations and problem areas can be identified. The input can be rough pessimistic estimates of traffic intensity, facade and ground reflection. If under these rough estimates noise regulations are met, then there is no need to refine input data and models. Otherwise the detailed fuzzy model has to be used.

Noise exposure maps are often used in public documents. Lay persons may not be able to relate the levels in dB to an actual loudness of noise and as such these noise maps are not easy to interpret. Based on scientific evidence, noise specialists may indicate what noise levels are unhealthy or unacceptable. In [1] it is argued that this knowledge can best be described by a set of two noise limits, a sustainable and an unacceptable limit, that are vague by nature. Vagueness emerges from imperfect knowledge on dosage-effect relationships but also from using concepts such as *acceptable effect*, *sustainability*, *sharing the burden amongst citizens in a just and fair way*, etc. Figure 16.8 shows a set of *fuzzy noise limits* together with the calculated exposure level. Evaluation is based on set containment. If the noise immission fuzzy set is within the unacceptable fuzzy limit, the situation is evaluated as bad and may be colored *red* in the maps. If the set lies within the sustainable limit, it is evaluated good and may be colored *green*. Partial containment may be expressed by varying the intensity of the color. For black and white presentation, containment in sustainable and unacceptable limit are shown in separate Figures 16.9 and 16.10. Grey areas show where the evaluation is not certain. Note that towards the left and right border of the figure, the evaluation is definitely *good* although the uncertainty on calculated noise levels is rather large. This results from the broad noise level fuzzy set being completely contained in the sustainable limit however.

## 16.6 Perception Modeling with Fuzzy Rule Bases

### 16.6.1 Noise Annoyance Modeling

Based on noise exposure maps, one can start modeling the impact on man. The degree of noise annoyance that people experience is generally considered

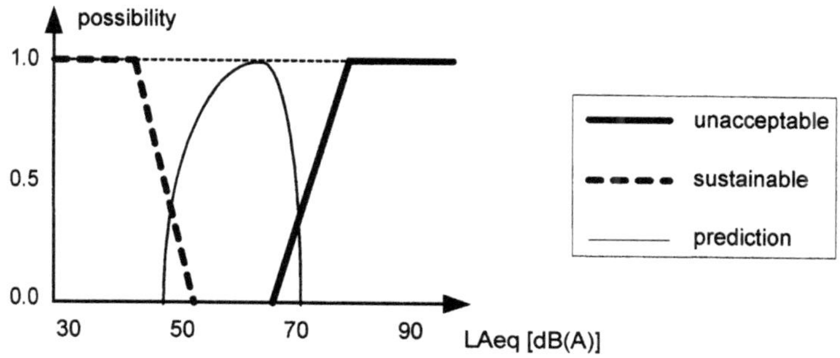


**Figure 16.8.** Fuzzy noise limit

**Figure 16.9.** Fuzzy noise limit map, sustainable

as the most important impact indicator. However, this modeling task is rather difficult because the construction of annoyance is a very complex psychological process. Not only noise exposure but also many other factors influence the way a person perceives noise in his or her environment. This whole environment in which noise occurs has been named the soundscape. Some of these factors are person related, e.g. age, gender, size of household, while other factors are non-acoustical features of the physical environment, e.g. land use. An overview of existing literature with variables that are known or assumed to have a direct or an indirect influence on noise annoyance has been compiled in [2].

Hence, difficulties arise in two areas. Firstly, an inherently vague concept such as noise annoyance has to be modeled. Secondly, the available knowledge is also vague and uncertain. A fuzzy rule based system which allows to incorporate knowledge expressed in almost natural language is a very appropriate way to handle such a modeling task.

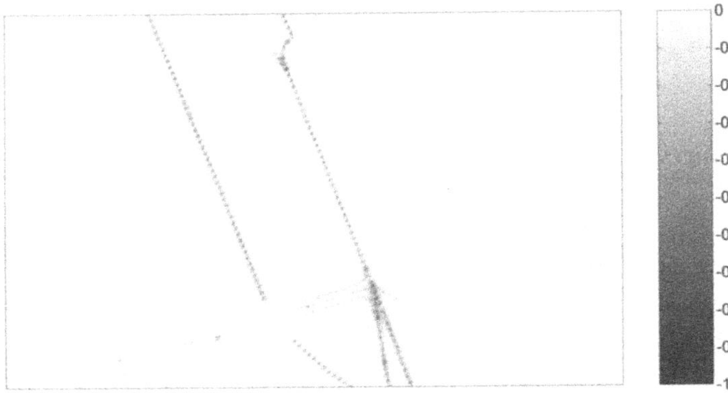


**Figure 16.10.** Fuzzy noise limit map, unacceptable

### 16.6.2 Technical Infrastructure

The technical infrastructure of the noise annoyance model is depicted in Figure 16.11.

Input data stems from two sources. A GIS is used to retrieve data that is available as maps of the area of interest. This is also the place where the fuzzy noise exposure map is stored. Additional variables that are relevant for noise annoyance modeling are collected through social surveys. Although these data could also be stored in a spatial database at the location of the dwelling of the survey participants, this data is mainly stored in a non-spatial database containing one record per person. Typically, such surveys also include a question that asks for the level of annoyance experienced by the respondent, providing a 4 or 5-point linguistical scale to choose from. This information is used to tune the model.

The knowledge base is the component where the domain-specific knowledge is stored. It is composed of a database and a rule base. The database contains the fuzzy set definitions of the linguistic terms that are used in the fuzzy rules. These definitions can be extracted from observations using fuzzy clustering or they can be forwarded by experts. Alternatively, they can be derived from studies specifically aimed at finding the meaning of words in a specified context. Such a study has been done internationally for the description of noise annoyance terms [9], fuzzy representations for these terms have been constructed in [18]. In general, the membership functions in the database are defined by piecewise linear functions or by more fluent sigmoidal and bell-shaped functions. The second part of the knowledge base, the rule base, contains the relationships between variables expressed as IF-THEN rules in natural language. A typical set of rules is shown in Table 16.1. Although all rules are stored in the rule base, the annoyance model can still be hierarchi-

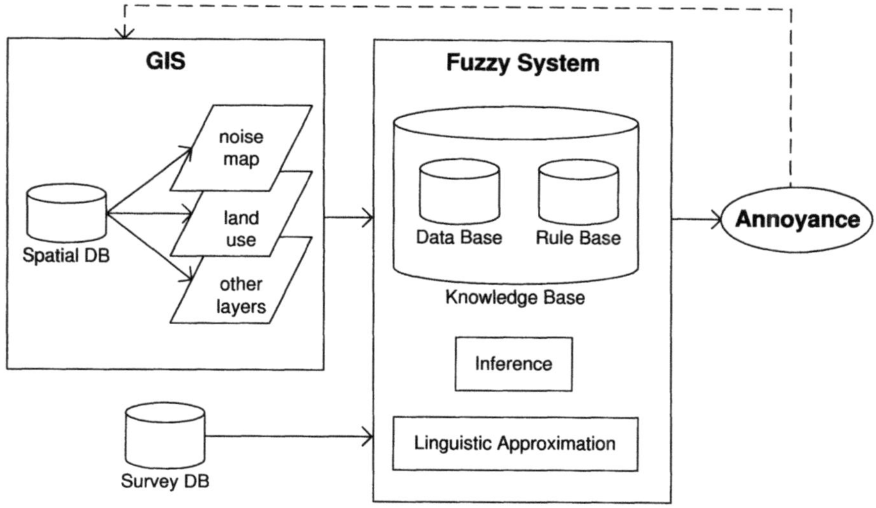


**Figure 16.11.** Fuzzy rule based system

cally structured. This allows the use of submodels to calculate intermediate variables. If these variables are themselves measurable (e.g. through a social survey) the submodels can be tuned separately, improving model quality. As an example of such submodel, consider the urbanization rules in Table 16.1. They allow a prediction of the degree of urbanization (country-side, suburb, city or city center) based on more objective variables: the population density and percentage of built-up and agriculture land use, which can be derived from appropriate maps in a GIS system. Also for the amount of traffic, a submodel based on traffic simulation data can be constructed.

The inference engine implements the compositional rule of inference to obtain the level of noise annoyance $H$, experienced by an individual. The fuzzy set $H$ can be shown in maps just as the noise exposure level. To be more in line with human evaluation we prefer however to express the level of annoyance using natural language. The outcome of the model, $H$, is compared with the set of annoyance terms $U_L$ that are stored in the database using a similarity measure. Let $L_i$ denote the membership function of the $i$-th label. The label $i$ will be considered a possible description of the annoyance of the subject if $L_i$ is contained within $H$. As the containment of two fuzzy sets is interpreted in a fuzzy way, the degree of containment [8], this procedure results in a possibility distribution over the available annoyance terms that can be calculated as

$$\pi_{U_L}(i) = L_i \subseteq H = \min_{u \in U}(\max(1 - L_i(u), H(u))) \qquad (16.7)$$

for each label $i$.

**Measures of Model Quality**

The model can be said to predict for each of the available linguistic labels the possibility that it is suitable for describing the noise annoyance situation of an individual. With this interpretation, the model is correct when the actual label word to report annoyance belongs to the fuzzy set of predicted labels. To evaluate model quality a fuzzy extension of false negative is used. It expresses to which degree the reported label $t$ is not considered as a possible outcome by the fuzzy rule base. Hence, it can easily be calculated as $1 - \pi_{U_L}(t)$.

Using only this false negative measure would favor a system that is indecisive; not excluding any label would be a good strategy. To make the system more useful, indecisiveness should be included as a measure of poor quality. This can be done by calculating the non-specificity of $\pi_{U_L}$ [12],

$$NS(\pi_{U_L}) = \sum_{i=2}^{N} \pi_{U_L}(i) log_2 \left( \frac{i}{i-1} \right) \tag{16.8}$$

where $\pi_{U_L}$ has been put in decreasing order.

### 16.6.3 Tuning

The model is tuned by modifying the sufficiency of every rule in such a way as to minimize a suitable prediction error on a database obtained from a social survey combined with spatial data. This optimization problem is solved using a genetic algorithm (GA) [10] because of the highly multimodal and non-continuous search space. Every individual in the population evolved by the GA is in this case an instance of the model and is completely represented by a string of real values that is formed by the sufficiency of every rule. The fitness of every individual —used by the GA in its imitation of natural selection— is the inverse of the prediction error of its model on the database.

The prediction error $e$, is defined as a combination of the two quality measures that have been described above [3],

$$e = \sum_{k=1}^{N} \frac{(1 - \pi_{U_L,k}(t_k))^2 + \alpha^2 (NS(\pi_{U_L,k}))^2}{f(t_k)} \ , \tag{16.9}$$

in which the index $k$ runs over all survey records and $f(t_k)$ is the frequency of occurrence of annoyance label $t_k$. Weighting with $f(t_k)$ is required to obtain a model that performs equally well for high annoyance levels that occur less often in the training set. The weight $\alpha$ is introduced to assign more importance to the non-specificity or to the false negative measures of the model, where the price to pay for decisiveness or specificity is more frequent false negatives.

The result after performing this tuning process for a road traffic noise annoyance model based on a survey conducted in Flanders is shown in Figure 16.12. The survey was part of the Investigation of the Environmental

Living Quality performed on behalf of the Flemish Environmental Administration (AMINAL) by the companies Deloitte & Touche and M.A.S. In the survey as well as in the model, five different annoyance labels have been used. A random sample of 2472 individuals was included in the test database.

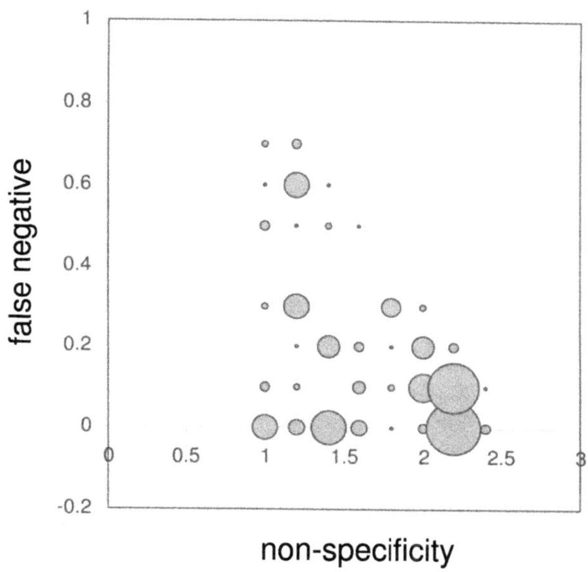


**Figure 16.12.** Distribution of the subjects over a false negative versus non-specificity plane that is tuned with $\alpha = 0.5$. The area of the bubbles is proportional to the number of subjects.

## 16.7 Representation of Perception Results using GIS

Finally, the results of the annoyance model can be visualized in a map. The result is the possibility distribution over the available annoyance terms. The non-specificity variable which is an indication of the uncertainty of the outcome, may already have been calculated for this fuzzy set.

Depending on the purpose of the map, these model results can be shown in several ways. One approach could be to create one map for every annoyance term and visualize its possibility degree by varying the color intensity. Although such maps would show all available data, it would be hard to interpret them. We suggest to use this representation only if colored surface maps are required.

Alternatively a map which depicts the possibility distributions in a more condensed form can be created. Of course, it is more difficult to represent a possibility distribution over five labels in a single point. A feasible solution is the use of a pie-chart as a mark. The greater the possibility of a label, the larger the piece of the pie that is assigned to it. In this case, the uncertainty is implicitly shown. If one label receives a dominant part of the pie, then the uncertainty will be small. However, if the pie is equally distributed among the labels, then there is high uncertainty about the result. An example of such a map is shown in Figure 16.13. It is evident that the more closely located to a large road, the larger fraction of the pie-chart is filled in black, indicating extreme annoyance.

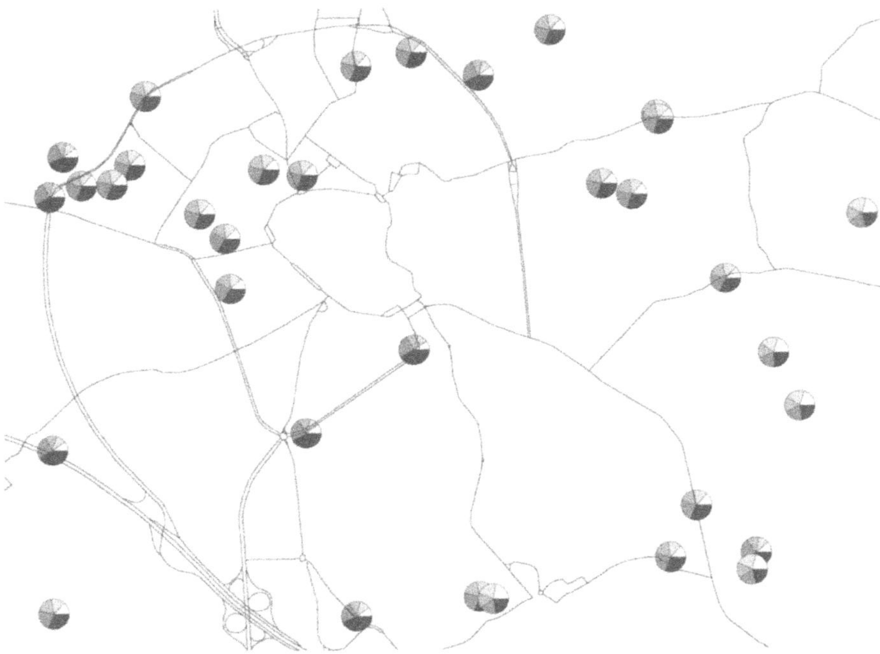

**Figure 16.13.** Map with possibility distribution represented as a pie-chart. White indicates the lowest level of annoyance ("not at all annoyed") while black indicates the highest level of annoyance ("extremely annoyed").

Another option is to visualize only the most plausible label, i.e. the annoyance term with the highest possibility. However, the model tends to peak only in extreme labels because it is far more easier for an expert to formulate clear rules for these extreme categories of annoyance. Hence, a significant amount of information on the almost equal plausibility of middle degrees of annoyance gets lost in this type of map. One way to compensate for this loss of information is to include the non-specificity as an indication of the uncertainty of the result, e.g. by varying the size of the mark. The resulting map of the same region is shown in Figure 16.14. Again, note the large black (certainly extremely annoyed) and occasionally small white points located near large roads.

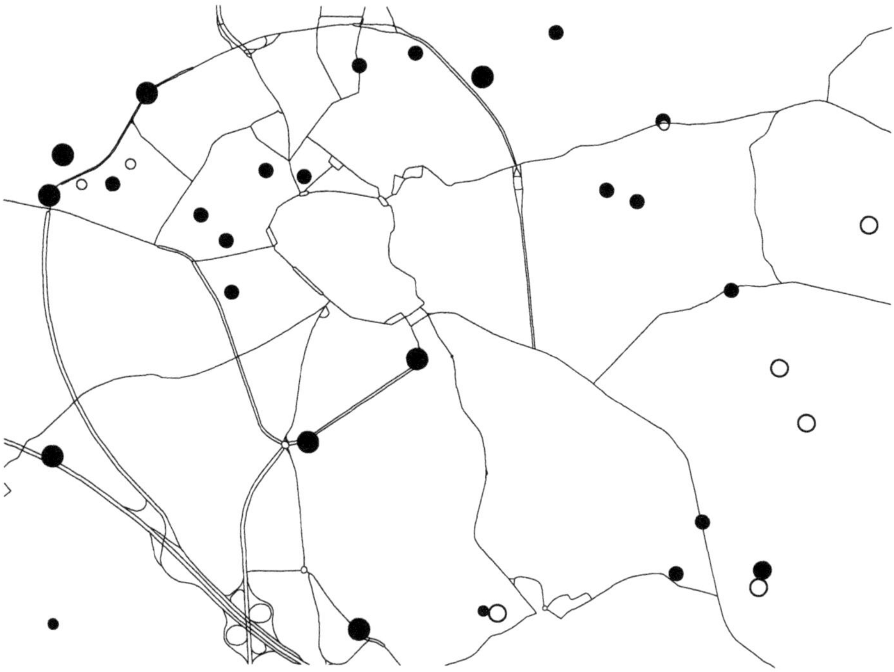

**Figure 16.14.** Map with most plausible label and the associated non-specificity. White indicates the lowest level of annoyance ("not at all annoyed") while black indicates the highest level of annoyance ("extremely annoyed"). The smaller the mark, the more uncertain.

## 16.8 Discussion and Conclusions

This chapter has demonstrated how fuzzy set theory, fuzzy number calculus, and fuzzy systems can be used to handle data imperfection, model imperfection and concept vagueness often encountered in the environmental impact assessment process. Different types of data imperfection are identified and representations that are optimally suited for working with these types of imperfection are presented. For the particular case of noise annoyance, a model, well rooted in the general DPSI(R) framework, is adapted to work with these imperfect data. It was shown that fuzzy number calculus (possibility theory), implemented at the lowest level of calculation (the so called microscopic level) yields qualitatively similar degrees of uncertainty in the resulting noise maps as a probability based model encapsulating the classical model it is based upon. An application based on fuzzy number calculus at microscopic level typically runs several orders of magnitude faster than its probability based macroscopic cousin, a feature that is widely appreciated in large geographic systems. Although possibility and probability both incorporate a different flavor of imperfection by nature, it can therefore be concluded that for practical purposes a fuzzy number based microscopic model is preferable. Vagueness, typically encountered in impact prediction and assessment, quite automatically lends itself to be handled using fuzzy sets. In this chapter we have shown how vague and uncertain knowledge about the construct and occurrence of perceptions such as noise annoyance can be molded into a fuzzy rule based system that is easily joined to the fuzzy noise map (considered here as an example of a state-layer) thus creating an impact-layer in the EIA-GIS. Regardless of the fact whether an impact-layer is constructed or not, the scientific knowledge on the effects together with vague society preferences and values such as sharing the burden in a fair and just way, can give rise to fuzzy limits. Contrasting these limits to uncertain impacts using fuzzy containment allows to present the importance of environmental impacts such as noise in a way that is easy to understand by the population at large. This chapter also gave some thought to presenting the fuzzy information in policy decision supporting maps. The obvious option of extracting a crisp number from the fuzzy sets, can be used for mapping continuous state-variables (e.g. noise levels in dB) but this is not a very creative solution. Maps using color shading (presented in black and white in this book) where color is used to represent a fact (e.g. exceeding a noise limit) and shade is used to indicate the certainty obtained by the model allow to include more information. Mapping the results of a perception based impact obtained on an ordinal scale (using linguistic labels) is even more difficult. Small pie-charts and bubbles are proposed for situations where results are available at a limited number of dwellings only.

## References

1. Botteldooren D. (2003) Fuzzy Noise Limits : A Fuzzy Set Theoretical Approach. Noise Control Engineering (to be published).
2. Botteldooren D., Verkeyn A., Lercher P. (2002) Noise annoyance modeling using fuzzy rule based systems. Journal of Noise and Health 15(4):1463–1741.
3. Botteldooren D., Verkeyn A., Lercher P. (2003) A Fuzzy Rule Based Framework for Noise Annoyance Modelling. Journal of the Acoustical Society of America 114(3):1487–1498.
4. De Muer T., Botteldooren D. (2003) Uncertainty in Noise Mapping: Comparing a Probabilistic and a Fuzzy Approach. Lecture Notes in Artificial Intelligence 2715.
5. Dubois D., Prade H. (1991) Fuzzy sets in approximate reasoning, part I: Inference with possibility distributions. Fuzzy sets and systems 40:143–202.
6. Dubois D., Prade H. (1996) What are fuzzy rules and how to use them. Fuzzy sets and systems 84:169–185.

7. Dubois D., Prade H., Sandri S. (1993) On Possibility/Probability Transformations. In: Lowen R., Roubens M. (Eds.) Fuzzy Logic, Kluwer Academic Publishers, Boston, pp. 103–112.
8. Dubois D., Foulloy L., Galichet S., Prade H. (1999) Performing approximate reasoning with words. In: Zadeh L.A., Kacprzyk J. (Eds.) Computing with words in information/intelligent systems I: foundations, Springer-Verlag, Studies in Fuzziness and Soft Computing.
9. Fields J.M., De Jong R.G., Gjestland T., Flindell I.H., Job R.F.S., Kurra S., Lercher P., Vallet M., Yano T., Guski R., Felscher-Suhr U., Schumer R. (2001) Standardized general-purpose noise reaction questions for community noise surveys: research and a recommendation. Journal of Sound and Vibration 242(4):641–679.
10. Fogel D.B. (2000) Evolutionary computating: Towards a new philosophy of machine intelligence. 2nd edition, IEEE Press.
11. European Committee (1996) Future noise policy. Green Paper, Brussels 1996; COM(96)540def, http://europa.eu.int/en/record/green/gp9611/noise.htm
12. Higashi M., Klir G. (1983) Measures of uncertainty and information based on possibility distributions. International Journal of General Systems 9(1):43–58.
13. Jonasson H.G., Storeheier S. (2001) Nord2000. New Nordic Prediction Method for Road Traffic Noise. Technical report, SP, SINTEF, SP Rapport 2001:10.
14. Kerre E.E. (1993) Introduction to the Basic Principles of Fuzzy Set Theory and Some of its Applications. Communication and Cognition, Ghent.
15. Medasani S., Kim J., Krishnaparum R. (1998) An overview of membership function generation techniques for pattern recognition. International Journal of Approximate Reasoning 19:391–417.
16. OECD core set of indicators for environmental performance reviews. (1993) Environment Monographs 83.
17. Smets P. (1996) Imperfect Information: Imprecision–Uncertainty. In: Uncertainty Management in Information Systems: From Needs to Solution, Kluwer Academic Publishers, Boston.
18. Verkeyn A., De Cock M., Botteldooren D., Kerre E.E. (2003) Generating membership functions for a noise annoyance model from experimental data. In: Reznik L., Kreinovich V. (Eds.) Soft computing in measurement and information acquisition, Springer-Verlag, Studies in Fuzziness and Soft Computing.
19. Zadeh L.A. (1975) The concept of a linguistic variable and its application to approximate reasoning I. Information Sciences 8:199–249.
20. Zadeh L.A. (1975) The concept of a linguistic variable and its application to approximate reasoning III. Information Sciences 9:43–80.
21. Zadeh L.A. (1978) Fuzzy sets as a basis for theory of possibility. Fuzzy Sets and Systems 1:3–28.
22. Zimmermann H.-J. (1996)Fuzzy Set Theory - And its applications. 3rd edition, Kluwer Academic Publishers, Boston.

**Table 16.1.** Overview of a typical set of rules. Used abbreviations: "at least" (at l.), "at most" (at m.), "at most very" (at m.v.)

| label | definition | label | definition |
|---|---|---|---|
| IF noise exposure (dB) | | THEN annoyance | |
| very low | Trap(0, 0, 52.5, 60) | not at all | 1-Sigm(0, 2) |
| low | Triang(52.5, 57.5, 65) | slightly | Bell(-3, 1.5, 5.5) |
| moderate | Triang(57.5, 65, 70) | moderately | Bell(1, 4, 9) |
| high | Triang(65, 70, 75) | very | Bell(5, 8, 11) |
| very high | Trap(70, 75, 80, 80) | extreme | Sigm(7, 10) |
| IF distance to highway (m) | | THEN annoyance | |
| close | Trap(0, 0, 500, 1500) | high | Sigm(5, 7.5) |
| far | Trap(500, 1500, ∞, ∞) | not high | 1-Sigm(3.5, 7.5) |
| IF distance to main road (m) | | THEN annoyance | |
| close | Trap(0, 0, 200, 600) | high | Sigm(5, 7.5) |
| far | Trap(200, 600, ∞, ∞) | not high | 1-Sigm(3.5, 7.5) |
| IF distance to through road (m) | | THEN annoyance | |
| close | Trap(0, 0, 60, 250) | high | Sigm(5, 7.5) |
| far | Trap(60, 250, ∞, ∞) | not high | 1-Sigm(3.5, 7.5) |
| IF age (year) | | THEN annoyance | |
| young | Trap(0, 0, 20, 30) | below moderate | 1-Sigm(4, 10) |
| old | Trap(50, 60, 100, 100) | below moderate | 1-Sigm(4, 10) |
| IF urbanization | | THEN annoyance | |
| country | Enum(0, 0, 0, 1) | low | 1-Sigm(1,8, 5.8) |
| suburb | Enum(0, 0, 1, 0) | below moderate | 1-Sigm(2.2, 6.6) |
| city | Enum(0, 1, 0, 0) | above moderate | Sigm(2.2, 7) |
| center | Enum(1, 0, 0, 0) | above moderate | Sigm(3, 6.2) |
| IF amount of traffic [very much, much, ...] | | THEN annoyance | |
| very much | Enum(1, 0, 0, 0, 0) | at l. strongly | Sigm(5, 6.6) |
| much | Enum(0, 1, 0, 0, 0) | at l. fairly | Sigm(3, 5) |
| few | Enum(0, 0, 0, 1, 0) | at m. v. slightly | 1-Sigm(3, 4.5) |
| very few | Enum(0, 0, 0, 0, 1) | at m. slightly | 1-Sigm(2, 3.5) |
| IF built up (% of area) | | THEN urbanization [country, ...] | |
| open | Trap(0, 0, 15, 25) | country | Enum(0, 0, 0, 1) |
| around 25 % | Triang(15, 25, 30) | suburb | Enum(0, 0, 1, 0) |
| around 35 % | Triang(25, 30, 35) | city | Enum(0, 1, 0, 0) |
| dense | Trap(30, 35, 100, 100) | center | Enum(1, 0, 0, 0) |
| IF population density (nr/km$^2$) | | THEN urbanization [country, ...] | |
| low | Trap(0, 0, 770, 1300) | country | Enum(0, 0, 0, 1) |
| below medium | Triang(1000, 2200, 3200) | suburb | Enum(0, 0, 1, 0) |
| above medium | Triang(1300, 2800, 3800) | city | Enum(0, 1, 0, 0) |
| high | Trap(2200, 3400, ∞, ∞) | center | Enum(1, 0, 0, 0) |
| IF agriculture land (% of area) | | THEN urbanization [country, ...] | |
| little | Trap(0, 0, 15, 25) | center | Enum(1, 0, 0, 0) |
| around 23 % | Triang(13, 23, 33) | city | Enum(0, 1, 0, 0) |
| around 30 % | Triang(20, 30, 50) | suburb | Enum(0, 0, 1, 0) |
| a lot | Trap(30, 50, 100, 100) | country | Enum(0, 0, 0, 1) |

www.ingramcontent.com/pod-product-compliance
Ingram Content Group UK Ltd.
Pitfield, Milton Keynes, MK11 3LW, UK
UKHW020833270526
471509UK00003B/41
* 9 7 8 3 5 4 0 2 2 2 1 4 9 *